George Payn Quackenbos

**A Natural Philosophy**

Embracing the Most Recent Discoveries in the Various Branches of Physics

George Payn Quackenbos

**A Natural Philosophy**
*Embracing the Most Recent Discoveries in the Various Branches of Physics*

ISBN/EAN: 9783337025373

Printed in Europe, USA, Canada, Australia, Japan

Cover: Foto ©berggeist007 / pixelio.de

More available books at **www.hansebooks.com**

A

# NATURAL PHILOSOPHY:

EMBRACING

## THE MOST RECENT DISCOVERIES

IN THE

### VARIOUS BRANCHES OF PHYSICS,

AND EXHIBITING

THE APPLICATION OF SCIENTIFIC PRINCIPLES IN EVERY-DAY LIFE.

*ADAPTED TO USE WITH OR WITHOUT APPARATUS, AND ACCOMPANIED WITH*
*FULL DESCRIPTIONS OF EXPERIMENTS, PRACTICAL EXERCISES,*
*AND NUMEROUS ILLUSTRATIONS.*

## BY G. P. QUACKENBOS, LL. D.,

PRINCIPAL OF "THE COLLEGIATE SCHOOL", N. Y.; AUTHOR OF "FIRST LESSONS IN
COMPOSITION", "ADVANCED COURSE OF COMPOSITION AND RHETORIC"
"ILLUSTRATED SCHOOL HISTORY OF THE UNITED STATES", ETC.

NEW YORK:
D. APPLETON AND COMPANY,
90, 92 & 94 GRAND STREET.
1870.

**By the same Author:**

FIRST LESSONS IN COMPOSITION: In which the Principles of the Art are developed in connection with the Principles of Grammar. 12mo, pp. 182.

ADVANCED COURSE OF COMPOSITION AND RHETORIC: A Series of Practical Lessons on the Origin, History, and Peculiarities of the English Language, Punctuation, Taste, Figures, Style and its Essential Properties, Criticism, and the various Departments of Prose and Poetical Composition. 12mo, pp. 451.

AN ENGLISH GRAMMAR: 12mo, pp. 288.

PRIMARY HISTORY OF THE UNITED STATES: Made easy and interesting for Beginners. Child's Quarto, splendidly illustrated, pp. 192.

ILLUSTRATED SCHOOL HISTORY OF THE UNITED STATES: Embracing a full Account of the Aborigines, Biographical Notices of Distinguished Men, numerous Maps, Plans of Battle-fields, and Pictorial Illustrations. 12mo, pp. 473.

# PREFACE.

THE importance of the physical sciences is now so generally admitted that there are few institutions of learning in which they are not made regular branches of study. And very properly,—for what can be more interesting and instructive, what more worthy of the attention of intelligent creatures, what more calculated to inspire their minds with a thirst for further knowledge, and fill their hearts with reverent gratitude to the Divine Being, than an acquaintance with the laws of the material world, the mysterious influences constantly at work in nature, and the principles by which atoms and worlds are alike controlled ?

It is in the hope of investing this subject with a lively interest, and bringing it home to the student by exhibiting the application of scientific principles in every-day life, that the Natural Philosophy here presented to the public has been prepared. The author has sought to render a subject, abstruse in some of its connections, easy of comprehension, by treating it in a clear style, taking its principles one at a time in their natural order, and illustrating them fully with the facts of our daily experience. The range of topics is comprehensive. By avoiding unnecessary repetitions, room has been found for chapters on Astronomy and Meteorology; one of which subjects, at least, has heretofore been invariably omitted in similar treatises, though a summary of both is important, as time is seldom found for pursuing these branches in separate volumes.

The incorrectness of many of the text-books on Natural Philosophy has been a subject of general complaint. Grave errors, both of theory and fact, have been handed down from one to another, and the results obtained by modern research have been too often overlooked. In preparing this volume, every effort has been made to ensure accuracy, the most recent authorities have been consulted, and it is believed that a faithful view is presented of the various sciences embraced, as far as they are at present developed. It is the intention of the author to keep his book up to the times by constant revision, and to make such alterations and additions as the progress of discovery may require.

Two styles of type are used in the text; a larger size for leading principles, a smaller size for descriptions of apparatus and experiments, explanatory illustrations, &c. By confining a class to the former when the saving of time is an object, a brief yet complete course may be taken. Questions at the bottom of each page will be found to facilitate the examiner's duty, and to afford the pupil a means of testing his preparation before reciting. At the end of such chapters as admit of it, easy practical examples have been introduced, to illustrate the rules and principles set forth.

An important feature of this work is its adaptation to use with or without apparatus. The majority of schools have few facilities for experimental illustration. The wants of these are here met by a free use of engravings, full descriptions of experiments, and explanations of their results. A number of these engravings have been furnished by BENJAMIN PIKE, jr., of 294 Broadway, New York, and are not mere fancy-sketches, but actual representations of instruments (the best and most modern of their kind) manufactured at his establishment. Mr. Pike's life has been devoted to this branch of industry; and it may not be improper to add that institutions desirous of procuring a set of apparatus, partial or complete, will find his assortment unsurpassed in variety and excellence.—For convenience of recitation, those cuts to which reference is made by letters are reproduced apart from the text, in the back of the book.

An Alphabetical Index closes the volume

NEW YORK, July 1, 1859.

# CONTENTS.

CHAPTER                    PAGE

I.—MATTER AND ITS FORMS . . . . . . . . 7

II.—PROPERTIES OF MATTER . . . . . . . . 12

III.—MECHANICS.
  Motion.—Momentum.—Striking Force . . . . . 26

IV.—MECHANICS (*continued*).
  Laws of Motion . . . . . . . . . 34

V.—MECHANICS (*continued*).
  Gravity . . . . . . . . . . 46

VI.—MECHANICS (*continued*).
  Centre of Gravity . . . . . . . . 70

VII.—MECHANICS (*continued*).
  The Motive Power.—The Resistance.—The Machine.—
  Strength of Materials . . . . . . . 81

VIII.—MECHANICS (*continued*).
  The Mechanical Powers . . . . . . . 94

IX.—MECHANICS (*continued*).
  Wheelwork.—Clock and Watchwork . . . . . 120

X.—MECHANICS (*continued*).
  Hydrostatics . . . . . . . . . 130

XI.—MECHANICS (*continued*).
  Hydraulics . . . . . . . . . 152

XII.—PNEUMATICS . . . . . . . . . 165

XIII.—PYRONOMICS . . . . . . . . . 192

  The Steam Engine . . . . . . . . 219

CHAPTER                                                          PAGE
  XIV—Optics . . . . . . . . . . . . 229

  XV.—Acoustics . . . . . . . . . . . 274

  XVI.—Electricity . . . . . . . . . . 289
          Frictional Electricity . . . . . . . 290
          Voltaic Electricity . . . . . . . . 316
          Thermo-electricity . . . . . . . . 332

  XVII.—Magnetism . . . . . . . . . . 333
          Electro-magnetism . . . . . . . . 349
          Magneto-electricity . . . . . . . . 366

XVIII.—Astronomy . . . . . . . . . . 368

  XIX.—Meteorology . . . . . . . . . 401

      Figures reproduced . . . . . . . . 407

      Index . . . . . . . . . . . 440

# NATURAL PHILOSOPHY.

———◆◆◆———

## CHAPTER I.

### MATTER AND ITS FORMS.

1. *Matter.*—Whatever occupies space, whatever we can see or touch, is known as Matter. Earth, water, air, are different forms of matter.

A distinct portion of matter is called a Body. The Earth, a ball, a rain-drop, are Bodies.

2. All matter, properly speaking, is Ponderable,—that is, has weight.

*Imponderable* means *without weight.* The term *Imponderable Matter* has been applied by some to heat, light, electricity, and magnetism. As late researches seem to indicate that these are forces or conditions of matter, and not themselves varieties of matter, they are now generally called Imponderable Agents.

3. *Forms of Ponderable Matter.*—Ponderable Matter exists in three forms; Solid, Liquid, and A-er'-i-form.

A body is said to be Solid when its particles cohere, so that they can not move among themselves; example, ice. Solid bodies are called Solids.

---

1. What is Matter? Give examples. What is a Body? Give examples. 2. What is said of all matter? What does *imponderable* mean? To what has the term *imponderable matter* been applied? What are heat, light, electricity, and magnetism generally called? 3. In how many forms does ponderable matter exist?

A body is said to be Liquid when its particles cohere so slightly that they can move freely among themselves; example, water. Liquid bodies are called Liquids.

Aëriform means *having the form of air*, and matter is said to exist in this state when its particles repel each other, tending to separate and spread out indefinitely; example, steam. Aëriform bodies are called Gases and Vapors.

Liquid and aëriform bodies are embraced under the general name of Fluids.

There are marked points of difference between solids and fluids. A solid has a permanent shape; a fluid accommodates its shape to that which contains it. A solid may often be moved by moving a portion of its particles; as a pitcher by its handle. The particles of a fluid, on the other hand, do not cohere, and therefore, when we move some of them, the rest are detached by their own weight; thus by dipping a tumbler into a pail of water, we can not remove all the fluid, but only as much as the tumbler contains. Again, a solid resists a force which seeks to penetrate it. A fluid, on the contrary, is easily divided; if we move slowly through the air, for instance, we feel no resistance.

The same substance may, under different circumstances, appear in all three of these forms. Thus water is a liquid; when frozen, it becomes ice, which is a solid; when exposed to a certain degree of heat, it is converted into steam, which is aëriform.

4. *Classes of Bodies.*—Bodies are distinguished as Simple and Compound.

A Simple Body consists of matter that can not be resolved into more than one element; as, gold.

A Compound Body consists of matter that can be resolved into two or more elements; as air, which is composed of two gases.

The simple bodies, or elements, of which every thing in the universe is composed, are sixty-two in number. Of these, fifty, distinguished by a peculiar lustre, are called Metals. The remaining twelve are known as Nonmetallic Elements.

Name them. When is a body said to be solid? What are solid bodies called? When is a body said to be liquid? What are liquid bodies called? What does *aëriform* mean? When is a body said to be aëriform? What are aëriform bodies called? What name is applied to both liquid and aëriform bodies? Mention some of the marked points of difference between solids and fluids. In how many forms may the same substance appear? Give an example. 4. Into how many classes are bodies divided? Name them. What is a Simple body? What is a Compound body? How many simple bodies are there? How are they divided? Name the principal met.

The principal metals are the seven known to the ancients,—gold, silver, iron, copper, mercury, lead, and tin; antimony, which was next discovered, in 1490; bismuth, zinc, arsenic, cobalt, plat'-i-num, nickel, manganese, &c. The twelve non-metallic elements are ox'-y-gen, hy'-dro-gen, ni'-tro-gen, chlorine [klo'-reen], iodine [i'-o-deen], bromine [bro'-meen], fluorine [flu'-o-reen], se-le'-ni-um, sulphur, phosphorus, carbon, and bo'-ron.

These simple substances are rarely found; nearly every body that we meet with, whether natural or artificial, is composed of two or more elements, and is therefore compound. Such is the case with air, which was anciently thought to be a simple substance, but was proved, towards the close of the eighteenth century, to be a mixture of 21 parts of oxygen and 79 parts of nitrogen. Water, also, has been found to be a compound substance, made up of oxygen and hydrogen combined in the proportion of 1 to 8. Of the sixty-two elements referred to above, twenty are so rare that their properties are not yet fully known; thirty more are comparatively seldom met with; the remainder constitute the great bulk of the globe and all that is thereon.

The consideration of the simple substances, with their properties and combinations, belongs to the science of CHEMISTRY. The force that causes them to combine and produce compound substances, is called Chemical Affinity. Oxygen and hydrogen combine and form water, in consequence of their chemical affinity.

Chemical affinity subsists only between certain substances. If sulphuric acid be poured on a piece of zinc, the two substances will combine and form a compound entirely different from either. Pour the acid on a lump of gold, and no such change will ensue, because there is no chemical affinity between them.

5. *Natural Philosophy.*—Natural Philosophy is the science that treats of the properties and laws of matter. It is also called PHYSICS.

Pythagoras was the first to use the term *philosophy.* From him and his followers it was borrowed by Socrates; who, when the other sages of his time called themselves *sophists,* or *wise men,* modestly declared himself a *philosopher,* or *lover of wisdom.*—Philosophy implies a search for truth; and Natural Philosophy, as distinguished from Moral and Intellectual Philosophy, searches for the truths connected with the material world.

---

als. Name the twelve non-metallic elements. What is said of the simple substances? What kind of substances are air and water? Of what is air composed? Of what, water? How many elements constitute the great bulk of the globe? What is said of the rest? To what science does the consideration of the simple substances belong? What causes the simple substances to combine? Give an instance of chemical affinity. Illustrate the fact that chemical affinity subsists only between certain substances. 5. What is Natural Philosophy? With whom did the term *philosophy*

6. *Modes of Investigation.*—We arrive at the facts of Natural Philosophy in two ways; by Observation and Experiment. Observation consists in watching such phenomena, or appearances, as occur in the course of nature. Experiment consists in causing such phenomena to occur when and where we wish, for the purpose of noting the attendant circumstances and results.

For example, we arrive at the fact that an unsupported body will descend to the earth's surface, when we see an apple fall from a bough; this is by Observation. We learn the same fact, when, with the view of ascertaining what it will do, we let an apple drop from our hands; this is by Experiment.

7. *Modes of Reasoning.*—Having obtained our facts in the two ways just described, and classified them, we next proceed from individual cases to deduce general laws. This is called Reasoning by Induction.

Thus, if we try the experiment with many different apples, and find that each, when let go, will fall to the ground, we lay down the general law that *all apples* will fall in like manner. If we find that not only apples do this, but also other objects with which we make the trial, we go a step further, and announce another law, that *all objects* left unsupported will fall to the ground.

It is by this process that most of the laws and principles of Natural Philosophy have been established. Archimedes [*ar-ke-me'-deez*], the Sicilian philosopher, used it over two thousand years ago. Gal-i-le'-o revived it in modern times, and it may be said to lie at the foundation of all the great discoveries of Newton.

When we have two similar phenomena and know that one proceeds from a certain cause, we attribute the other to the same cause. This is called Reasoning by Analogy.

Such reasoning is employed in the case of all bodies that are beyond our reach. From what is near, we draw conclusions respecting what is remote. It is thus, for example, that the philosopher explains the motions of the heavenly bodies, extending to them, by analogous reasoning, the same principles that govern the motion of bodies on the earth.

8. *Division of the Subject.*—Natural Philosophy, hav-

originate? Who borrowed it from Pythagoras? What does philosophy imply? What is the particular province of Natural Philosophy? 6. How do we arrive at the facts of Natural Philosophy? In what does Observation consist? In what, Experiment? Illustrate these definitions. 7. What is meant by *reasoning by induction*? Give an example. By what philosophers has this mode of reasoning been employed? What is meant by *reasoning by analogy*? Give an example. 8. What

ing to treat of matter in all its forms, embraces the following distinct sciences:—

Mechanics, which treats of forces and their application in machines.   To Mechanics belong

Hy-dro-stat'-ics, which treats of liquids at rest;

Hy-drau'-lics, which treats of liquids in motion.

Pneumatics [*nu-mat'-ics*], which treats of gases and vapors.

Pyr-o-nom'-ics, which treats of heat.

Optics, which treats of light and vision.

Acoustics [*a-cow'-stics*], which treats of sound.

Electricity, which treats of the electric fluid.   To Electricity belong

Galvanism, which treats of electricity produced by chemical action;

Thermo-electricity, which treats of electricity developed by heat;

Magneto-electricity, which treats of electricity developed by magnetism.

Magnetism, which treats of magnets and the forces they develop.   To Magnetism belongs

Electro-magnetism, which treats of magnetism developed by electricity.

Astronomy, which treats of the heavenly bodies.

Me-te-o-rol'-o-gy, which treats of the phenomena of the atmosphere.

---

branches does Natural Philosophy embrace?  Of what does Mechanics treat?  Hydrostatics?  Hydraulics?  Pneumatics?  Pyronomics?  Optics?  Acoustics?  Electricity?  Galvanism?  Thermo-electricity?  Magneto-electricity?  Magnetism?  Electro-magnetism?  Astronomy?  Meteorology?

# CHAPTER II.

## PROPERTIES OF MATTER.

9. EVERY distinct portion of matter possesses certain properties. Some of these belong in common to all bodies, solid, liquid, and aëriform, and are called Universal Properties of matter. Others, again, are found only in certain substances, and these are known as Accessory Properties.

The Universal Properties of matter are Extension, Figure, Impenetrability, Indestructibility, Inertia [in-er'-sha], Divisibility, Porosity, Compressibility, Expansibility, Mobility, and Gravitation.

The principal Accessory Properties are Cohesion, Adhesion, Hardness, Tenacity, Elasticity, Brittleness, Malleability, and Ductility.

We proceed to consider these properties in turn.

10. EXTENSION.—Extension is that property by which a body occupies a certain portion of space. The portion of space thus occupied is called its Place.

In other words, every body, however small, must have some size, or a certain length, breadth, and thickness, which are called its Dimensions. The greatest of these three dimensions is its Length; the next greatest, its Breadth, or Width; the least, its Thickness. But, instead of any of these terms, we use the word *height* to denote distance from bottom to top in the case of objects towering above us, and *depth* to denote distance from top to bottom in the case of objects extending below us.

11. FIGURE.—Figure is that property by which a body has a certain shape.

This property necessarily follows from Extension; for since every body must have length, breadth, and thickness, it must also have some definite

9. What is meant by Universal Properties of matter? What is meant by Accessory Properties? Enumerate the universal properties. Mention the principal accessory properties. 10. What is Extension? What is meant by the dimensions of a body? What is Length? Breadth? Thickness? When are the terms *height* and *depth* used? 11. What is Figure? From what does figure follow? What is the

shape. While this is true of all bodies, it must be remembered that the form of solids is permanent, while that of fluids varies, to adapt itself to every new surface with which it comes in contact. A bullet keeps the same shape, wherever it is placed; whereas a quantity of water, poured from a tumbler into a pail, visibly changes its form.

12. IMPENETRABILITY.—Impenetrability is that property by which a body occupies a certain portion of space, to the exclusion for the time of all other bodies.

Impenetrability may be illustrated with a variety of simple experiments. Fill a tumbler to the brim with water, and drop in a bullet; the water will at once overflow. Fill a bottle with water, and try to put the cork in; the cork will not enter till it has displaced some of the water: if it fit so closely that the water can not escape, and a hard pressure be exerted, the bottle will burst.

Fig. 1.

The impenetrability of air is shown with the apparatus represented in Figure 1. A is a glass jar fitted with an air-tight cork, through which a funnel, B, enters the jar. C is a bent tube, one end of which also passes through the cork into the jar, while the other is received in a glass of water, D. Let water be poured into the funnel; as it descends, drop by drop, into the jar, air passes out through the bent tube, and escapes through the water in D in the form of bubbles. Thus it is shown that water and air can not occupy the same space at the same time.

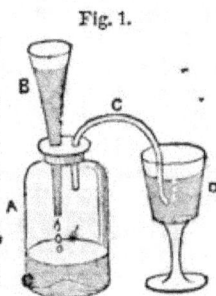

13. Impenetrability belongs to all substances, though in some cases it may appear to be wanting. A nail, for instance, is driven into a piece of wood without increasing its size; but it effects an entrance by forcing together the fibres of the wood, not by occupying their space at the same time with them. In like manner, a certain amount of salt and sugar may be successively dropped into a tumbler brim-full of water without causing it to overflow. The particles of water, which are supposed to be globular, do not everywhere touch each other, and the particles of salt are accommodated in the interstices between them. These in turn leave minute spaces, into which the still smaller particles of sugar find their way. Fig. 2 exhibits such an arrangement. To illustrate it familiarly, we may fill a vessel with as many oranges as it will hold, and then pour on a quantity of peas, shaking the vessel slightly so that they may settle in the empty spaces.

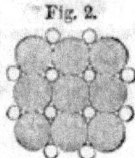

Fig. 2.

difference between solids and fluids as regards figure? 12. What is Impenetrability? Give some familiar illustrations of this property. Describe the experiment with the apparatus represented in Fig. 1. 13. What is said of those cases in which impenetrability appears to be wanting? Illustrate this with the nail. Explain how salt and sugar may be dropped into a tumbler full of water without causing it to over-

When the vessel will receive no more peas, repeat the process with fine gravel, and it will be found that a considerable quantity will lodge between the oranges and peas.

14. Indestructibility.—Indestructibility is that property which renders a body incapable of being destroyed.

Matter may be made to assume a new form and new properties, but it can not cease to exist. The quantity of matter now in the world is precisely the same as when it was first called into being, and it will continue undiminished till the end of time. The Deity alone created, and it is only He that can destroy.

15. To this universal law we have some apparent exceptions; but, when closely examined, it will be found that they are exceptions in appearance only. Water, for instance, exposed to the air in a shallow dish, will at length disappear by evaporation; but it is not destroyed. Assuming the form of vapor, it ascends, becomes incorporated with clouds, is condensed into rain, and falls,—to go through the same process again.—The oil in a burning lamp gradually gets lower and lower till at last it is all gone, and we say it is *burned up;* but the process of combustion, or burning, only changes it into invisible gases,—not one particle of its substance is lost. In like manner, when fuel of any kind is consumed, there is only a change of form, not a destruction of the least portion of matter.

Such changes are constantly going on in the operations of nature. One body perishes, and of the materials that composed it another is formed. Our own frames may contain particles that were in the bodies of Adam, Noah, or Socrates; or, if they do not now, may do so to-morrow, for they are constantly parting with portions of their substance, the place of which is as constantly supplied by new matter. It is supposed that the whole body, including even the innermost parts of its hardest bones, is completely renewed every seven years. Yet, amid all the countless transitions of nature, not a single particle of matter is destroyed or lost.

16. It was by a knowledge of the indestructibility of matter that Sir Walter Raleigh is said to have won a wager of Queen Elizabeth. Having weighed out a sufficient quantity of tobacco to fill his pipe, he came into the queen's presence, and as the wreaths of smoke curled up offered to bet her Majesty that he could tell their weight. Elizabeth accepted the bet, and Sir Walter quietly finished his pipe; then, having shaken out the ashes, he weighed them, and, subtracting the amount from that of the tobacco originally put

---

flow. 14. What is Indestructibility? What can be done to matter, and what not? 15. What is said of the apparent exceptions to this law? What becomes of water exposed to the air? What becomes of the oil in a burning lamp? What is said of the changes of nature? What is said of the changes in the human body? 16. How did Sir Walter Raleigh teach Queen Elizabeth that matter is indestructible? 17. What

iu, told the queen the exact weight of the smoke. Elizabeth paid the wager, and thus learned to her cost that *matter is indestructible.*

17. INERTIA.—Inertia is that property which renders a body incapable of putting itself in motion when at rest, or coming to rest when in motion.

When a stationary body begins to move, or a moving body comes to rest, it is not through any power of its own, but because it is acted on by some external agency, which we call a Force.

That no inanimate body can put itself in motion, is evident from our daily experience. The rocks that we saw on the earth's surface ten years ago are to-day in precisely the same place as they then were, and there they will remain forever unless some force removes them.

It is equally true, though not so obvious, that a body once in motion can not of itself cease to move. The earth revolves on its axis, the heavenly bodies move in their orbits, just as they did at the time of the Creation; they have no power to stop. It is true that on the surface of the earth a moving body gradually comes to rest, when the force which put it in motion ceases to act; but this is owing to the resistance of the air and a force which draws it towards the centre of the earth—not to any agency of its own. Remove all external forces, and its inertia would keep it moving on in a straight line forever.

18. *Familiar Examples.*—It is in consequence of inertia that a horse has to strain hard at first to move a load, which, when it is once in motion, he can draw with ease. A car, through its inertia, continues moving after the locomotive is detached. Through inertia, a person standing erect in a stationary boat or wagon is thrown backward if it suddenly starts: his feet, touching the bottom, are carried forward with it, while his body by its inertia does not partake of the onward motion and falls backward. So, a person standing erect in a boat or wagon that is moving rapidly, is thrown forward if it suddenly stops; his feet cease to move at once, while his body continues in motion in consequence of its inertia, and falls forward.

Fig. 3.

19. An interesting experiment to illustrate inertia may be performed with the apparatus represented in Fig. 3. On the top of a short pillar is placed a card, and on the card a brass ball. Beside the pillar is fixed a steel spring, with an apparatus for drawing it back. If the

spring is drawn back and then suddenly released, it will drive the card from the top of the pillar, while the ball in consequence of its inertia will retain its place.

Those who have not the above apparatus may balance a card with a penny placed upon it on the tip of one of the fingers of the left hand, and strike it

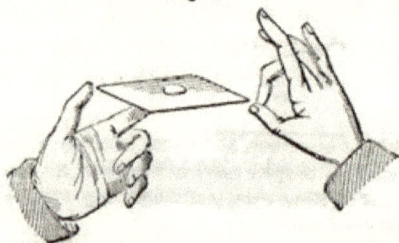

Fig. 4.

suddenly with the middle finger of the right hand, as represented in Fig. 4. If properly balanced and evenly struck, the card will fly away, and the penny will be left on the finger.

In these cases, there is not sufficient time for the card to overcome the inertia of the ball and the penny, and impart to them its own motion. When, however, motion has once been communicated by one body to another resting on it, the inertia of the latter keeps it in motion. A person riding in a carriage partakes of its motion, and if he jumps from it runs the risk of being thrown down, because his feet cease to move the instant they strike the ground, while the inertia of his body carries it forward. The circus-rider

Fig. 5.

takes advantage of this fact. While his horse is going at full speed, he jumps over a rope extended across the ring (see Fig. 5), and regains his footing on the saddle without difficulty. To do this, he has only to leap straight up as he comes to the rope, for his inertia bears him along in the same direction as his horse.

A bullet thrown at a pane of glass breaks it into many pieces, but, fired at it from a rifle, merely makes a circular hole. In the latter case, all the particles of glass, on account of their inertia, can not immediately acquire the rapid motion of the bullet; and consequently only that portion which is struck is carried onward. On the same principle, a thin stick resting on two wine-glasses (see Fig. 6) may be broken by a quick blow with a poker in its centre, without injury to its brittle supports.

---

the experiment with the card and penny. What is the effect of inertia, when motion has once been communicated to a body? Why is a person who jumps from a carriage in motion thrown down? Explain the leap of the circus-rider. What is the effect of throwing a bullet against a pane of glass, and what of firing it? What causes the difference? What experiment may be performed to illustrate this point? 20. To

20. The heavier a body is, the greater is its inertia; the more strongly does it resist forces that would set it in motion, change its motion, or stop its motion.

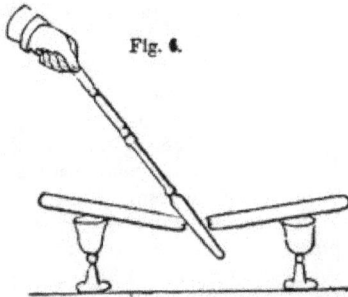

Fig. 6.

Instinct teaches this fact. A child, when nearly overtaken by a man, will suddenly turn, or " dodge" as he calls it, thus gaining ground, inasmuch as the greater weight and inertia of the man compel him to make a longer turn. So a hare, in making for a cover, often escapes a hound by making a number of quick turns. The greater inertia of the hound carries him too far, and thus obliges him to pass over a greater space, as seen in Fig. 7, in which the continuous line shows the hare's path and the dotted line the hound's.

Fig. 7.

21. DIVISIBILITY. — Divisibility is that property which renders a body capable of being divided.

*Atomic Theory.* — Practically, there is no limit to the divisibility of matter. Most philosophers, however, hold what is called the Atomic Theory,—that if we had more acute senses and instruments sufficiently delicate, we would at last, in dividing and subdividing matter, arrive at exceedingly small particles, incapable of further division. Such particles they call ATOMS, a term derived from a Greek word meaning *indivisible*.

According to this theory, different kinds of matter are made up of different kinds of atoms; but in the same substance the atoms are always the same in shape and nature. It must be remembered, however, that no particle has yet been arrived at that can not be divided.

22. *Instances of Divisibility.*—Matter has been divided into parts incredi-

bly minute. With the proper instrument, ten thousand distinct parallel lines can be drawn on a smooth surface an inch in width. So minute are these lines that they can not be seen without a microscope, not even a scratch being visible to the naked eye.

A grain of musk will diffuse a perceptible odor through an apartment for twenty years. It does this by filling the air with particles of its substance; but so inconceivably minute are these particles, that, if the musk is weighed at the end of the twenty years, no loss of weight can be detected.

A grain of copper dissolved in nitric acid will impart a blue color to three pints of water. Each separable particle of water must contain a portion of the grain of copper,—which is thus, it has been computed, divided into no less than 100,000,000 parts.

23. Nature affords many striking examples of the divisibility of matter. The spider's web is so attenuated that a sufficient quantity of it to go around the earth would weigh only eight ounces; and yet this minute thread consists of about a thousand separate filaments.

Blood is composed of small red globules floating in a colorless liquid. Of these globules, every drop of human blood contains at least a million. Minute as they are, they may be divided into globules much more minute. As we descend in the scale of creation, we come to animals whose whole bodies are no larger than these little globules of human blood, yet possess all the organs necessary to life. How inconceivably small are the vessels through which the fluids of their bodies must circulate!

The microscope reveals to us wonders of animal life that are almost incredible. It shows us in duck-weed animalcules so small that it would take ten thousand millions of them to equal the size of a hemp-seed. In a single drop of ditch-water, it exhibits myriads of moving creatures. The mineral called tripoli is formed of these animalcules fossilized or turned into stone; and it has been shown that the fortieth part of a cubic inch of this mineral contains the bodies of no less than a thousand million animalcules—or more than all the human beings on the globe.

24. POROSITY.—What shape the atoms of different bodies are, we have no means of determining. By reason of their shape, however, or from some other cause, they do not everywhere touch each other, but are separated by interstices, to which we give the name of Pores. Pores are often visible to the naked eye, as in sponge and pumice-stone; in other cases, as in gold and granite, they are too minute to be detected even with the microscope.

25. Porosity is the property of having pores. It belongs to all bodies.

26. That water is porous, is proved by the fact that a vessel filled with it will receive considerable quantities of salt and sugar without overflowing. What can become of these substances, unless, as shown in Fig. 2, their particles lodge in the interstices between the particles of water? It is on this principle that hot water receives more salt and sugar without overflowing than cold. Heat expands water,—that is, forces its particles further apart,— and thus enables a greater quantity of salt and sugar to lodge between them.

That granite is porous, is shown by placing a piece of it in a vessel of water under the receiver of an air-pump (described on page 178), and removing the air. Little bubbles will soon be seen rising through the water. These bubbles are the air contained in the invisible pores of the granite.

A piece of iron is made smaller by hammering. This proves its porosity. Its particles could not be brought into closer contact, if there were no interstices between them.

27. An experiment performed some years ago at Florence, Italy, to ascertain whether water could be compressed, proved that gold is porous. A violent pressure was brought to bear on a hollow sphere of gold filled with water. The water made its way through the gold and appeared on the outside of the sphere. Water will thus pass through pores not more than one half of the millionth of an inch in diameter.

28. *Density and Rarity.*—The fewer and smaller the pores in a body, the more compact are its particles, and the greater is the weight of a given bulk. Bodies whose particles are close together are called *Dense;* those with large or numerous pores are called *Rare.*

29. COMPRESSIBILITY AND EXPANSIBILITY.—These two properties are the opposites of each other. Compressibility is that property which renders a body capable of being reduced in size. Expansibility is that property which renders a body capable of being increased in size.

Compressibility and Expansibility follow from porosity. Since the particles of bodies do not everywhere touch each other, the application of a sufficient force will bring them closer together, and the size of the bodies will thus be re-

of the pores? 25. What is Porosity? 26. How is water proved to be porous? Why does hot water receive more salt and sugar than cold? How may it be proved that granite is porous? How is the porosity of iron proved? 27. Give an account of the experiment by which the porosity of gold was proved. How small pores will water pass through? 28. What bodies are called *dense?* What bodies are called *rare?* 29. What is Compressibility? What is Expansibility? Show how these properties

duced. A sponge, for instance, by the simple pressure of the hand, can be reduced to one-tenth of its natural size. In like manner, if the pores of a body are made larger by any agency (as they are by heat), its size is proportionately increased.

Fig. 8.

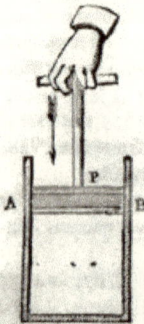

30. All bodies possess these properties. A rod of iron, too large to enter a certain opening, may be so compressed by hammering as to pass through it, and then so expanded by heat as to render its entrance again impossible. Liquids, which were long considered incompressible, are now known to yield to a high degree of pressure; their expansibility is illustrated by the rise of mercury in the thermometer.

The compressibility and expansibility of air are shown by the apparatus represented in Fig. 8. Let P be a piston, fitted, air-tight, to the cylinder A B. As the piston is driven down, the air, unable to escape, is compressed; as it is drawn back, the air expands.

Aëriform bodies are more easily compressed and expanded than any others.

31. MOBILITY.—Mobility is that property which renders a body capable of being moved.

Though the inertia of bodies prevents them from moving themselves, yet there is no body that can not be moved by the application of a proper force.

32. GRAVITATION.—Gravitation (or Gravity, as it is called when acting at short distances) is the tendency

Fig. 9.

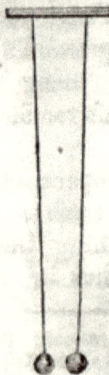

which one body has to approach another, under the influence of the latter's attraction. A cannon ball dropped from the hand falls to the earth by reason of its gravity. The earth at the same time moves towards the cannon ball, but through a space inconceivably small in consequence of its vast superiority in size over the ball.

That the cannon ball is capable of attracting as well as being attracted, may be proved by suspending two balls close to each other by very long cords. In consequence of the attraction of the balls, the cords will not hang parallel, but will incline towards each other as they descend, as shown in Fig. 9.

follow from porosity. 30. How may compressibility and expansibility be illustrated with an iron rod? What is said of these properties in liquids? How may the compressibility and expansibility of air be shown? What bodies are most easily com-

We now proceed to the Accessory Properties, which are confined to certain bodies.

33. COHESION.—Cohesion is that property by which the particles of a body cling to each other. As particles are also called *mol'-e-cules*, Cohesion has received from some authors the name of Mo-lec'-u-lar Attraction.

Cohesion belongs particularly to solids, and is in fact the cause of their solidity. In some it is stronger than in others, rendering them harder or more tenacious. Liquids have so little cohesion that their weight alone overcomes it, and causes a separation of particles. In aëriform fluids cohesion is entirely wanting, its place being supplied by a Repulsive Force, which tends to make their particles spread out from each other.

34. ADHESION.—Adhesion is that property by which the surfaces of two different bodies placed in contact cling together.

The bodies in question may be of the same kind of matter. This is proved by an experiment with two glass plates ground perfectly even. Let these be pressed together, and it will be found, on attempting to pull them apart by their handles, that considerable force will be required. The larger the surfaces of the plates, the harder it will be to separate them. A pair of Adhesion Plates is represented in Fig. 10.

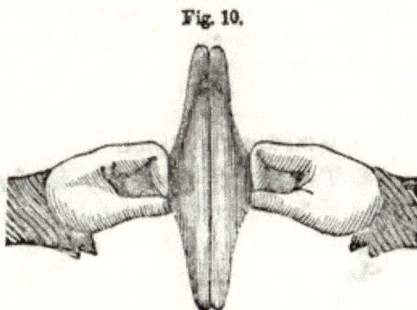

Fig. 10.

ADHESION PLATES.

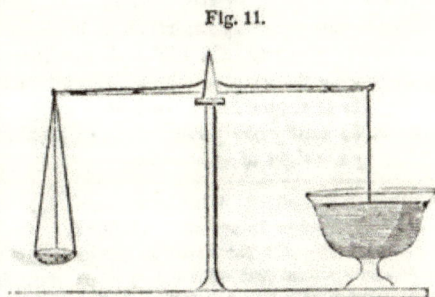

Fig. 11.

Adhesion also operates between the surfaces of solids and liquids. Suspend a piece of copper-plate from one side of a pair of scales, in such a way that its under surface may be parallel to the floor, and balance it with weights placed in the scale on the other side. Then, without disturbing the copper, place a vessel beneath it, as in Fig. 11, and pour in water till the liquid just reaches the plate. The adhesion between the solid and the liquid is now so strong that additional weights (more or less, according to the extent of surface) may be put in the scale on the other side without causing them to separate.

35. HARDNESS.—Hardness is that property by which a body resists any foreign substance that attempts to force a passage between its particles.

The hardness of a body depends on the degree of firmness with which its particles cohere. It is therefore entirely distinct from density, which depends on the number of particles in a given bulk. Thus lead is *dense*, but not *hard*.

Neither liquids nor aëriform fluids possess this property; and even in some solids, for instance butter and wax, it is almost entirely wanting.

Of two bodies, that is the harder which will scratch the surface of the other. By trying the experiment with different substances, it is found that the precious stones are harder than any other class of bodies, the diamond standing first, and the ruby, sapphire, topaz, and emerald following in order. Rhodium and iridium are among the hardest metals, on which account they are used for the tips of gold pens.

36. TENACITY.—Tenacity is that property by which a body resists a force that tends to pull it into pieces.

Both hardness and tenacity are the result of cohesion; but they must not be confounded. Of several rods equally thick, that which will support the greatest weight without breaking is the *most tenacious ;* that which it is most difficult to cut into, is the *hardest.*

The metals generally are remarkable for their tenacity. Some, however, possess this property in a higher degree than others. This may be shown by comparing the weights which different metallic wires of the same size are capable of supporting. An iron wire one-tenth of an inch in diameter will sustain nearly 550 pounds without breaking, while one of lead will be broken by a weight of 28 pounds.

---

33. What is Cohesion? What other name has been given to cohesion? What is said of cohesion in solids? In liquids? In aëriform fluids? 34. What is Adhesion? Describe the experiment with adhesion plates. Describe the experiment which proves that adhesion operates between solids and liquids. 35. What is Hardness? What is the difference between hardness and density? In what is hardness wanting? How may it be determined which of two bodies is the harder? What bodies are the hardest as a class? Mention the order in which they rank. What two metals are distinguished for their hardness? 36. What is Tenacity? Of what are both hardness and tenacity the result? Show the differ-

Iron is the most tenacious of the metals. A cable of this material, composed of wires one-thirtieth of an inch across, will support the enormous weight of 60 tons for each square inch in its transverse section. In consequence of this great tenacity, such cables are used for the support of suspension bridges.

37. *Tenacity of Different Substances.*—It is important in building and other arts to know the relative tenacity of different woods and metals. To determine this, experiments have been made. Their results do not precisely agree, inasmuch as there are differences in different trees of the same kind and different pieces of the same metal; yet we may take the following as the average weights that can be supported by the several materials mentioned,— taking in each case a rod of given length with a transverse section of a square inch.

| | POUNDS. | | POUNDS. |
|---|---|---|---|
| *Metals.*—Cast Steel, | 134,250 | *Woods.*—Ash, | 14,000 |
| Swedish Iron, | 72,000 | Teak, | 13,000 |
| English Iron, | 55,800 | Oak, | 12,000 |
| Cast Iron, | 19,000 | Fir, | 11,000 |
| Cast Copper, | 19,000 | Maple, | 8,000 |
| Cast Tin, | 4,700 | *Rope*, one inch around, | 1,000 |
| Cast Lead, | 1,825 | *Rope*, three inches around, | 5,600 |

It is a curious fact that a composition of two metals may be more tenacious than either of them separately. Thus brass, which is made of zinc and copper, has more tenacity than either of those metals.

38. The liquids have comparatively little tenacity, yet there is a difference in them in this respect. Milk, for instance, is more tenacious than water; this makes it boil over more readily, inasmuch as its bubbles do not break, but accumulate, climbing one upon another till they overtop the vessel. In like manner, it is on account of their superior tenacity that soap-suds will make a lather while pure water will not.

39. BRITTLENESS.—Brittleness is that property which renders a body capable of being easily broken.

ence between them. What is said of the tenacity of the metals? How may their relative tenacity be shown? Compare iron and lead in this respect. What is said of the tenacity of iron? 87. Explain the fact that experiments for determining the tenacity of different substances show different results. What does the table show? Of the metals mentioned in the table, which has the greatest tenacity? Which, the least? Of the woods mentioned, which is the most tenacious? Which, the least? What curious fact is mentioned respecting a composition of two metals? 88. What is said of the tenacity of liquids? How do milk and water compare in tenacity?

Brittleness is the opposite of tenacity, but often characterizes hard bodies. Glass, which is so hard that it will scratch the surface of polished steel, is remarkable for its brittleness.

A substance naturally tenacious may be so treated as to become brittle. Thus a bar of iron raised to a high degree of heat, if allowed to cool gradually, retains its tenacity, and bends rather than breaks; but, if suddenly cooled by being plunged into cold water, it is made brittle.

40. ELASTICITY.—Elasticity is that property by which a body, compressed, dilated, or bent by an external force, resumes its form when that force has ceased to act.

Stretch a piece of india rubber; when you let go the ends, they will fly back. Bend a bow; when the string is released, the bow will at once return to its former curve. These are familiar examples of elasticity.

41. The force with which a body resumes its form is called the Force of Restitution. Those bodies whose force of restitution brings them back, under all circumstances, exactly to their original form, are said to be *perfectly elastic.* The only perfectly elastic substances are the aëriform bodies. A body of air may be kept compressed for years; yet, on being freed from the compressing force, it will immediately expand to its former dimensions.

42. Many of the hard and dense solids are highly elastic; for example, steel, marble, and ivory. The soft solids generally, such as butter, putty, &c., have little or no elasticity; there are a few, however, that exhibit it, among which are india rubber and silk thread.

43. The elasticity of steel is increased by making it suddenly contract when expanded by heat. This is called *tempering,* and is effected by raising the steel to an intense heat, plunging it in cold water, and keeping it there for a certain time. The process is a nice one. At Damascus, in Syria, and Toledo, in Spain, it was long performed with peculiar skill, so that the sword-blades of those two cities were considered superior to all others. At the World's Fair in London, a Toledo sword was exhibited, of such exquisite temper that it could be bent into a circle, yet on being released sprung back and became as straight as ever.

44. A compound of two metals may possess a higher degree of elasticity

Soap-suds and water? 39. What is Brittleness? Of what is brittleness the opposite? What is said of glass? How may iron be made brittle? 40. What is Elasticity? Give some familiar examples. 41. What is meant by the Force of Restitution? When is a body said to be perfectly elastic? What are the only perfectly elastic substances? 42. What solids are for the most part elastic, and what not? 43. How is the elasticity of steel increased? What is this process called? Describe the process of tempering. Where was it long done with peculiar skill? Give an account of the Toledo blade exhibited at the World's Fair. 44. What is said of a compound of two

than either of them separately. Thus bell-metal is much more elastic than either the tin or the copper of which it is composed.

45. An elastic body, thrown against any hard substance, rebounds. An india rubber ball bounds back from a wall, to a distance proportioned to the force with which it is thrown. In such cases, the ball is flattened at the point of contact, but instantly resumes its former shape with such force as to drive the ball back.

To prove this, take two ivory balls (Fig. 12), smear one of them with printer's ink, and suspend them near each other by strings of equal length. Bring them gently in contact, and a few particles of ink will adhere to the surface of the clean ball: strike them violently together, and a larger spot of ink will be found there. This could not happen if the two balls were not flattened at the moment of striking.

Fig. 12.

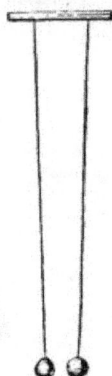

46. There is a limit to the elasticity of most bodies, beyond which, if compressed, dilated, or bent, they will fail to regain their original form. An iron wire, if slightly bent, springs back, so that no change of form can be detected; but not so, if violently bent. A continued application of the compressing, dilating, or bending force, has the same effect. A bow, if kept bent for a long time, will lose its elasticity. For this reason, an archer, before putting his bow away, is careful to un-string it.

47. The liquids have but little elasticity. They are therefore called Non-elastic Fluids; while aëriform bodies, which possess this property in a higher degree than any others, are known as Elastic Fluids.

48. MALLEABILITY.—Malleability is that property which renders a body capable of being rolled out or hammered into sheets.

From a piece of copper, a workman with no other instrument than his hammer will make a hollow vessel without joint or seam, the malleability of the metal preventing it from giving way under his blows. Dough, which can be made into very thin sheets under the rolling-pin, affords a familiar illustration of malleability.

Malleability belongs chiefly to the metals, yet in some of them, such as antimony and bismuth, it is wanting. It is strikingly exhibited in silver,

metals? Give an example. 45. What does an elastic body do, when thrown against a hard substance? In such cases, what takes place? Prove this by an experiment. 46. What is said of the limit of elasticity? Give examples. 47. What names have been given to liquids and aëriform bodies? Why? 48. What is Malleability? Give

2

platinum, iron, and copper, but most of all in gold. A cubic inch of this metal may be beaten out till it covers 282,000 square inches, which makes the leaf only $\frac{1}{282000}$ of an inch thick. In other words, it would take 282,000 strips of such gold leaf, lying on each other, to make the thickness of an inch.

49. DUCTILITY.—Ductility is that property which renders a body capable of being drawn out into wire.

The malleable metals are for the most part ductile, but not always in the same degree. Thus gold exceeds all the other metals in ductility as well as in malleability; but tin, which can readily be beaten into very thin sheets, can not be drawn out into small wire.

Gold wire has been made so attenuated that fifty miles of it would weigh but an ounce. Platinum, which is nearly as ductile as gold, has been drawn into wire only $\frac{1}{30000}$ of an inch in diameter and invisible to the naked eye. Glass, when softened by fire, becomes exceedingly ductile, and may be spun out into flexible and elastic threads scarcely larger than the thread of the silk-worm.

---

# CHAPTER III.

## MECHANICS.

50. MECHANICS is that branch of Natural Philosophy which treats of forces and their application in machines.

51. FORCE AND RESISTANCE.—When we see a body begin to move, cease to move, or change its motion, since it can do neither of itself, we know that it has been acted on by some external agency, which we call a Force. The elasticity of a bow which sends an arrow through the air, is a force; the wind, which changes its direction, is a force; gravity, which brings it to the earth and helps to stop its motion, is a force.

---

examples. To what does malleability chiefly belong? Show the extreme malleability of gold. 49. What is Ductility? What substances are for the most part ductile? What is the most ductile substance known? What facts are stated, illustrating the ductility of gold, platinum, and glass?

50. What is Mechanics? 51. What is a Force? Give illustrations. What is the

That which opposes a force is called the Resistance. In the above example, the inertia of the arrow is the resistance.

Forces may act on bodies so as to produce either Motion or Rest.

## Motion.

52. Motion is a change of place.

53. Motion is either Absolute or Relative.

Absolute Motion is a change of place with reference to a fixed point. Relative Motion is a change of place with reference to a point that is itself moving.

Two balls are rolled on the floor. The motion of each, as regards the point from which it was thrown, is absolute; their motion with reference to each other is relative.

54. REST.—Rest is the opposite of motion, and implies continuance in the same place.

Like motion, Rest is either Absolute or Relative. A man sitting on a steamer that is moving forward five feet in a second, is at rest *relatively* to the other objects on board. To be at rest *absolutely*, he must walk five feet every second towards the stern of the boat.

Strictly speaking, there is no such thing as absolute rest in any of the objects that surround us; for the earth moves round the sun at the rate of nearly 99,000 feet in a second, and carries with it every thing on its surface. Hills, trees, and houses, therefore, though they occupy the same place with respect to each other, are really travelling through space with immense rapidity. Yet as this is the case with ourselves, with the atmosphere, and all things about us, we regard an object as absolutely at rest if it has no other motion than this.

55. VELOCITY.—The Velocity of a body is the rate at which it moves.

This rate is determined by the space it passes over in a given time. The greater the space, the greater the velocity. Thus, if A walks two miles an hour, and B four, B's velocity is twice as great as A's.

Resistance? What may the action of forces on bodies produce? 52. What is Motion? 53. How is motion distinguished? What is Absolute Motion? What is Relative Motion? Illustrate these definitions. 54. What is Rest? Illustrate Absolute and Relative Rest. Show that there is really no such thing as absolute rest. 55. What is

56. The relation between the space passed over, the time employed, and the velocity, is such, that when two are given, we can find the third.

*Rule* 1.—To find the velocity of a body, divide the space passed over by the time.

*Example.* A locomotive goes 120 miles in 4 hours; what is its velocity?—Dividing 120 by 4, we get 30; *answer*, 30 miles an hour.

*Rule* 2.—To find the time, divide the space by the velocity.

*Example.* A locomotive goes 120 miles at the rate of 30 miles an hour; how long is it on the way?—Dividing 120 by 30, we get 4; *answer*, 4 hours.

*Rule* 3.—To find the space, multiply the velocity by the time.

*Example.* A locomotive goes 4 hours at the rate of 30 miles an hour; how far does it travel?—Multiplying 30 by 4, we get 120; *answer*, 120 miles.

57. *Table of Velocities.*—It may not be uninteresting to compare the average velocities of the following moving objects:—

| | Miles per hour. | | Miles per hour. |
|---|---|---|---|
| A man walking | 3 | A hurricane | 80 |
| A horse trotting | 7 | Sound | 764 |
| A slow river | 3 | A musket-ball, when first | |
| A rapid river | 7 | discharged | 850 |
| A fast sailing vessel | 10 | A rifle-ball | 1,000 |
| A fast steamboat | 18 | A 24-lb. cannon-ball | 1,600 |
| A railroad train | 25 | Earth in its orbit | 68,040 |
| A moderate wind | 7 | Light | 691,200,000 |
| A storm | 50 | Electric Fluid | 1,036,800,000 |

58. KINDS OF MOTION.—There are three kinds of motion; Uniform, Accelerated, and Retarded.

59. Uniform Motion is that of a body which moves over equal spaces in equal times.

Uniform motion would be produced by a force acting once and then

Velocity? How is it determined? 56. What is said of the relation between the space, the time, and the velocity? Give the rule for the velocity, and example. Give the rule for the time, and example. Give the rule for the space, and example. 57. What is the velocity of a slow river? A rapid river? A moderate wind? A hurricane? Sound? Light? The electric fluid? A rifle-ball? The earth in its orbit? 58. Name the three kinds of motion. 59. What is Uniform Mo-

ceasing to act, if the moving body were free from all other influences, for its inertia would keep it moving at the same rate. Gravity and the resistance of the air, however, constantly retard a moving body; and, therefore, to keep up a uniform motion, a force just sufficient to nullify these retarding agencies must continue acting. There are very few cases of uniform motion either in nature or art.

60. Accelerated Motion is that of a body whose velocity keeps increasing as it moves. It is produced by the continued action of a force.

A ball dropped from a height is a familiar instance of accelerated motion. The moment it is let go, the attraction of gravitation causes it to descend. Were this force and every other then suspended, the ball would fall to the earth with a uniform motion; but gravity, continuing to act, forces it along faster and faster, and thus imparts to it an accelerated motion.

A body is said to have a Uniformly Accelerated Motion, when its velocity keeps increasing at the same rate; when, for instance, it moves two feet in the first second, four in the next, eight in the third, &c.

61. Retarded Motion is that of a body whose velocity keeps diminishing as it moves. It is produced by the continued action of some resistance on a moving body.

A ball rolled over the ground, under the continued action of gravity and the resistance of the air, moves more and more slowly, till finally it comes to rest. This is an example of retarded motion.

A body is said to have a Uniformly Retarded Motion, when its velocity keeps diminishing at the same rate; when, for instance, it moves eight feet in the first second, four in the next, and two in the third.

### Momentum.

62. The Momentum (plural, *momenta*) of a body is its quantity of motion.

A ten-pound ball, moving at the rate of 400 feet in a second, may be supposed to be divided into ten pieces, each weighing one pound. Each piece has a motion of 400 feet in a second; and the quantity of motion, or momen-

---

tion? Theoretically, how is uniform motion produced? Practically, how is it produced? 60. What is Accelerated Motion? How is it produced? Give an example of accelerated motion. When is a body said to have a Uniformly Accelerated Motion? 61. What is Retarded Motion? How is it produced? Give an example. When is a body said to have a Uniformly Retarded Motion? 62. What is Momentum? Give

tum, of all ten, that is, of the whole ball, will be ten times 400, or 4,000. Hence the following rule:—

63. *Rule.*—To find the momentum of a moving body, multiply its velocity by its weight.

*Example.* What is the momentum of a ten-pound ball, moving at the rate of 400 feet in a second?—Multiplying 400 by 10, we get 4,000; *answer*, 4,000.

64. When the momenta of different objects are to be compared, their weight and velocity must be expressed in units of the same denomination: if the weight of one is given in pounds, that of the other must be in pounds; if the velocity of one is so many feet per second, that of the other must be expressed in feet per second. If different denominations are given, reduce them to the same denomination.

Thus: A weighs 50 pounds, and has a velocity of 7,200 miles an hour; B weighs 100 pounds, and has a velocity of 4 miles a second. Which has the greater momentum?

3,600 seconds make an hour; and if A's velocity is 7,200 miles an hour, in a second it will be $\frac{1}{3600}$ of 7,200 miles, or 2 miles.

A's weight 50 multiplied by A's velocity 2 gives A's momentum 100.

B's weight 100 multiplied by B's velocity 4 gives B's momentum 400. Therefore B's momentum is 4 times as great as A's.

65. Two bodies of the same weight have momenta proportioned to their velocities. Thus, if two balls weighing 5 pounds each, move respectively at the rate of 20 and 10 miles an hour, then their momenta will be in the proportion of 20 to 10, or two to one.

Two bodies moving with the same velocity, have momenta proportioned to their weight. Thus, if two balls moving at the rate of 5 miles an hour, weigh 20 and 10 pounds respectively, then their momenta will be in the proportion of 20 to 10, or two to one.

66. Since momentum depends on velocity as well as weight, it is obvious that, by increasing its velocity sufficiently, a small body may be made to have a greater momentum than a large one. Thus, a bullet fired from a gun has a greater momentum than a stone many times larger thrown from the hand.

On the same principle, a very heavy body, though its motion may be hardly perceptible, may have an immense momentum. This is the case with icebergs, rendering them fatal to objects with which they come in collision.

---

an example. 63. Repeat the rule for finding a body's momentum. Give an example. 64. When the momenta of different objects are to be compared, what is essential? Give an example. 65. When two bodies have the same weight, to what are their momenta proportioned? Give an example. When two bodies have the same velocity, to what are their momenta proportioned? Give an example. 66. How may a greater momentum be given to a small body than a large one? Illustrate this. How do you account for the great momenta of icebergs, notwithstanding their slow mo-

## Striking Force.

67. The Striking or Living Force of a moving body is the force with which it strikes a resisting substance.

Striking Force is sometimes confounded with momentum, but improperly, inasmuch as it is the product of the weight into *the square of the velocity*. Two moving bodies may have the same momentum, but differ greatly in their striking force.

Thus, the ball A, weighing 200 pounds and moving 2 miles a minute, has a momentum of 200 multiplied by 2, or 400. The ball B, weighing 20 pounds and moving 20 miles a minute, also has a momentum of 400 (20 multiplied by 20). How do they compare in striking force? That of A is equal to its weight 200 multiplied by the square of its velocity, 4,—or 800. That of B is equal to its weight 20 multiplied by the square of its velocity 400,—or 8,000. Therefore, though the momenta of the two balls are equal, the striking force of B is 10 times as great as that of A; if both were fired into a bank of moist clay, B would penetrate ten times as far as A.

68. As the velocity of a body increases, its striking force increases also, but in a higher degree.

If, for instance, a train of cars be moving 50 miles an hour, and another train of the same weight 10 miles an hour, the striking force of the former will not be to that of the latter as 50 to 10, but as the square of 50 is to the square of 10, or as 2500 is to 100. The former train would therefore do 25 times as much damage as the latter to any object with which it came in collision, or to itself in case of being thrown from the track. This result is borne out by facts.

69. *Rule.*—To find the striking force of a moving body, multiply its weight into the square of its velocity.

If the striking force of one body is to be compared with that of another, see that their weight and velocity are in units of the same denomination.

*Example.* The stone A, weighing 1 pound, is thrown at the rate of 20 ft.

---

tion? 67. What is meant by the Striking or Living Force of a moving body? What is the difference between a body's striking force and its momentum? Exemplify this difference. 68. How does a body's striking force increase, compared with its velocity? Give an example. How is this result borne out? 69. Give the rule for finding the striking force of a moving body. When bodies are to be compared with respect to their striking force, how must their weight and velocity be expressed? Solve the example under the rule.

a second. The stone B, weighing 3 pounds, is thrown at the rate of 2,400 ft.
a minute. Which will penetrate further into a snow-bank?

20 times 20 is 400 = square of A's velocity.

400 × 1 (A's weight) = 400, A's striking force.

Reduce B's velocity to the same denomination as A's. If B move 2,400
feet in a minute, in a second it will move $\frac{1}{60}$ of 2,400 feet, or 40 feet.

40 times 40 is 1,600 = square of B's velocity.

1,600 × 3 (B's weight) = 4,800, B's striking force.

*Ans.*—A's striking force being 400, and B's 4,800, B will penetrate into
the snow-bank 12 times as far as A.


### EXAMPLES FOR PRACTICE.

1. (*See Rule* 1, § 56.)　A fox-hound will run 30 miles in three hours. What
   is its velocity?
2. At the battle of Brandywine, Gen. Greene's detachment marched 4 miles
   in 42 minutes, to relieve Gen. Sullivan. With what velocity did they
   move?
3. At the most flourishing period of its history, ancient Athens was 25 miles
   in circumference. With what velocity would an Athenian have had to
   move, in order to walk round the city in 5 hours?
4. A pigeon will fly 100 miles in 2 hours. What is its velocity?
5. P walks 2 miles in 30 minutes; Q walks 4 miles in 2 hours. Which has
   the greater velocity?

   REMARK.—*When different denominations are used, they must be reduced to
   the same denomination, as shown in* § 64.

6. The current of a rapid river runs 1,200 feet in 2 minutes; a horse at a moderate trot passes over 30 feet in 3 seconds. Which moves with the
   greater velocity?
7. (*See Rule* 2, § 56.)　Strabo tells us that ancient Nineveh was 47 miles in
   circumference; in what time could a person have walked around it, at
   the rate of 10 miles a day?
8. The bombardment of Ostend, on the coast of Holland, was heard in London, a distance of 70 miles. There are 5,280 feet in a mile, and sound
   travels at the rate of 1,120 feet in a second. How many seconds after a
   cannon was fired at Ostend, was the report heard in London?
9. From the base of the Pyramid of Cheops to its top is 704 feet; how long
   will it take a person to ascend it, walking at the rate of 4 feet per second?
10. A rifle-ball moves at the rate of 1,000 miles an hour. If it could maintain
    the same speed, how long would it be in crossing the Atlantic Ocean,
    which is 3,000 miles broad?
11. Light moves 192,000 miles in a second, electricity 288,000 miles in the
    same time. How long before we could see a flash of lightning in a cloud
    2 miles off, and how long before the lightning could strike an object by
    our side?
12. In the year 1804, the French philosopher Gay Lussac ascended in a bal-

loon to the height of 4½ miles. He came down at the rate of 660 feet in a minute; how long was he in making the descent?

13. (*See Rule* 3, § 56.) Some of the Alpine glaciers move 25 feet annually. How far would they move in 4 years?

14. The comet observed by Newton in 1680 moved 880,000 miles an hour. How far at this rate would it move in a day?

15. Which will pass over the greater space—a hurricane, moving at the rate of 80 miles an hour in 4 hours, or a locomotive, going 30 miles an hour, in 10 hours?

16. If the earth moves in its orbit 68,040 miles an hour, and is 365 days, 6 hours, in completing its revolution, how long is its orbit?

17. If a ray of light travels 691,200,000 miles in an hour, how far will it go in a day?

18. (*See Rules*, §§ 63, 69.) A 24-pound cannon-ball moves at the rate of 1,000 miles an hour. A battering-ram weighing 10,000 pounds moves at the rate of 10 miles an hour. How do their momenta compare?—*Ans. As 24 to* 100; *that is, the cannon-ball has a little less than one-fourth of the momentum of the battering-ram.*

How does the striking force of the above cannon-ball compare with that of the battering-ram; that is, what would be their comparative effect on the wall of a fortress?—*Ans. That of the ball would be* 24 *times as great as that of the battering-ram.*

19. An iceberg weighing 50,000 tons moves at the rate of 2 miles an hour. An avalanche of 10,000 tons of snow descends with a velocity of 10 miles an hour. How do their momenta compare?

How do they compare in striking force?

20. How does the momentum of a 32-pound ball with a velocity of 2,000 miles an hour, compare with that of a 16-pound ball with a velocity of 1,000 miles an hour?

Which would penetrate further into a bank of moist clay?

21. A locomotive weighing 20 tons moves with a velocity of 40 feet a second. Another locomotive weighing 25 tons moves at the rate of 4,800 feet in a minute. How do their velocities compare?

How do they compare in momentum?

If the one with the less striking force penetrate 10 feet into a snow-bank, how far will the other penetrate?

22. A stone weighing 15 ounces is thrown from the hand with a velocity of 1,320 feet in a minute. A rifle-ball weighing 3 ounces is discharged at the rate of 15 miles a minute. How do their velocities compare?

How do they compare in momentum?

How many times greater is the striking force of the rifle-ball than that of the stone?

2*

# CHAPTER IV.

## MECHANICS (CONTINUED).

### LAWS OF MOTION.

### Mathematical Definitions.

70. BEFORE treating of the laws of motion, it is necessary to define the mathematical terms used in connection with them.

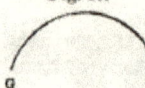

Fig. 13.

A——————B

1. A Right or Straight Line is one that has the same direction throughout its whole extent; as, A B.

Fig. 14.

C————D
E————F

2. Parallel Lines are those which have the same direction; as, C D, E F.

Fig. 15.

3. A Curve Line, or Curve, is one that changes its direction at every point; as, G H.

G        H

4. A Circle is a figure bounded by a curve, every point of which is equally distant from a point within, called the Centre. Fig. 16 represents a circle, and E its centre.

Fig. 16.

5. The Circumference of a circle is the curve that bounds it; as, A C F B D.

6. Any part of the circumference is called an Arc; as, A C, C F.

7. A Diameter of a circle is a straight line drawn through the centre, terminating at both ends in the circumference; as, A B. Every circle has an infinite number of diameters, all equal to each other.

8. A Radius (plural, *radii*) of a circle is a straight line drawn from the centre to the circumference; as, E D, E C, E F, E A, E B. Every circle has an infinite number of radii, all equal to each other. The radius of a circle is just half its diameter.

9. A Tangent of a circle is a straight line that touches the circumference

in a single point, without cutting it at either end when produced; as, A B, C D.

10. The circumference of every circle is divided into 360 equal parts, called Degrees. One fourth of the circumference contains 90 degrees, and is called a Quadrant.

11. An Angle is the difference in direction of two straight lines that meet or cross each other.

12. The Vertex (plural, *vertices*) of an angle is the point at which its sides meet; as, D in Fig. 18.

An angle is named from the letter at its vertex, if but one angle is formed there. Otherwise, it is named from the letters on each side and at the vertex, that at the vertex being placed in the middle. Thus the angle in Fig. 18 is called D; if more than one angle were formed there, it would be distinguished as C D B or B D C.

The size of an angle does not depend on the length of its sides, but simply on their difference of direction. We may extend the lines D C, D B, as far as we choose, without making the angle D any larger.

13. When a straight line meets another straight line in such a way as to make the two adjacent angles equal, that is, so as to incline no more to one side than the other, it is said to be Perpendicular to the latter; and the angle which it makes on either side is called a Right Angle. Thus, F E B and F E A (being equal) are Right Angles, and the line F E is Perpendicular to the line A B.

— A right angle, it will be seen, is measured by one fourth of the circumference of a circle, or 90 degrees.

14. An Obtuse Angle is one that is greater than a right angle; as, F E D in Fig. 19.

15. An Acute Angle is one that is less than a right angle; as, F E C in Fig. 19.

16. A Triangle is a figure bounded by three straight lines; as, A B C, Fig. 20.

17. A Quadrilateral is a figure bounded by four straight lines; as, A B C D, Fig. 21.

18. A Diagonal of a quadrilateral is a straight line which joins the vertices of two opposite angles; as, A C, D B, in Fig. 21.

19. A Parallelogram is a quadrilateral whose opposite sides are parallel; as, A B C D, Fig. 21.

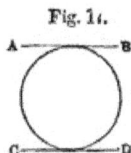

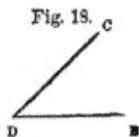

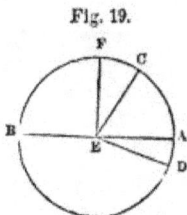

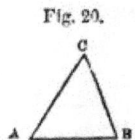

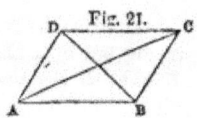

Fig. 16.

Fig. 18.

Fig. 19.

Fig. 20.

Fig. 21.

diameter? What is a Tangent of a circle? How is the circumference of every circle divided? What is a Quadrant? What is an Angle? What is the Vertex of an angle? How is an angle named? On what alone does the size of an angle depend? When is one line said to be Perpendicular to another? What is a Right Angle? By what is a right angle measured? What is an Obtuse Angle? What is an Acute An-

Fig. 22.          20. A Rectangle is a quadrilateral whose angles are
all right angles; as, E F G H, Fig. 22.

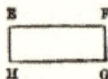

21. A Square is a rectangle whose sides are equal;
as, I J K L, Fig. 23.

Fig. 23.

22. A Sphere is a solid bounded by a curved surface,
all the points of which are equally distant from a point within called
the centre; as, A B C D, Fig. 24.

Fig. 24.

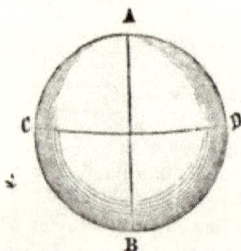

23. The Axis of a sphere is a straight line
passing through its centre and terminating in its
surface, round which it revolves; as, the straight
line connecting A and B, in Fig. 24.

24. The Poles of a sphere are the extremities
of its axis; as, the points A, B, in Fig. 24.

25. The Equator of a sphere is a great circle
which we imagine to be drawn round it on its
surface, midway between the poles; as, the circle C D, in Fig. 24.

26. An Oblate Spheroid is a figure which differs from a sphere only in being flattened at its
poles, like an orange.

27. A Prolate Spheroid is a figure which differs from a sphere only in being lengthened out at its poles, like a lemon.

28. A Cylinder is a circular body of uniform diameter, the ends of which
form equal and parallel circles. A lead-pencil, before it is sharpened, is a
cylinder; a stove-pipe is a hollow cylinder.

71. By investigating the principles of motion, Newton
arrived at three great laws, which have ever since been
received.

## First Law of Motion.

72. *A body at rest remains at rest, a body in motion
moves in a straight line with uniform velocity, unless acted
on by some external force.*

This law follows from inertia. No body has power of itself to move, to
cease moving, or to change its direction or velocity.

73. The air is a powerful agent in stopping motion. This is shown by
causing a wheel to revolve on a pivot, first in the air, and then under a glass

---

gle? What is a Triangle? What is a Quadrilateral? What is a Diagonal of a quadrilateral? What is a Parallelogram? What is a Rectangle? What is a Square?
What is a Sphere? What is the Axis of a sphere? What are the Poles of a sphere?
What is the Equator of a sphere? What is an Oblate Spheroid? What is a Prolate
Spheroid? What is a Cylinder? 71. How many laws of motion did Newton arrive
at? 72. What is the First Law of Motion? From what does this law follow?
73. How may it be shown that the air is a powerful agent in stopping motion?

receiver from which the air has been removed with an air-pump. In the former case, the wheel soon ceases to move; in the latter, it retains its motion for a long time. A pendulum (see § 138) will vibrate nearly a day in an exhausted receiver.

74. Friction is the resistance with which a body meets from the surface on which it moves. The rougher the surfaces brought in contact, the greater the friction, and the sooner the moving body will come to rest. A ball rolled over a stony road is soon stopped by the obstacles it encounters; on a level pavement it goes much farther, and farther still on a smooth sheet of ice. This is because the friction becomes less in proportion as the surface on which the ball rolls becomes smoother.

75. According to this law, every body left free to obey the force that set it in motion will move in a straight line. We observe few such motions in nature. The planets in their orbits, rivers in their channels, rolling waves, and ascending smoke, all move in curves, in consequence of their being acted on by other forces, besides those that set them in motion. The tendency of the moving body, however, is always to continue in a straight line, even when from overruling causes it moves in a circle.

Attach a ball, for instance, to a cord; and, fastening the end of the cord at a point, O, give a quick impulse to the ball. It will be found to move in a circle, A B C D, because the cord keeps it within a certain distance of the centre. Were it not for this, it would move in a straight line. Thus, let the cord be cut when the ball is at A, and it will be found to move to E in a tangent to the circle A B C D. In like manner, at B it will fly off in a tangent to F, and so at C, D, or any other point.

Fig. 25.

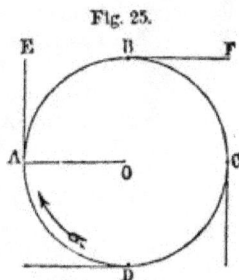

76. THE CENTRIFUGAL FORCE.—The force which tends to make a body fly from the centre round which it revolves, is called the Cen-trif'-u-gal Force.

The opposite force, which draws a body towards the centre round which it revolves, is called the Cen-trip'-e-tal Force.

Magnificent examples of these two forces are exhibited

74. What is Friction? On what kind of surfaces does a moving body encounter the most friction? Exemplify this. 75. What is said of the motions that we find in nature? Give some instances. What is the tendency of the moving body? Illustrate this with a ball and cord. 76. What is the Centrifugal Force? What is the Centrip-

by the planets revolving round the sun in space. At each successive point of their orbits, in obedience to the Centrifugal Force, they tend to fly off in tangents, disturbing the harmony of the universe and carrying desolation in their path. They are constantly restrained, however, by a Centripetal Force equally powerful, the sun's attraction; and the result is that they revolve in curves.

Fig. 26.

77. *Familiar Examples.*—Whirl a wet mop rapidly round, and drops of water, propelled by this force, will fly off from it in straight lines.

Suspend a glass vessel containing some colored water, by a cord passed round the rim, as shown in Fig. 26. Turn the vessel round till the cords become tightly twisted, and then suddenly let it go. It will rapidly revolve, and the centrifugal force will give the water an impulse away from the centre. As it can not escape, it will spread up the sides. Should there be water enough, it will rise above the top of the vessel, and fly off in straight lines.

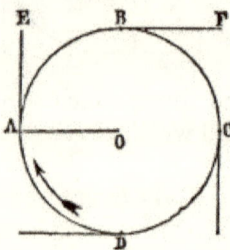

We take advantage of the centrifugal force in discharging a stone from a sling. The stone is whirled quickly round the hand as a centre, which it is prevented from leaving by two strings connected with the strap on which it rests. The instant one of the strings is let go, the centrifugal force carries off the stone in a tangent to the circle it was describing. Its direction varies according to the point at which the string is let go, as will appear from Fig. 27. Great velocity may be communicated to the stone with this simple apparatus. In the hands of the Persians, the Rhodians, and other ancient nations, the sling was a formidable weapon.

Fig. 27.

When a wagon turns a corner rapidly, it is liable to be upset in consequence of the centrifugal force. A person sitting in it feels his body sway outward, and one who is on his feet has to grasp the wagon to avoid being thrown from his place. To counteract the effects of the centrifugal force in curves on railroads, the outer rail is laid higher than the inner one, as represented in Fig. 28. Were it not for this precau-

etal Force? What examples of these two forces does nature furnish us? 77. How may a mop be made to illustrate the centrifugal force? How does the apparatus represented in Fig. 26 illustrate the Centrifugal Force? Describe the mode in which a stone is discharged from a sling, and explain the principle. What is the effect of the centrifugal force, when a wagon turns a corner rapidly? How is this effect counter-

tion, trains moving swiftly round a curve would often be thrown from the track.

Instinct teaches a horse running rapidly round a small circle, to incline his body inward, that he may counteract the centrifugal force. For the same reason, a circus-rider, going swiftly round the ring, has to lean towards the centre.

Jugglers take advantage of the centrifugal force to astonish their audiences with a striking experiment, represented in Fig. 29. A B is a wheel with a broad rim, or felly. A wine-glass partly filled with water is placed on the inner surface of the felly, and the wheel is then made to revolve rapidly round the axle O. If the proper amount of motion be communicated to the wheel, not only will the wine-glass keep its place on the felly, but the water also will remain in it, not a drop being spilled, even when the glass is at W. Gravity, which, if the wheel were stationary, would at once cause both glass and water to fall, is completely nullified by the centrifugal force.

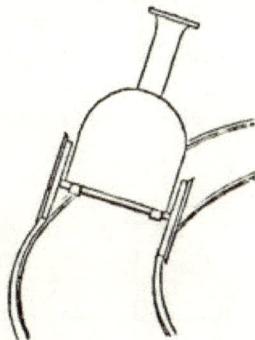

Fig. 28.

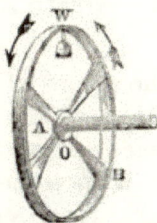

Fig. 29.

78. *Law of the Centrifugal Force.*—The centrifugal force of a revolving body increases according to the square of its velocity. If, therefore, the earth revolved round the sun twice as fast as it now does, its centrifugal force would be 4 times as great; if 3 times as fast, 9 times as great; if 4 times as fast, 16 times as great, &c.

This explains why a cord with which a stone is whirled round, as in a sling, is more apt to break under a rapid motion than a slow one. Every time the velocity is doubled, the strain on the cord is increased fourfold.

79. *Effect of the Centrifugal Force on Revolving Bodies.*—The centrifugal force acts, not only on bodies moving in curves, but also on fixed bodies revolving on their own axes.

When large wheels are turned rapidly by machinery, the centrifugal force at the circumference becomes an agent

acted in railroads? How does instinct teach a horse to counteract the centrifugal force? Describe the juggler's trick performed with the aid of the centrifugal force. 78. What is the law of the centrifugal force? When is the cord of a sling most apt to break, and why? 79. On what, besides bodies moving in curves, does the centrifugal

of tremendous power. Unless such wheels are made of very strong materials, their cohesion will be overcome by the centrifugal force, and they will fly into fragments. Ponderous grindstones sometimes burst, with the most disastrous effects, when too great a velocity is imparted to them.

Fig. 30.

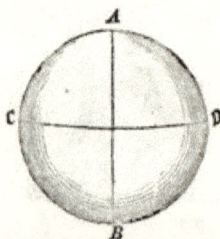

Fig. 30 represents a sphere revolving on its axis. All parts of the surface have to complete their revolution in exactly the same time; therefore, as the parts lying on the equator CD are further from the axis, and have a greater distance to go, they must travel faster than the rest. Now we have seen that the centrifugal force increases with the square of the velocity; and, therefore, at the equator CD it will be stronger than at any other part of the surface.

Hence the general law:—On a revolving sphere, the centrifugal force is greatest at the equator, and diminishes from that point till at the poles it wholly disappears.

Fig. 31.

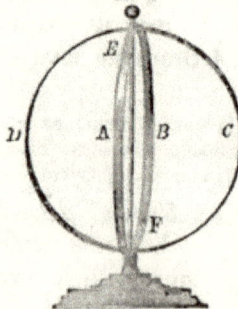

80. This difference of intensity in the centrifugal force at different points is shown when a sphere of moist clay is made to revolve rapidly, as on a potter's wheel. The tendency of particles on and near the equator to fly off is so great that in those parts the sphere bulges out, becoming proportionately flattened at the poles.

A similar result is produced in the apparatus represented in Fig. 31. Two thin and flexible metal hoops are fixed, at right angles to each other, on the axis E F,—fastened at the end F, but loose at E, so as to admit of their moving freely up and down the rod E F. A rapid rotary motion being communicated to the hoops, they will assume an oval form, bulging out more and more as their velocity is increased. When allowed to come to rest, they will rise to their original position at E.

force act? What is sometimes its effect on large wheels moved by machinery? What is the law of the centrifugal force in the case of revolving spheres? Explain the reason of this. 80. What is the effect of the centrifugal force on a sphere of moist clay made to revolve rapidly? Describe the experiment with the apparatus represented

81. The centrifugal force, acting as just described, is supposed to have given the earth its present form. The matter of which our planet is composed seems at one time to have been soft, and under a rapid rotary motion, before becoming solid, it swelled out at the equator and became depressed at the poles. The earth thus became an oblate spheroid, the distance from pole to pole being about 26 miles less than the equatorial diameter.

## Second Law of Motion.

82. *A given force always produces the same effect, whether the body on which it acts is in motion or at rest; whether it is acted on by that force alone or by others at the same time.*

The earth, as it turns on its axis, carries all things on its surface with great velocity from west to east; yet a force acting on any object on the surface causes it to move in the same direction, and with the same rapidity, as if the earth were at rest.

Let a stone be dropped from the mast-head of a vessel, and it will fall at the bottom of the mast, whether the vessel moves or is at rest.

A person sitting in a wagon throws up an orange and catches it in his hand, whether the wagon is moving or not.

Fig. 32.

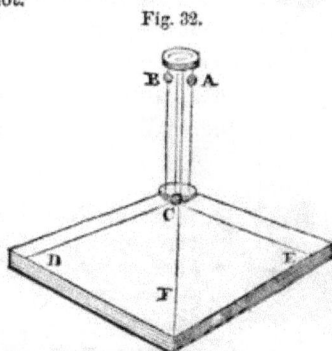

83. SIMPLE MOTION.—Motion produced by a single force is called Simple Motion.

84. RESULTANT MOTION.—Motion produced by the joint action of more than one force is called Resultant Motion.

Resultant motion is illustrated with the apparatus represented in Fig. 32. The ball C is placed on a square frame between two upright wires, on each of which a ball slides so as to strike C when it descends. Let the ball A drop, and it will drive C to D; this is an example of simple motion. Let the ball B drop, and it will drive C to E; this, also, is simple motion. Let A and B

in Fig. 31. 81. What is supposed to have been the effect of the centrifugal force on the form of the earth? How does the equatorial diameter of the earth compare with the distance from pole to pole? 82. What is the Second Law of Motion? Give some familiar illustrations of this law. 83. What is Simple Motion? 84. What is Resultant Motion? Describe the apparatus with which resultant motion is illustrated.

drop at the same instant, and they will drive C to F; this is resultant motion.

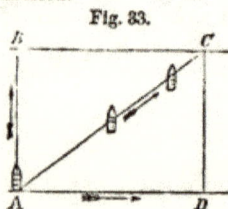

Fig. 33.

85. We have an example of resultant motion in a boat (see Fig. 33) which a person attempts to row north across a river, while the tide carries it to the east. Each force produces the same effect as if it acted alone; and the boatman, when he has crossed the river, will find himself neither due north nor due east of the point from which he started, but north-east of it.

If, in addition to the boatman's efforts and the tide, the wind should blow, this also will produce its full effect; and the boat will exhibit a resultant motion produced by the joint action of the three forces.

86. THE PARALLELOGRAM OF MOTION.—If Figures 32 and 33 be examined, it will be seen that a body acted on by two forces moves in a diagonal direction, between the lines in which they would separately propel it.

In Fig. 33, the boatman, starting at A, would row his boat to B; the tide in the same time would carry it to D. When both act, to get the direction of the boat and the point it would reach, we must draw the other sides of the parallelogram, B C, D C; the diagonal A C will then show the course of the boat, and its extremity C the point it would reach.

87. If the two forces are equal, the body will move in the diagonal of a square, that is, directly between the lines in which they would carry it. If one is greater than the other, the parallelogram must be constructed accordingly.

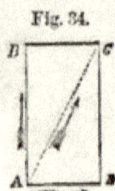

Fig. 34.

Let, for instance, the force used by the boatman be twice as great as that of the tide. Then by the time he would reach B, the tide would have carried his boat one-half of that distance, to D. Completing the parallelogram, as in Fig. 34, and drawing the diagonal A C, we find that under the joint action of these forces the boat would reach C.

## Third Law of Motion.

88. *Action* is the force which one body exerts on another subjected to its operation.

---

*Reaction* is the counter-force which the body acted upon exerts on the body acting.

The third Law of Motion is as follows :—*Reaction is always equal to Action, and opposite to it in direction.*

89. *Examples of Action and Reaction.*—We strike an egg against a table; the table reacts on the egg with the same force and in the contrary direction, breaking its shell. We push a wagon forward, and feel the reaction in the resistance it offers. A bird, when flying, strikes the air downward blows with its wings; the air reacts upward and supports the bird. A rower pulls his oar against the water; the water reacts and drives the boat in the opposite direction. A boy fires a gun; the exploding powder carries forward the ball, but the air thus struck reacts on the gun and causes it to recoil against the boy's shoulder. Two boats of equal weight, A and B, are connected with a rope : a man in A pulls the rope; action and reaction being equal, not only will the boat B move towards him, but the boat A, which he is in, will move with the same velocity towards B.

90. It is reaction that kills a person who falls from a height on a hard pavement. Another, falling the same distance, lights on a feather bed, and receives little or no injury; not because there is *less* reaction, but because the reaction is *more gradual*, and therefore his body does not receive so great a shock. On the same principle, if a steamboat in making her landing is likely to strike violently against the dock, the force of the collision is deadened and the boat saved from damage by interposing a coil of rope, or some other substance softer than wood.

Hence also a bullet, which would penetrate a board, will not go through a soft cushion, its motion being gradually and not instantaneously opposed by the reaction of the cushion. A person may catch a very heavy stone without being hurt, if he allows his hand, the instant he catches it, to be carried in the direction in which the stone was moving, and thus makes the reaction gradual.

91. Reaction often nullifies action. This was the case with the man who tried to raise himself over a fence by pulling at the straps of his boots. Tug as he might, he found that all the upward impulse he could give himself was counterbalanced by an equally strong downward impulse, and that his utmost efforts could not reverse the law

What is the Third Law of Motion? 89. Give some familiar illustrations of the third law of motion. 90. What is the effect of reaction on a person falling from a height on a hard pavement? What is the effect, if the person falling lights on a feather bed? What causes the difference? Give another instance of gradual reaction. How may a person catch a very heavy stone without being hurt? 91. What is often the effect

Fig. 35.

of nature — that action and reaction are equal in force and opposite in direction.

We read of another man no less ingenious, who rigged a huge bellows in the stern of his sail-boat, that he might always be able to make a fair wind. On trying the experiment, he found that with all his blowing he could not move the boat an inch; for the reaction of the air on the bellows kept her back as much as its action on the sail tended to move her forward.

92. ACTION AND REACTION IN NON-ELASTIC AND ELASTIC BODIES.—Action and reaction are always equal, but they are exhibited differently in non-elastic and elastic bodies. This difference is shown with suspended balls of soft clay and ivory, the latter of which are elastic, while· the former are the reverse.

Fig. 36.

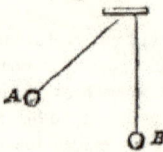

Fig. 37.

Fig. 36 represents two clay or non-elastic balls. A is raised and allowed to fall. If it met with no resistance, it would rise to about the same height on the opposite side. But, encountering B, it imparts to it a portion of its motion, and both move on together, as shown in Fig. 37, though only half as far as A would have gone alone. The reaction of B is clearly equal to the action of A ; for the latter loses just as much motion as the former gains.

If the two balls be of ivory, or any other highly elastic substance, A will impart the whole of its motion to B, and remain stationary after striking; while B, as

of reaction ? What humorous instance is given of the nullifying effect of reaction ? State the case of the man with the sail-boat. 92. In what two classes of bodies are action and reaction differently exhibited ? How is this difference shown ? What does Fig. 36 represent ? Show the effect of action and reaction in these non-elastic

shown in Fig. 38, will swing to the same height that A would have reached if unresisted. Here again the reaction of B, which brings A to rest, is evidently equal to the action of A, which sets B in motion.

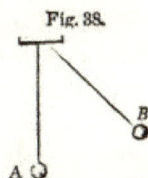

Fig. 38.

93. Fig. 39 affords a further illustration of action and reaction in elastic bodies. Five ivory balls are suspended by strings of equal length, so as to fall in front of a graduated arc, with the aid of which the distance they move can be observed. Let the first, A, be drawn out and allowed to fall. It will impart all its motion to the second, and by the reaction of the latter will be brought to rest. In like manner, the second imparts its motion to the third, and is kept at rest by reaction; and so with the third and the fourth. The fifth, B, finally receives the motion; and, there being in this case no reaction to stop it, it flies off to the same height from which A started.

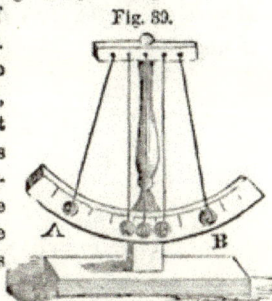

Fig. 39.

94. REFLECTED MOTION.—Reflected Motion is the motion of a body turned from its course by the reaction of another body against which it strikes. A ball rebounding from a wall against which it has been thrown, affords an example of Reflected Motion.

If a body possessing little or no elasticity be thrown against a wall, it will rebound but a short distance, if at all. We find the most striking instances of reflected motion in the most elastic bodies. Every boy knows that an india rubber ball will bound higher than one made of yarn, and that a yarn ball will bound higher than one stuffed with cotton.

95. When a ball is thrown perpendicularly against another body, it rebounds in the same line towards the hand from which it was thrown. Thus, in Fig. 40, if a ball be thrown from F against the surface B C so as to strike it perpendicularly at A, it will return in the line A F. If thrown from D, however, it will glance off on the other side of the perpendicular, at the same angle, to E. If D were nearer the perpendicular, the line A E would also be nearer to it; if it were farther from the perpendicular, A E would be farther in proportion.

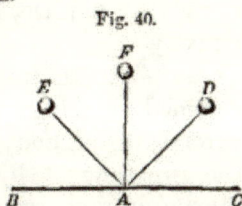

Fig. 40.

balls. What does Fig. 38 represent? 93. Describe the apparatus represented in Fig. 39, and tell how it operates. 94. What is Reflected Motion? Give an example. What bodies exhibit reflected motion most strikingly? 95. When a ball is thrown perpendicularly against another body, how does it rebound? When thrown so as to make an angle with the perpendicular, how will it rebound? Illustrate this with

96. The angle D A F in Fig. 40, made by the body in its forward course with the perpendicular at the point of contact, is called the Angle of Incidence.

The angle E A F, made by the body in its backward course with the same perpendicular, is called the Angle of Reflection.

The great law of reflected motion is as follows:—*The Angle of Reflection is always equal to the Angle of Incidence.*

---

# CHAPTER V.

## MECHANICS (CONTINUED).

### GRAVITY.

97. TERRESTRIAL GRAVITY.—When a stone is let go, we all know that it does not fly up in the air or move sideways, but falls to the ground. This is owing, as already mentioned, to a universal property of matter. The stone and the earth mutually attract each other; but the earth, being vastly superior in size, draws the stone to itself, or in other words, causes it to fall.

The tendency of bodies, when unsupported, to approach the earth's surface, is called Terrestrial Gravity, or simply Gravity.

98. GRAVITATION.—Attraction is universal. It is not confined to things on and about the earth's surface, but extends throughout space, millions of miles, and is in fact the great agent by which the heavenly bodies are kept moving in their respective spheres. The earth as certainly attracts the planet Uranus, at the vast distance of 1,828,000,000 miles, as it does the falling stone.

---

Figure 40. 96. What is the Angle of Incidence? What is the Angle of Reflection? What is the great law of reflected motion?

97. When a stone is let go, what does it do? To what is this owing? What is meant by Terrestrial Gravity? 98. What is Gravitation? How far does gravitation

The attraction subsisting between the heavenly bodies is called Gravitation.

To Sir Isaac Newton the world owes the great discovery of the law of Universal Gravitation. Galileo had investigated the subject of terrestrial gravity (A. D. 1590), but he did not imagine that any similar force existed beyond the neighborhood of the earth. Kepler advanced a step nearer the truth, and spoke of gravitation as acting from planet to planet; still he did not conceive of its having any effect on the planetary motions. This discovery, one of the most important that modern science has achieved, was reserved for the mighty genius of Newton. Sitting in his orchard one day (A. D. 1666), he observed an apple fall from a bough. This simple circumstance awakened a train of thought. Gravity, he knew, was not confined to the immediate surface of the earth. It extended to the greatest heights with which man was acquainted; why might it not reach out into space? Why not affect the moon? Why not actually cause her to revolve around the earth? To test these speculations, Newton at once undertook a series of laborious calculations, which proved that the attraction of gravitation is universal; that it determines the orbits and velocities of the planets, causes the inequalities observed in their motions, produces tides, and has given its present shape to the earth.

99. Three facts have been established respecting gravitation :—

1. Gravitation acts instantaneously. Were a new body created in space 1,000 miles from the earth, its attraction would be felt at the sun just as soon as at the earth, though the one would be 95,000,000 miles off, and the other only 1,000.

2. Gravitation is not lessened by the interposition of any substance. The densest bodies offer no obstacle to its free action. Were a body placed on the other side of the moon, it would be attracted by the earth just as much as if the moon were not between them.

3. Gravitation is entirely independent of the nature of matter. All substances that contain equal amounts of matter attract and are attracted by any given body with equal

extend? Give an example. By whom was the law of Universal Gravitation discovered? What advance had been made towards it by Galileo? What, by Kepler? Give an account of the circumstances and reasoning that led Newton to this discovery. What was proved by his calculations? 99. What is the first fact that has been established respecting gravitation? Give an example. What is the second fact? Give an example. What is the third fact? What evidence is there of this? 100. What

force. The action of the sun is found to be the same on all the heavenly bodies.

100. DIRECTION OF GRAVITY.—If a piece of lead suspended by a string be left free to move, it will point towards the earth. This is the case in all parts of the globe. Now, as the earth is round, it follows that at two opposite points of its surface, the plummet, or plumb-line (as this suspended lead is called), will point in opposite directions. This will be seen from the relative positions of A and B, C and D, in Fig. 41. The lead, therefore, has no tendency to fall in any particular direction as such, but takes all directions according to the part of the earth's surface which it is near. The universal law is, that *it must point towards the centre of the earth.*

Fig. 41.

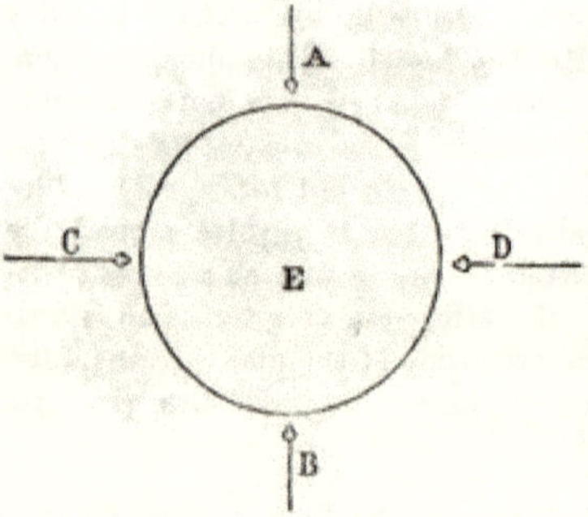

It is not because any peculiar attractive power resides in the centre that a falling body tends towards that point; but because, in a sphere, this is the result of the attraction of all the particles. The particles on one side attract the falling body as much as those on the other; and consequently it seeks a point between them.

No two plummets suspended in different places have exactly the same direction, for the lines in which they hang would meet at the centre of the earth. At short distances, however, the difference of direction is so slight as to be imperceptible, and the plummets seem to point the same way.

101. It follows that *up* and *down* are relative and not absolute terms. What is *up* to a person in New York, is down to a ship a few miles south-west of Australia. If a person in a standing position at New York were to be carried in a straight line through the earth to its centre, and on in the same direction to the opposite side of the earth, he would come out in the Indian Ocean south-west of Australia, but would find himself on his head instead of his feet. His head, which at New York pointed up, would now point down.

is a piece of lead suspended by a string called? How does the plummet always point? On what does the absolute position of the plummet depend? Why does a falling body tend towards the *centre* of the earth? What is said of the difference of direction in plummets suspended in different places? 101. What is said of the terms *up* and *down*? Exemplify this. What is the real meaning of *up* and *down*? Why

*Down*, therefore, simply means towards the centre of the earth, and *up* away from the centre.

This explains what the unreflecting are sometimes puzzled to account for, —why persons and things on the side of the earth opposite to them do not fall off. Regarding themselves as on the *upper* side, they can not see what keeps those on the *under* side from being precipitated into space. But really there is no *under* side. All things are alike drawn towards the centre; all are kept on the earth's surface by the same force of gravity.

102. LAWS FOR THE FORCE OF GRAVITY.—The force of gravity (and the term is here used in its widest sense, including gravitation) depends on two things,—1. Amount of matter; 2. Distance,—according to the following laws:

1. *The force of gravity increases as the amount of matter increases.*
2. *The force of gravity decreases as the square of the distance increases.*

103. According to the first law, if the sun contained twice as much matter as it now does, it would attract the earth with twice its present force; if it contained three times as much matter, with three times its present force; &c. Observe, we say *if it contained twice as much matter*, not *if it were twice as large;* for it might be twice its present size, and yet so rare as to contain less matter and attract less strongly than it now does. If there were two heavenly bodies, the one of iron and the other of cork, the latter, though twice as large as the former, would have less attraction because it would contain less matter.

As already remarked, the earth is so much larger than the bodies near its surface that it is not perceptibly affected by their attraction. Even if a ball 500 feet in diameter were placed in the atmosphere 500 feet from the earth's surface, the earth, being 580 million million times greater than the ball, would draw the latter to itself, while it would advance to meet it, less than one ninety-six-thousand-millionth of an inch—a distance so small that it can not be appreciated.

The sun is 800 times greater than all the planets put together. It is on account of this enormous amount of matter that its attraction is felt by the most remote bodies of the solar system at a distance of many millions of miles.

---

do not objects on the *under* side of the earth fall off? 102. On what does the force of gravity depend? Repeat the two laws of gravity. 103. Explain the first law. Why is not the earth perceptibly affected by the attraction of bodies near its surface? Give an example. Why is the attraction of the sun so great? What would be its effect

A man carried to the surface of the sun would be so strongly attracted by its immense mass that he would be literally crushed by his own weight.

104. According to the second law, if the sun were twice as far from the earth as it now is, it would attract the latter with but $\frac{1}{4}$ of its present force; if three times as far, with $\frac{1}{9}$; if four times as far, with $\frac{1}{16}$, &c. So, if two equal masses were situated respectively 5,000 miles and 10,000 miles from the earth's centre, the nearer would be attracted not twice, but 4 times, as strongly as the more distant.

105. All bodies on the earth's surface, however small, attract each other with greater or less force according to their masses and distance. This attraction, in most cases, is absorbed in the far greater attraction of the earth, and consequently can not be perceived. In the case of mountains, however, it is so strong as to have a sensible effect on plummets suspended at their base. Instead of pointing directly towards the centre of the earth, a plumb-line in such a position is found to incline slightly towards the mountain.

106. WEIGHT.—When a body is supported or prevented from following the impulse of gravity, it presses on that which supports it, more or less strongly according to the force with which it is attracted. This downward pressure is called its Weight.

Weight is simply the measure of a body's gravity, and is proportioned to the amount of matter contained. A ball of iron is heavier than a ball of cork of equal size, because it contains more matter.

Weight being nothing more than the measure of the force with which bodies are drawn towards the earth, it follows that, if the earth contained twice as much matter as it now does, they would have twice their present weight; if it contained three times as much matter, three times their present weight, &c.

107. *Weight above and below the Earth's Surface.*— Since the weight of a body is the measure of its gravity, and since gravity decreases as the square of the distance from the earth's centre increases, it follows that bodies be-

on a man carried to its surface? 104. Illustrate the second law with an example. 105. Why is not attraction exhibited between small bodies on the earth's surface? How is a plummet suspended near the base of a mountain affected? 106. What is Weight? To what is weight proportioned? If the earth contained twice as much matter as it now does, how would the weight of objects on its surface compare with

come lighter in the same proportion as they are taken up from the earth's surface. A mass of iron which at the earth's surface weighs a thousand pounds, taken up to a height of 4,000 miles, would weigh only 250 of such pounds, or one-fourth as much as before.

The reason of this is clear. The earth being about 8,000 miles through, from its centre to its surface is 4,000 miles; and from its centre to a point 4,000 miles above its surface, is 8,000 miles. 4,000 is to 8,000 as 1 to 2; but the weight at the surface would not be to the weight 4,000 miles above the surface as 2 to 1, but as the squares of these numbers, 4 to 1. Hence, if it would weigh 1,000 pounds at the surface, it would weigh only $1/4$ as much, 4,000 miles above the surface. For the same reason, it would weigh $1/9$ of 1,000 pounds at a distance of 8,000 miles from the surface; $1/16$, at a distance of 12,000 miles; $1/25$, at a distance of 16,000 miles, &c. These results are exhibited in Fig. 42.

At small elevations, the weight which an object loses amounts to but little. Four miles above the earth's surface, a body weighing 1,000 pounds would become only two pounds lighter. Raised to a height of 240,000 miles, the distance of the moon from the earth, its weight would be reduced to less than five ounces.

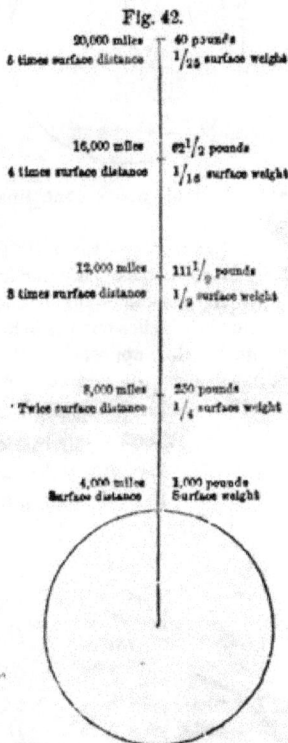

Fig. 42.

20,000 miles / 5 times surface distance — 40 pounds / $1/25$ surface weight

16,000 miles / 4 times surface distance — $62\,1/2$ pounds / $1/16$ surface weight

12,000 miles / 3 times surface distance — $111\,1/9$ pounds / $1/9$ surface weight

8,000 miles / Twice surface distance — 250 pounds / $1/4$ surface weight

4,000 miles / Surface distance — 1,000 pounds / Surface weight

108. If we could go from the surface of the earth to the centre, we should find a given object weigh less and less as we advanced. The moment we descended beneath the surface, we would leave particles of matter behind us, and the attraction of these would act in a direction exactly opposite to gravity.

their present weight? 107. What is said of the weight of bodies taken up from the earth's surface? What would 1,000 pounds of iron weigh, 4,000 miles above the earth's surface? Show the reason of this. What is said of the loss of weight at small elevations? Four miles above the surface, how much would a body weighing 1,000 pounds lose? What would be its weight, 240,000 miles from the earth? 108. If we

Thus, in Fig. 43, let C represent the centre of the earth, and O any object beneath the surface. All the particles below the line A B attract O down-

Fig. 43.               Fig. 44.

ward, but all above that line attract it upward, and thus diminish its weight.

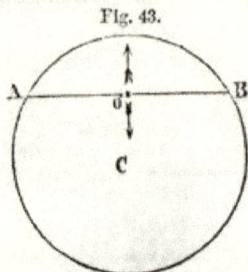

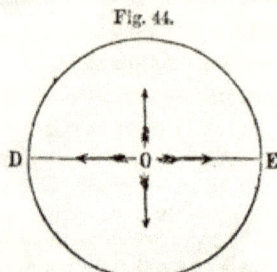

At the centre of the earth (see Fig. 44) no object would weigh any thing. There would be as many particles above the line D E as below it; and O, being equally attracted on all sides, would have no weight.

109. All bodies carried below the earth's surface would, therefore, become lighter as they approached the centre. Their weight at any given number of miles below the surface may be found as follows:—

For 1 mile below, take $\frac{3999}{4000}$ of the surface weight.

For    2 miles, take $\frac{3998}{4000}$ of the surface weight.

For   100 miles, take $\frac{3900}{4000}$ of the surface weight.

For 1,000 miles, take $\frac{3000}{4000}$ of the surface weight, &c.

Fig. 45.

| 8,000 miles | $\frac{1}{4}$, 250 p. |
| 7,000 miles | $\frac{16}{49}$, $326\frac{26}{49}$ p. |
| 6,000 miles | $\frac{4}{9}$, $444\frac{4}{9}$ p. |
| 5,000 miles | $\frac{16}{25}$, 640 p. |
| 4,000 miles | 1,000 pounds |
| 3,000 m. | 750 p. |
| 2,000 m. | 500 p. |
| 1,000 m. | 250 p. |
| centre | 0 pounds. |

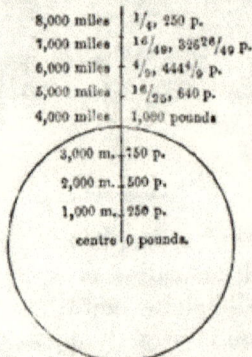

110. *Law of Weight.*—From the above principles the following law of weight is deduced:—*All objects weigh the most at the surface of the earth: ascending from the surface, their weight diminishes as the square of their distance from the centre increases; descending towards the centre, their weight diminishes as their distance from the surface increases.*

Fig. 45 shows the operation of this law in the case of an object weighing 1,000 pounds at the earth's surface.

could go from the surface of the earth to the centre, what would we find respecting the weight of a given body? What is the reason of this decrease? Illustrate this with Fig. 43. What would all objects weigh at the centre? Show the reason of this with Fig. 44. 109. How may we find the weight of a given body one mile below the

111. *Weight at different Parts of the Earth's Surface.*
—The weight of a body differs at different parts of the
earth's surface. A mass of lead, for instance, that weighs
1,000 pounds at the poles, will weigh only 995 such pounds
at the equator.

112. This is owing to two causes :—

1. The equatorial diameter is about 26½ miles longer
than the polar diameter; and therefore an object at the
equator is farther from the centre and less strongly attracted than at any other point.

2. The centrifugal force, as shown in § 79, is greatest
at the equator, and therefore counterbalances more of the
downward attraction there than at any other part of the
surface, making the weight less. It has been computed,
that, if the earth revolved 17 times as fast as it now does,
the centrifugal force at the equator would counterbalance
gravity entirely, and thus deprive all bodies of weight. If
the earth's velocity were further increased, all things at the
equator would be thrown off into space.

113. The general effect of gravity is
to draw bodies towards the earth; but
sometimes it causes them to rise. A
balloon, for instance, mounts to the
clouds. This is because it contains less
matter than a mass of air of the same
bulk, or, as we say briefly, it is lighter
than air. Hence the air, acted on more
strongly by gravity than the balloon, is
drawn towards the earth under the latter, which is thus caused to rise.

For the same reason, smoke ascends.
So, if a flask of oil be uncorked at the

Fig. 46.

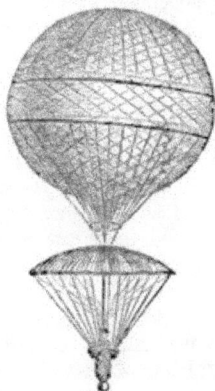

A BALLOON.

earth's surface? Two miles? A hundred miles? A thousand miles? 110. Repeat
the law of weight. 111. What is said of the weight of a body at different parts of the
earth's surface? Give an example. 112. To what causes is this owing? What would
be the result, if the earth revolved on its axis with seventeen times its present velocity? 113. Show how gravity sometimes causes a body to rise. Give some illustra-

bottom of a pail of water, the water will be drawn down below the oil, and force the latter to the top.

## Falling Bodies.

114. VELOCITY OF FALLING BODIES.—If a feather and a cent be dropped from a height at the same time, the cent will reach the ground some seconds before the feather. This fact Aristotle and his successors explained by teaching that the velocity of falling bodies is proportioned to their weight; that a body of two pounds, for instance, would reach the ground in just half the time required by a body weighing one pound. Galileo was the first to correct this error (about A. D. 1590). He held that the velocity of falling bodies is independent of their weight, and that, if no other force than gravity acted on them, all objects dropped at the same time from the same height would reach the ground at the same instant.

So startling a proposition was at once condemned by the learned men of the day; but Galileo, convinced of the truth of his position, challenged his opponents to a trial.

The leaning tower of Pisa [pe'-zah], Italy, was chosen as the scene of the experiment, and multitudes flocked to witness it. Two balls were produced, one of which weighed exactly twice as much as the other, and after being examined, to prevent the possibility of deception, at a given signal they were dropped. In breathless anxiety the crowd awaited the result, doubting not that it would confound the bold youth of six-and-twenty years, who had dared to oppose not only the sages of his own time, but also the established opinion of centuries and the great master Aristotle himself. To their amazement, the bold youth was right; the balls reached the earth at the same instant. Unable to credit their own senses, again and again they repeated the experiment, but each time with the same result. This triumph, though it awakened the jealousy of his defeated rivals, and cost Galileo his place as professor of mathematics in the university of Pisa, established the fact that *gravity causes all bodies to descend with equal rapidity, without reference to their weight*, and that all apparent differences are caused by some other agency.

115. RESISTANCE OF THE AIR.—The cause of the differ-

tions. 114. If a feather and a cent be dropped at the same time, which will reach the ground first? How did Aristotle explain this fact? What was Galileo's opinion on the subject? How was his theory received by the learned men of the day? Give an account of the trial that was made at Pisa. What fact was established by the experi-

ence of velocity in a falling feather and a falling cent is the Resistance of the Air.

This resistance is proportioned to the extent of surface which the falling body presents to the air. The surface, indeed, may be so extended that gravity can hardly overcome the air's resistance; thus, gold may be beaten into a leaf so thin that it will be exceedingly slow in its descent, floating for a time in the air.

Fig. 47.

116. That the resistance of the air causes the difference of velocity exhibited by falling bodies, may be proved in two ways:—

1. A piece of paper, a sheet of gold-leaf, or a feather, with its surface extended, floats slowly downward; roll it into a compact mass, and it will descend rapidly like a stone.

2. Remove the air from a high glass tube (see Fig. 47) by means of an instrument called the air-pump, to be described hereafter. Then, from an apparatus provided for the purpose, drop a feather and a cent simultaneously, and they will reach the bottom at precisely the same instant. Let in the air and drop them, and the feather will be several seconds longer than the cent in reaching the bottom.

117. *The Parachute.*—It is the resistance of the air that enables a person to descend in safety from a balloon at great heights above the earth's surface. A parachute, which spreads open like a large umbrella, is suspended beneath the balloon. Having taken his position in the basket-shaped car hanging beneath, the aërial voyager fearlessly detaches himself from the balloon; for, though he is borne downward by gravity, the force of his fall is so broken by the resistance which the air offers to the extended surface of the parachute that he incurs

Fig. 48.

A PARACHUTE.

ment? What was its result to Galileo? 115. What causes the difference of velocity in a falling feather and a falling cent? To what is the resistance of the air proportioned? How may the air's resistance almost be made to counterbalance gravity? Give an illustration. 116. Prove in two ways that the resistance of the air causes the difference of velocity in falling bodies. 117. How is a person enabled to descend safely

little danger. To ensure the safety of a common-sized man, a parachute must be at least 22 feet across. Fig. 48 represents a parachute; Fig. 46 shows it attached to a balloon.

118. LAW OF FALLING BODIES.—We have found that all bodies acted on solely by gravity fall to the earth with the same velocity. It is evidently an accelerated velocity; for gravity, which first causes the motion, continues acting. In other words, gravity gives a falling body a certain velocity in the first second of its descent; still forcing it downward, it increases that velocity in the following second; and so on till it reaches the earth.

To find the exact spaces passed over in successive seconds, and the velocity at any given point of the descent, was formerly exceedingly difficult, on account of the rapidity with which falling bodies move, and the want of conveniences for experimenting on them. Even the greatest perpendicular heights were inadequate to the purpose, as a falling body would reach their base in a few seconds. These difficulties are now removed by an ingenious apparatus, called, after its inventor, Atwood's Machine.

119. *Atwood's Machine.*—Atwood's Machine is represented in Fig. 49. It consists of a pillar, G, about six feet high, surmounted by a horizontal plate, J K; from which to the base of the stand extends a perpendicular graduated scale, C L, divided into feet, inches, and tenths of an inch. The plate J K supports a vertical wheel, D, the axis of which, that it may revolve as far as possible without friction, rests on four other wheels, *a, b, c, d* (*d*, being behind the rest, is not seen in the figure). A and B are equal weights, connected by a cord, which passes over the wheel D. F is a pendulum which vibrates once in a second; and I is a dial-plate and index (like the face and hand of a clock) for marking seconds.

B, having exactly the same weight as A, just counterbalances it. Now attach to A a small weight equal to one sixty-third of the combined weight of A and B. This slight addition causes A to descend; but as A descends, B of course ascends; and as neither A nor B, being counterbalanced

from a balloon at a great height? Describe the process. How large must a parachute be for a common-sized man? 118. With what sort of velocity must falling bodies descend? Why so? What made it difficult formerly to ascertain the velocity, &c., of falling bodies? What apparatus is now employed for this purpose? 119. Describe Atwood's Machine from the plate. Show its mode of operation. How does this ma-

each by the other, has any gravity, the gravity of the small weight attached to A, which sets them in motion, must be divided into 64 equal parts. Hence A with the added weight is 64 times as long in descending as it would be if dropped freely in the air, and the experimenter thus has an opportunity of observing its velocity at different points, and ascertaining the relative distances passed over during the successive beats of the pendulum. The distances passed over in the first, the second, the third, and the fourth second, &c., bear the same relation to each other, as if the bodies were falling freely in space. The velocity, moreover, having been greatly diminished, the resistance of the air becomes so slight that it need not be taken into calculation.

120. It is found with Atwood's Machine, that, calling the distance traversed in the 1st second 1, that traversed in the 2d will be 3; that in the 3d, 5; that in the 4th, 7; and so on in the series of odd numbers. The velocity at the end of the 1st second will be a mean between 1 and 3, or 2; at the end of the 2d, it will be a mean between 3 and 5, or 4; at the end of the 3d, 6; at the end of the 4th, 8; and so on in the series of even numbers.

In 1 second a falling body descends $16\frac{1}{12}$ feet; therefore, according to the results obtained with Atwood's Machine, it has a velocity at the end of the 1st second of twice $16\frac{1}{12}$ feet, or $32\frac{1}{6}$

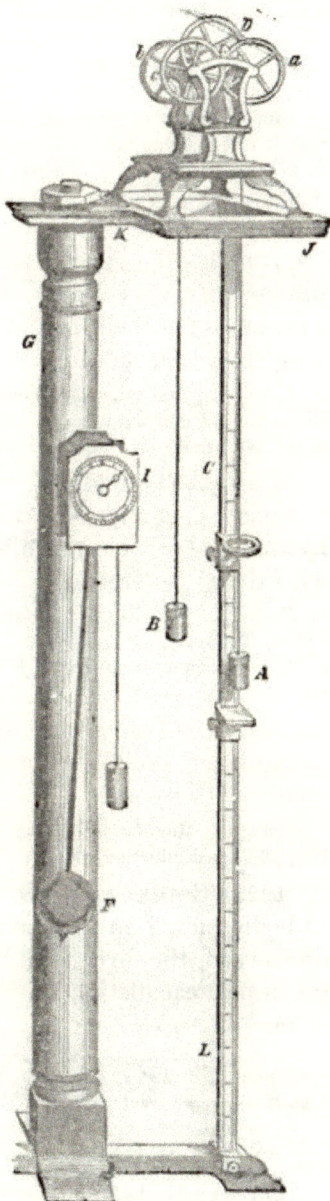

chine aid the experimenter? 120. What is found with Atwood's Machine, respecting the distances traversed in successive seconds? What is the relative velocity at the end of successive seconds? How far does a body fall in the first second? According

feet, per second. In the second second it descends 3 times $16^1/_{12}$ feet, or $48^1/_4$ feet, and at its termination has a velocity of 4 times $16^1/_{12}$ feet, or $64^1/_3$ feet, per second. In the third second, it descends 5 times $16^1/_{12}$ feet, or $80^5/_{12}$ feet, and at its termination has a velocity of 6 times $16^1/_{12}$, or $96^1/_2$ feet, per second, &c.

Now, as to the whole space passed through in any given time. In 1 second, it will be $16^1/_{12}$ feet; in 2 seconds, by addition $(16^1/_{12}+48^1/_4)$, $64^1/_2$ feet; in 3 seconds, $(16^1/_{12}+48^1/_4+80^5/_{12})$ $144^3/_4$ feet; in 4 seconds, $(16^1/_{12}+48^1/_4+80^5/_{12}+112^7/_{12})$ $257^1/_3$, and so on.

121. These results are summed up in the following rules:—

*Rule* 1.—To find the space through which a falling body passes during any second of its descent, multiply $16\frac{1}{12}$ feet by that one in the series of odd numbers which corresponds with the given second.

*Example.* How far will a stone fall in the tenth second of its descent?—The series of odd numbers is 1, 3, 5, 7, 9, 11, 13, 15, 17, 19, &c. The tenth is 19; $16^1/_{12}$ multiplied by 19 gives $305^7/_{12}$.—*Answer*, $305^7/_{12}$ feet.

*Rule* 2.—To find the velocity of a falling body at the termination of any second of its descent, multiply $16\frac{1}{12}$ feet by that one in the series of even numbers which corresponds with the given second.

*Example.* What is the velocity of a stone that has been falling ten seconds?—The series of even numbers is 2, 4, 6, 8, 10, 12, 14, 16, 18, 20. The tenth is 20; $16^1/_{12}$ multiplied by 20 gives $321^2/_3$.—*Answer*, $321^2/_3$ feet per second.

*Rule* 3.—To find the whole space passed through by a falling body, multiply $16\frac{1}{12}$ feet by the square of the given number of seconds.

*Example.* How far will a stone fall in 10 seconds?—Squaring 10 gives 100; $16^1/_{12}$ multiplied by 100 gives $1,608^1/_3$.—*Answer*, $1,608^1/_3$ feet.

122.—BODIES THROWN DOWNWARD.—These rules apply to bodies acted on by gravity alone. If a body is thrown downward, the force with which it is thrown must also be taken into calculation.

Thus, if a stone be cast from a height with a force that would propel it 50

_____

to the results obtained with Atwood's Machine, how far will it fall in successive seconds, and what will be its velocity at the end of each? 121. Repeat Rule 1, for finding the space traversed by a falling body during any second of its descent. Apply this rule in the given example. Repeat Rule 2, for finding the velocity of a falling body. Apply this rule in an example. Repeat Rule 3, for finding the whole distance traversed by a falling body. Give an example. 122. To what bodies do these rules

feet in a second, then in the tenth second, instead of falling $305^7/_{12}$ feet, as in the example under Rule 1, it would fall 50 feet farther,—that is $355^7/_{12}$ feet. Its velocity at the end of the tenth second would likewise be obtained by adding 50 feet per second to the velocity obtained in the example under Rule 2: $321^2/_3 + 50 = 371^2/_3$.—To obtain the whole space passed through, add to the result obtained by Rule 3, the distance traversed in consequence of the velocity originally imparted. A body thrown downward with a velocity of 50 feet per second, would, without any aid from gravity, pass through 500 feet in 10 seconds. Adding this to $1,608^1/_3$ feet, the distance through which gravity alone causes a body to fall in 10 seconds, we have $2,108^1/_3$ feet for the whole distance traversed in that time by a body thrown downward with a velocity of 50 feet per second.

123. In the above examples, no allowance is made for the resistance of the air. But even the bodies most favorably shaped for falling feel the effects of this resistance. Experiments in St. Paul's Cathedral, London, show that in $4\frac{1}{2}$ seconds a body falls 272 feet; whereas, according to the principles stated above, it should fall 325 feet. This difference, which amounts to nearly one-sixth of the whole distance, is owing principally to the resistance of the air.

124. As the velocity of a falling body increases $32\frac{1}{6}$ feet every second, it does not take long for it to acquire a tremendous speed; and, as the striking force is proportioned to the weight multiplied into the square of the velocity, it is clear that even a small body, falling any considerable distance, may become a very powerful agent. Hence the disastrous effects of hail-stones, which have been known to injure cattle and break through the roofs of houses, and which prove so destructive to the vineyards in parts of Southern Europe that the fields have to be protected from their visitations.

125. ASCENDING BODIES.—As a falling body increases in velocity $32\frac{1}{6}$ feet every second of its descent, so an ascending body, being acted on by the same force, loses a

apply? If a body is thrown from a height, what must enter into the calculation? If a stone were thrown down with a force that would propel it 50 feet in a second, how far would it fall in the tenth second? What would be its velocity at the end of the tenth second? What would be the whole distance traversed in ten seconds? 123. For what must allowance be made in applying these rules? How great a difference does the resistance of the air occasion? 124. How are the disastrous effects of hail-stones accounted for? 125. What is said of the velocity of an ascending body? How may

like amount, and will at last be brought to rest. The number of seconds during which it will continue to rise is found by dividing the number of feet per second with which it starts by 32¼.

The height, therefore, which an ascending body reaches, depends on the force with which it is projected upward; and, were there no air to resist its progress, it would always reach such a height as it would have to fall from in order to acquire the velocity with which it started. The spaces traversed and the velocities attained during successive seconds would be the same in the ascent as in the descent, only reversed in order.

Thus, if projected upward with a velocity of 321²⁄₃ feet per second, a ball unresisted by the air would continue to rise 10 seconds; because, to attain a velocity of 321²⁄₃ feet from a state of rest, it would have to fall 10 seconds. In the tenth second of its ascent, it would pass through the same distance as in the first second of its descent, 16¹⁄₁₂ feet; in the ninth second of its ascent, the same as in the second second of its descent, 48¹⁄₄ feet; in the eighth second of its ascent, the same as in the third of its descent, &c.

126. According to the principle just stated, a rifle-ball, shot vertically upward, would descend on whatever it struck with the same force that it had when originally discharged. But it does not do so, on account of the resistance of the air. This resistance prevents the ball from rising as high as it otherwise would do by about one-sixth of the whole distance (see § 123), and in its descent it again loses nearly one-sixth. The whole loss thus amounts to nearly one-third of the velocity, leaving a little over two-thirds remaining. Now, to find the proportion between the striking force of the ball when originally projected and its striking force on returning to the same point, we must square two-thirds; and this gives four-ninths; and thus we find that the ball, on returning to the surface, strikes an object with less than half the effect which it has immediately on being discharged—a result borne out by facts.

## Projectiles.

127. A Projectile is a body thrown through the air. An arrow discharged from a bow, a bullet from a gun, a stone from the hand, are all Projectiles.

we find the number of seconds that an ascending body will continue to rise? Were it not for the resistance of the air, how great a height would a body projected upward attain? What is said of the spaces traversed and the velocities attained during successive seconds? Exemplify this in the case of a ball thrown upward with a velocity of 321¼ feet per second. 126. According to this principle, with what force would a ball shot vertically upward descend on an object? Does it do so? Explain the rea-

Every projectile is acted on by three forces:—

1. The force by which it was thrown.
2. Gravity, which constantly impels it towards the earth.
3. The resistance of the air, which tends to bring it to rest.

128. PATH OF A PROJECTILE.—A projectile may be thrown with such force as to be borne some distance in a straight line, without having its direction sensibly altered by gravity or the air's resistance; as in the case of a cannon-ball. When, however, its velocity diminishes, the joint action of these forces causes it to move in a line more or less resembling the curve called the *pa-rab'-o-la*. The less the projectile force, the sooner does the body deviate from a straight line to a curve.

Fig. 50.

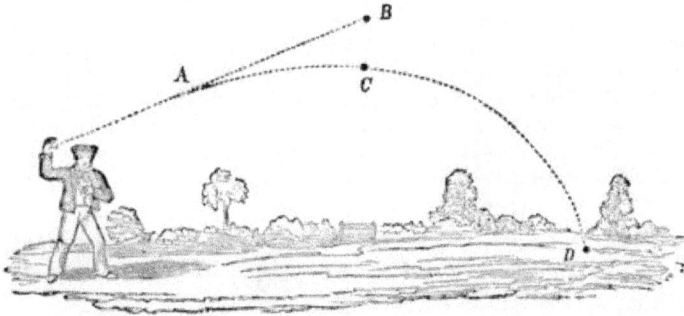

Fig. 50 shows the path of a stone thrown obliquely from the hand. The propelling force sends it in a straight line to A, and would take it on in the same direction to B, were it not that, as soon as its velocity becomes sufficiently diminished, gravity and the air's resistance give it a circular motion to C, and finally bring it to the earth at D.

129. If thrown straight up, a projectile will descend in the same line in which it ascended. If discharged horizontally from a height, it will describe a curve which varies

in form according to the velocity originally imparted. The greater this velocity, the greater the distance the projectile will pass through; but, whatever the distance traversed, it will always reach the ground in precisely the same time that it would take to fall to the earth from the height at which it was discharged.

Fig. 51.

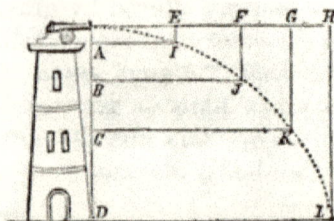

Thus, in Fig. 51, we have a cannon planted on a tower at such a height that it would take four seconds for a ball to fall from it to the ground. Dropped from the cannon's mouth, in the first second a ball would reach A; in the next, B: in the third, C; and in the fourth, D. Fired from the cannon, and acted on by the projectile force alone, it would in one, two, three, and four seconds, successively reach E, F, G, and H. When both forces act, the ball will move in the dotted line, reaching at the end of the successive seconds the points I, J, K, and L. The ball *fired* from the cannon will touch the ground at L at precisely the same instant that the ball *dropped* from it will strike the ground at D.

130. The resistance of the air, which is but slight when a body moves slowly through it, becomes a powerful agent as the velocity of the body increases. A cannon-ball, fired with a velocity of 2,000 feet in a second, would go 24 miles before gravity alone would stop it; whereas, when opposed by the air's resistance, as well as gravity, it goes but 3.

131. A projectile reaches a greater height and remains longer in the air, when thrown straight upward, than when thrown in any other direction.

132. RANDOM.—The Random, or Range, of a projectile is the distance in a straight line between the points at which it begins and ceases to move.

When thrown perpendicularly upward, a projectile returns to the point from which it started, and the random is *naught*. The more its course deviates from the perpendicular the greater the random becomes, until it is thrown

descend? If discharged horizontally from a height, what kind of a line does a projectile describe? What projectiles, so discharged, will traverse the greatest distance? How long will it take projectiles discharged horizontally from a height to reach the ground? Explain these principles with Fig. 51. 130. In what case does the resistance of the air become a very powerful agent? Show this in the case of a cannon ball. 131. In what direction must a projectile be thrown, to attain the greatest

at an angle of somewhat less than 40 degrees, from which point it again diminishes. Were it not for the resistance of the air, a projectile would have the greatest random when thrown at an angle of 45 degrees.

Figure 52 shows the course of projectiles thrown at different angles. The ball which leaves the cannon's mouth at an angle of about 37 degrees will be the only one to hit the vessel. The two balls fired at a greater and a less angle will fall short of it.

Fig. 52.

133. GUNNERY.—The laws relating to projectiles form the basis of the science of Gunnery. The artilleryman must know just at what angle to elevate his gun, and how great an allowance to make for gravity and the air's resistance.

134. Military projectiles are discharged with the aid of gunpowder. This is a solid, which by the application of a spark is instantaneously converted into a highly elastic fluid, and in that form expands to many times its previous bulk. This sudden expansion, confined within a cannon, finds vent at its mouth, and with such force as to impart great velocity to a ball or other missile.

Who invented gunpowder can not be ascertained. It was known many centuries before the Christian era to the Chinese, who used it for levelling hills, blasting rocks, and also, as the remains of ancient pieces of ordnance indicate, for military purposes. Other eastern nations appear to have been acquainted with its use at an early date. Roger Bacon, the celebrated English philosopher, in a work written about 1270 A. D., alludes to it as a well known composition. Fifty years later, Berthold Schwartz, a Prussian monk,

height? 132. What is the Random of a projectile? What is the random of a projectile thrown perpendicularly upward? At what angle must a projectile be discharged, to have the greatest random? What would be the angle, were it not for the resistance of the air? Explain Fig. 52. 133. What science is based on the laws of projectiles? 134. How are military projectiles discharged? Explain the mode in which a projectile is discharged with gunpowder. By whom was gunpowder invented? To whom was it early known? What English philosopher alluded to it, and when? What Prussian monk investigated its properties? Where and when were

investigated its properties; he has by some been called its inventor, as Bacon has by others. The first that we hear of cannon's being used in war is at the battle of Cressy, between the French and English, A. D. 1346.

135. As the striking force of a body increases with the square of its velocity, the pieces of ordnance used in attacking a fort are so charged as to give the balls the greatest possible velocity. In naval engagements, on the other hand, no greater velocity is desired than will just plant the balls in the enemy's hull; for thus, imparting the whole of their motion to the ship, they give it a greater shock, and do more damage by splintering its timbers, than if they have sufficient velocity to carry them completely through.

Fig. 53.

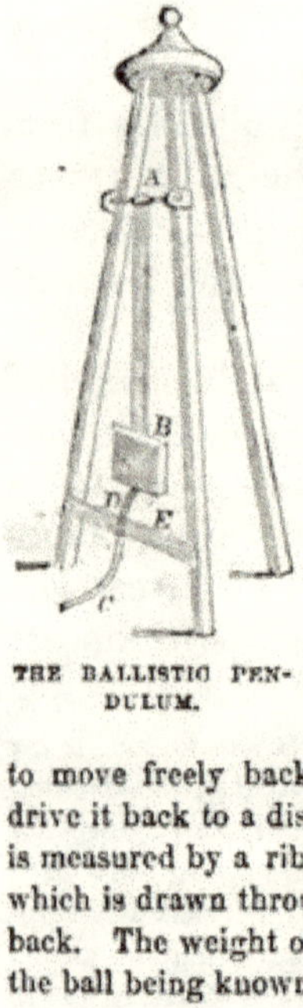


THE BALLISTIC PENDULUM.

136. THE BALLISTIC PENDULUM.—Several methods have been tried for measuring the velocity of cannon and musket balls. One is to suspend the piece from which the ball is fired and measure its recoil; action and reaction being equal, this recoil is proportioned to the force with which the ball is discharged. Another method is by means of an instrument called the Ballistic Pendulum, represented in Fig. 53.

From a cross-piece, A, on a stout framework, a heavy block of wood, B, is suspended, in such a way as to move freely backward and forward. A ball fired into this block will drive it back to a distance proportioned to the ball's velocity. This distance is measured by a ribbon, C D, attached to the lower end of the pendulum, which is drawn through an orifice in the cross-piece E as the block is carried back. The weight of the block, the distance it is driven, and the weight of the ball being known, the velocity of the ball can be determined.

137. It is found by experiments with the ballistic pendulum that the greatest velocity that can be given to a cannon-ball is a little over 2,000 feet in a second. To make

cannon first used in war? 135. How are pieces of ordnance charged for attacking a fort? How in naval engagements, and with what object? 136. What methods have been tried for measuring the velocity of balls? Describe the Ballistic Pendulum. 137. What is the greatest velocity that can be given to a cannon-ball? What is said a piece carry the greatest distance, it must be charged with a certain amount of powder, which is not uniform, but varies even in different pieces of the same size. A larger charge is not only useless, but dangerous, as it may burst the gun.

The longer the barrel of a gun, the greater is the velocity imparted to the ball; but its random is thus only slightly increased, and, for various reasons, great length is now regarded as a positive disadvantage.

## The Pendulum.

138. A Pendulum consists of a heavy ball suspended in such a way as to swing to and fro. Fig. 54 represents a Pendulum.

If raised on one side and let go, the ball of the pendulum, B, will be carried down by gravity with such force as to rise by its inertia to the same height on the opposite side. From this point it will again fall and rise on the other side; and, if no other force than gravity operated, it would keep on rising and falling for-

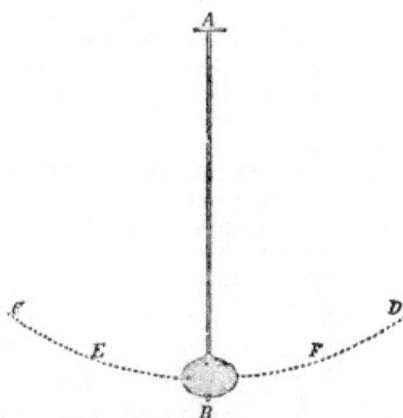

Fig. 54.

THE PENDULUM.

ever. The friction at the point of suspension, however, and the resistance of the air, are constantly tending to check its motion; and the consequence is that it swings each time a less distance, and finally comes to rest.

139. When swinging to and fro, a pendulum is said to *vibrate ;* and the portion of a circle through which it moves is called its *arc.* In Fig. 54, C D is the arc of the pendulum A B.

140. LAWS OF VIBRATION.—*First Law.—The vibra-*

*tions of a given pendulum are performed in very nearly the same time, whether it moves through longer or shorter arcs.*

Thus, in Fig. 54, if the pendulum A B were raised only to E, it would be as long in swinging from E to F as from C to D. The shorter the arc, therefore, the slower its motion. It is on this principle that a swing, when first set in motion, goes very slowly, but increases in velocity as it is pushed higher and higher.

141. *Second Law.—The vibrations of pendulums of different length are performed in different times ; and their lengths are proportioned to the squares of their times of vibration.*

One pendulum vibrates in 2 seconds, another in 4. Then the latter will be four times as long as the former ; because they will be to each other as the square of 2 is to the square of 4,—that is, as 4 is to 16. Hence, to have its time of vibration doubled, a pendulum must be made 4 times as long ; to have it tripled, 9 times as long ; to have it quadrupled, 16 times as long, &c. A pendulum, to vibrate only once in a minute, would have to be 60 times 60, that is 3,600, times as long as one that vibrates once in a second,—or a little over 2 miles.

Conversely, the times in which different pendulums vibrate are to each other as the square roots of their length. If one pendulum be 16 feet long and another 4, the former will be twice as long in vibrating as the latter ; for their times of vibration are to each other as the square root of 16 is to the square root of 4,—or as 4 to 2.

142. *Third Law.—The vibrations of the same pendulum are not performed in the same time at all parts of the earth's surface ; but, being caused by gravity, differ slightly, like gravity, according to the distance from the earth's centre.*

On the top of a mountain five miles high, for instance, a pendulum vibrating seconds would make 10 less vibrations in an hour than at the level of the sea, because it would be farther from the earth's centre. At either pole, a second-pendulum would make 13 more vibrations in an hour than at the equator, because it is nearer the centre, the earth being flattened at the poles. Hence the vibrations of the pendulum afford a means of measuring heights.

---

140. What is the first law relating to the pendulum? Illustrate this with Fig. 56. 141. What is the second law? Apply this law in an example. When the lengths of different pendulums are known, how can we find the relative times of vibration ? If we have two pendulums, 16 and 4 feet long, how will their times of vibration compare? 142. What is the third law? What is the difference in the number of vibrations in a second-pendulum at the level of the sea and at an elevation of five miles ? How would the number of vibrations at the pole compare with those at the equator ?

They also confirm what we have learned, that the polar diameter of the earth is 26¼ miles shorter than its equatorial diameter.

In the latitude of New York, a pendulum, to vibrate seconds, must be about 39¹/₁₀ inches long; whereas at Spitzbergen, in the far North, it must be a little over 39¹/₅, and at the equator exactly 39 inches.

143. APPLICATION OF THE PENDULUM TO CLOCK-WORK. —Galileo, to whom science owes so much, was the first to think of turning the pendulum to a practical use. Observing that a chandelier suspended from the ceiling of a church in Pisa, when moved by the wind, vibrated in exactly the same time, whether carried to a greater or less distance, he at once saw that a similar instrument might be employed in measuring small intervals of time in astronomical observations.

To adapt it to this use, it was necessary to invent some way of counterbalancing the constant loss of motion caused by friction and the air's resistance. This was done by the Dutch astronomer Huygens [hi'-genz], who in the year 1656 first applied the pendulum to clock-work. To this great invention modern astronomy owes its precision of observation, and consequently much of the progress it has made.

144. As a pendulum vibrating seconds, which is over 39 inches long, would be inconvenient in clocks, it is customary to use one that vibrates half-seconds; which, according to the principles laid down in § 141, is one-fourth as long, or a little less than 10 inches.

145. At the same distance from the equator, the same elevation above the sea, and the same temperature, a pendulum of given length will always vibrate in exactly the same time, and a clock regulated by a pendulum will keep uniform time. If taken from the equator towards the poles, the pendulum will vibrate more rapidly, and the clock

---

What is the length of a second-pendulum at New York? At Spitzbergen? At the equator? 143. Who first thought of turning the pendulum to a practical use? Relate the circumstance that led him to do so. To enable it to measure small intervals of time, what was first necessary? Who did this, and thus first applied the pendulum to clock-work? 144. What is the length of the pendulums generally used in clocks? 145. Under what circumstances will a pendulum always vibrate in the same

will go too fast. If taken up a mountain, the pendulum will vibrate less rapidly, and the clock will go too slow. If expanded by the heat of summer (for such we shall hereafter learn is the effect of heat), the pendulum will also vibrate less rapidly, and the clock will go too slow.

146. THE GRIDIRON PENDULUM.—To prevent a clock from being affected by heat and cold, the Compensation Pendulum is used.

Fig. 55.

GRIDIRON
PENDULUM.

One form of the Compensation Pendulum, known as the Gridiron Pendulum, is represented in Fig. 55. It consists of a frame of nine bars, alternately of steel and brass. These are so arranged that the steel bars, being fastened at the top, have to expand downward; while the brass ones, fastened at the bottom, expand upward. The expansive power of brass is to that of steel as 100 to 61; therefore, if the length of the steel bars is made $^{100}/_{61}$ the length of the brass bars, the expansion of the one metal counterbalances that of the other, and the pendulum always remains of the same length. The steel bars in the figure are represented by heavy black lines; the brass ones, by close parallel lines.

147. A clock is regulated by lengthening or shortening its pendulum. This is done by screwing the ball up or down on the rod. The ball is lowered when the clock goes too fast, and raised when it goes too slow.

### EXAMPLES FOR PRACTICE.

1. (See Fig. 45, and §§ 107, 109.) What would be the weight (that is, the measure of the earth's attraction) of an iceberg containing 40,000 tons of ice, if raised to a height of 1,000 miles above the earth's surface?

   What would it weigh 1,000 miles beneath the earth's surface?

2. A horse at the earth's surface weighs 1,200 pounds; what would he weigh 4,000 miles above the surface?

   How far beneath the surface would he have to be sunk, to have the same weight?

3. A Turkish porter will carry 800 pounds; how many such pounds could he carry, if he were placed half way between the surface and the centre of the earth, and retained the same strength?—Ans. 1,600.

   How many such pounds could he carry, if elevated 4,000 miles above the surface with the same strength?

time? What will cause it to vibrate more rapidly, and what less? 146. To prevent a clock from being affected by heat and cold, what is used? Describe the Gridiron Pendulum. 147. How is a clock regulated?

4. What would a body weighing 100 pounds at the earth's surface weigh 1,000 miles above the surface ?
   What would it weigh 1,000 miles below the surface?

5. Would an 18-pound cannon-ball weigh more or less, 2,000 miles above the earth's surface, than 2,000 miles below it,—and how much?

6. At the centre of the earth, what would be the difference of weight between a man weighing 200 pounds at the surface and one weighing 100 pounds? Four thousand miles above the surface, what would be the difference in their weight?

7. (*See Rule* 1, § 121.—*In the examples that follow, no allowance is made for the resistance of the air.*) A man falls from a church steeple; how many feet will he pass through in the third second of his descent?

8. How far will a stone fall in the twelfth second of its descent?

9. (*See Rule* 2, § 121.) How great a velocity does a falling stone attain in 7 seconds ?

10. A hail-stone has been falling one-third of a minute; what is its velocity?

11. (*See Rule* 3, § 121.) How far will a stone fall in 10 seconds?

12. How far will a hail-stone fall in one-third of a minute?

13. I drop a pebble into an empty well, and hear it strike the bottom in exactly two seconds. How deep is the well?
    How many feet does the pebble fall in the first second of its descent? How many, in the second?
    What velocity has the pebble at the moment of striking?

14. A musket-ball dropped from a balloon continues falling half a minute before it reaches the earth ; how high is the balloon, and what is the velocity of the ball when it reaches the earth?

15. What is the velocity of a stone dropped into a mine, after it has been falling 7 seconds, and how far has it descended ?

16. (*See* § 122.) What would be the velocity of the same stone at the end of the seventh second, if thrown into the mine with a velocity of 20 feet in a second, and how far would it have descended ?

17. An arrow falls from a balloon for 9 seconds. How far does it fall altogether, how far in the last second, and what velocity does it attain?
    What would these three answers be, if the arrow were discharged from the balloon with a velocity of 10 feet per second?

18. (*See* § 125.) How long will a ball projected upwards with a velocity of $125\frac{2}{3}$ feet per second, continue to ascend ?
    How great a height will it attain?
    What will be its velocity, after it has been ascending one second? After two seconds? After three seconds?

19. How many seconds will a musket-ball, shot upward with a velocity of $225\frac{1}{3}$ feet in a second, continue to ascend ?
    How many feet will it rise ?

20. A stone thrown up into the air rises two seconds; with what velocity was it thrown?

21. (*See* § 141.) How many times longer must a pendulum be, to vibrate only once in a second, than to vibrate three times in a second?

22. Two pendulums at the Cape of Good Hope vibrate respectively in 40 seconds and 10 seconds; how many times longer is the one than the other?
23. Two pendulums at New Orleans vibrate in 40 seconds and 10 seconds; how many times longer is one than the other?
24. In the latitude of New York, a pendulum vibrating seconds is $39\frac{1}{10}$ inches in length; how long must one be, to vibrate once in 10 seconds? —*Ans.* 3,910 *inches.*

How long must one be, to vibrate 4 times in a second at the same place? —*Ans.* $2\frac{71}{100}$ *inches.*
25. At the equator, a pendulum 39 inches long vibrates once in a second; how long must a pendulum be, to vibrate once in half an hour at the same place?

How long must one be, to vibrate 10 times in a second?
26. At Trinidad, a seconds-pendulum must be about $39\frac{1}{40}$ inches long; what would be the length of one that would vibrate 3 times in a second?

What would be the length of one that would vibrate 3 times in a minute?

What would be the length of one that would vibrate 3 times in an hour?

---

# CHAPTER VI.

## MECHANICS (CONTINUED).

### CENTRE OF GRAVITY.

148. The Centre of Gravity of a body is that point about which all its parts are balanced.

The centre of gravity is nothing more than the centre of weight. Cut a body of uniform density in two, by a plane passed in any direction through its centre of gravity, and the parts thus formed will weigh exactly the same. The whole weight of a body may be regarded as concentrated in its centre of gravity.

149. The Centre of Gravity must be carefully distinguished from the Centre of Magnitude and the Centre of Motion.

---

148. What is the Centre of Gravity? How may we divide a body of uniform density into two parts of equal weight? Where may we regard the whole weight of a body as concentrated? 149. From what must the centre of gravity be carefully

150. The Centre of Magnitude (or, as we briefly call it, the Centre) of a body, is a point equally distant from its opposite sides.

151. The Centre of Motion in a revolving surface is a point which remains at rest, while all the other points of the surface are in motion.

In all revolving bodies, a number of points remain at rest. The line connecting them is called the Axis of Motion, or briefly, the Axis of the body.

152. The centre of gravity may coincide with the centre of magnitude and lie in the axis of motion, but need not do so. In Fig. 56, A represents a wheel entirely of wood of uniform density; here the centre of gravity coincides with the centre of magnitude, C, and both lie in the axis of motion. B represents the same wheel

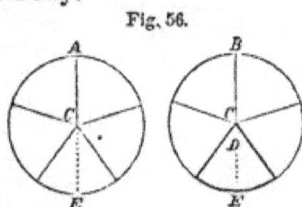

Fig. 56.

with its two lower spokes and part of the felly of lead. The centre of magnitude, C, still lies in the axis, but the centre of gravity has fallen to D.

When a body is of uniform density, its centre of gravity coincides with its centre of magnitude. When one part of a body is heavier than another, the centre of gravity lies nearer the heavier part.

153. A line drawn perpendicularly downward from the centre of gravity is called the Line of Direction. In Fig. 56, C E and D E are the Lines of Direction.

154. HOW TO FIND THE CENTRE OF GRAVITY.—The part of a body in which the centre of gravity is situated, may be found, in some cases, by balancing it on a point. Thus the centre of gravity of the poker represented in Fig. 57 lies directly over the point on which it is balanced.

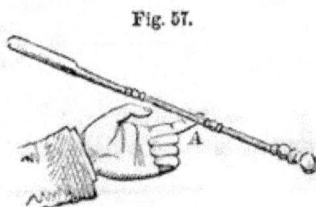

Fig. 57.

155. In a solid of regular

distinguished? 150. What is the Centre of Magnitude? 151. What is the Centre of Motion? What is the Axis of a revolving sphere? 152. Show, with Fig. 56, how the centre of gravity may not coincide with the centre of magnitude, or lie in the axis. When does a body's centre of gravity coincide with its centre of magnitude? When one part is heavier than another, where does the centre of gravity lie? 153. What

shape and uniform thickness and density, so thin that it may be regarded as a mere surface, such as a piece of pasteboard, the centre of gravity may be found by ascertaining any two straight lines drawn from side to side that will divide it into two equal parts. The point at which these lines intersect is the centre of gravity. Thus, in a parallelogram, the centre of gravity is the point at which its two diagonals intersect.

Fig. 58.

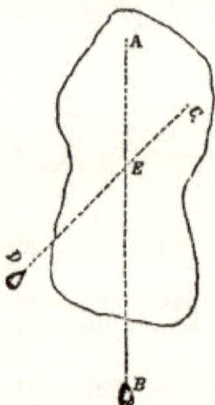

When such a surface is irregular in shape, suspend it at any point, so that it may move freely, and when it has come to rest, drop a plumb-line from the point of suspension and mark its direction on the surface. Do the same at any other point, and the centre of gravity will lie where the two lines intersect.

Thus, suspend the irregular body represented in Fig. 58 at the point A; and, dropping the plumb-line A B, mark its direction on the surface. Then suspend it at C; drop the plumb-line C D, and mark its direction. The lines cross at E, and there will be the centre of gravity.

156. When two bodies of equal weight are connected by a rod, the centre of gravity will be in the centre of the rod. When two bodies of unequal weight are so connected, the centre of gravity will be nearer to the heavier one. These principles are illustrated in Fig. 59, in which C represents the centre of gravity.

Fig. 59.

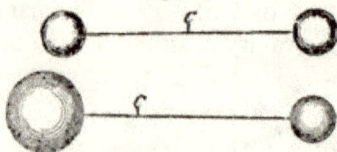

157. STABILITY OF BODIES.—The Base of a body is its lowest side. When a body is supported on legs, like a

chair, its base is formed by lines connecting the bottoms of its legs.

158. When the line of direction falls within the base, a body stands; when not, it falls.

In Fig. 60, G is the centre of gravity; since the line of direction, G P, falls within the base, the body will stand. In Fig. 61, the line of direction falls exactly at one extremity of the base, and the body will be overturned by the slightest force. In Fig. 62, the line of direction falls outside of the base, and the body will fall.

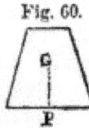

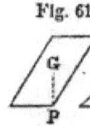

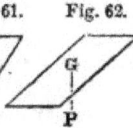

Fig. 60.    Fig. 61.    Fig. 62.

A man carrying a load on his back naturally bends forward, to bring his line of direction within the base formed by his feet. Otherwise, the line of direction falls outside of the base, as shown in Fig. 63; and the load, if heavy, may pull him over backward.

Fig. 63.

159. Of different bodies of the same height, that which has the broadest base is the hardest to overturn, because its line of direction must be moved the farthest to fall outside of its

Fig. 64.

EGYPTIAN PYRAMIDS.

base. Hence a pyramid is the most stable of all figures; and, of different pyramids of the same height, that which has the broadest base is the most stable. The pyramids of Egypt have withstood the storms of more than three thousand years.

The stability of stone walls is increased by making them broader at the base than at the top. Candlesticks and inkstands generally spread out at the bottom that they may not be easily upset. For the same reason, the legs of chairs bend outward as they approach the floor. A three-legged stool or table has a smaller base than one that has four legs, and is therefore more easily upset. Hence, also, the ease with which a man standing on one leg is overturned.

160. A ball of uniform density has its centre of gravity at its centre of magnitude. When such a ball rests on a level surface, the line of direction falls on the point of support, and it therefore remains in any position in which it is placed. But, as the base of a ball consists of a single point, —the point in which it touches a level surface,—a slight push throws the line of direction beyond the base, and causes the ball to roll.

Fig. 65.

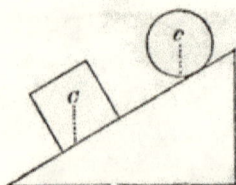

161. When a ball is placed on a sloping surface, the line of direction falls outside of the base, and the ball begins to roll. A cube placed on the same sloping surface maintains its position, because the line of direction falls within its base. See Fig. 65, in which C represents the centre of gravity.

162. Of different bodies with bases equally large, the lowest is the hardest to overturn, because its line of direction is least liable to fall outside of its base.

Why? What is the most stable of all figures? How long have the pyramids of Egypt stood? Give some familiar instances in which the base of a body is made larger than the top, to increase its stability. Why are three-legged chairs and tables easily overturned? 160. In a ball of uniform density, where is the centre of gravity? What is said of the stability of such a ball, when resting on a level surface? 161. When such a ball is placed on a sloping surface, what takes place? Compare it, in this respect, with a cube. 162. Of different bodies with bases equally large, which is the

This is apparent from Figs. 66 and 67. The unfinished tower, though leaning far over, maintains its upright

Fig. 66.

position, the line of direction falling within the base. When made higher by the addition of several stories, as shown in Fig. 67, it will fall, because the centre of gravity has been raised, and the line of direction now falls outside of the base.

Fig. 67.

High chairs for children are unsafe, unless their legs spread at the bottom. A coach with heavy baggage piled on its top is in danger of upsetting on a rough road. On the same principle, a load of stone may pass safely over a hill-side, on which a load of hay would be overturned. Fig. 68 shows that the line of direction in the one case would fall within the base, while in the other it would fall outside of it.

Fig. 68.

163. The lower its centre of gravity, the more stable a body is. Those, therefore, who pack goods in wagons or vessels, should place the heaviest lowest.

This principle has been turned to account in building leaning towers. The tower of Pisa, which is the most remarkable of these structures, with a height of 150 feet, leans to such a degree that its top overhangs its base more than 12 feet; yet it has stood firm for centuries. In this case, the centre of gravity has been brought lower than it would otherwise have been, by the use of heavy materials at the bottom and lighter ones higher up. The lower stories are of dense volcanic rock, the middle stories of brick, and the upper ones of

hardest to overturn? Why? Illustrate this point with Figs. 66 and 67. Give some familiar applications of this principle. 163. Why do those who pack goods in wagons place the heaviest lowest? In what has this principle been turned to account? De-

an exceedingly porous stone. Thus built, the tower is much less liable to fall, than if the same material had been used throughout.

**164.** When the centre of gravity is brought beneath the point of support, the stability of a body is still further increased.

Fig. 69.

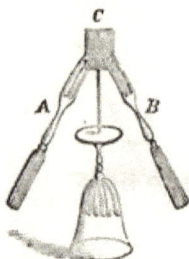

This is shown in Fig. 69. To balance a needle on its point is next to impossible, on account of the smallness of the base, and the height of the centre of gravity. It may be done, however, by running the head of the needle into a piece of cork, C, and sticking into opposite sides of this cork two forks, A, B, at equal angles. The whole may then be poised upon the needle's point on the bottom of a wine-glass. In this case, the heavy handles of the forks bring the centre of gravity below the point of support, in the stem of the glass.

Fig. 70.

ROCKING-HORSE.

The common toy known as the Rocking Horse, represented in Fig. 70, is made on this principle. To a horse of any light material, bearing a trooper or some other figure, is attached a wire to which a ball may be fastened. When the hind legs of the horse are placed on the stand without the ball, the line of direction falls outside of the base, and the horse and his rider fall. When the ball is attached, however, the centre of gravity is brought below the point of support; the horse will then maintain its upright position, and by moving the ball may be made to rock up and down.

**165. Effect of Rotary Motion.**—Rotary Motion, that is, motion round an axis, may keep a body from falling, even when its line of direction falls outside of its base. Thus, if a boy tries to balance his top on its point, he finds it impossible; but, when he spins it, it stands as long as the rotary motion continues. The centre of gravity is not over the point of support all the time the top is spinning, but is

scribe the tower of Pisa, and the materials of which it is built. 164. How is the stability of a body further increased? Show how a needle may be balanced on its point by applying this principle. Describe the Rocking Horse, and explain the principle involved. 165. What is meant by Rotary Motion? What is one of its effects? Why does a top fall over when we try to balance it on its point, but not fall when spinning?

constantly moving round the axis of motion; and, before the top can fall in consequence of its being on one side of the axis, it reaches the other side, and thus counteracts the previous impulse. Hence, the faster the top revolves, the steadier it is; as its motion slackens, it gradually reels more and more, and finally falls.

166. CENTRE OF GRAVITY IN MAN.—The centre of gravity in the body of a man lies between the hips; the base is formed by lines connecting the extremities of the feet. A man enlarges this base, and therefore stands more firm, when he turns his toes out and places his feet a short distance apart. When old and infirm, he enlarges his base and increases his stability still further by using a cane.

When attempting to rise from a sitting position, a man must either bend his body forward or draw his feet backward, in order to bring his centre of gravity over his base; otherwise, he will fall back in making the attempt. So, a person who keeps his heels against a wall, can not stoop without falling, because he has no room to throw the middle of his body far enough back to keep the line of direction within the base.

Fig. 71.

Fig. 72.

Nature teaches a man when descending a height to lean backward, and when ascending to lean forward, as shown in Fig. 71. In like manner, when carrying a weight on one side, we sway our body to the other, like the man with the watering-pot, in Fig. 72. We find it easier to carry a pail of water in each hand than to carry but one, because the weights balance each other, and no effort is necessary to keep the line of direction within the base.

An infant that has not learned to balance itself in a standing position creeps on all fours without danger, because it thus brings its centre of gravity lower and enlarges its base. In order to walk, it must know how to preserve its balance; and, as some practice is necessary for this, the child in its first efforts is likely to fall. The same is the case with a dizzy or an intoxicated person, who for the time loses the power of preserving his balance—that is, of keeping his line of direction within his base.

167. When a person slips on one side, he naturally throws out his arm on the other. He thus seeks to bring back his centre of gravity over his base, and, when he can do so, he saves himself from falling. A person skating has to use his arms constantly for this purpose. Rope-dancers, in performing

Fig. 73.

SHEPHERDS OF LANDES.

their feats, have to shift their centre of gravity from point to point with great rapidity; and, finding their arms insufficient for maintaining their balance on the rope, they use a long pole, with a slight motion of which they can throw the centre of gravity into any desired position.

168. The shepherds of Landes [lond], in the south-west of France, have turned the art of balancing to good account. Having to tend their sheep in a region covered with marsh in winter and hot sand in summer, they mount on stilts about four feet high. Though the centre of gravity is raised, and their liability to fall thus increased, by practising from an early age they become exceedingly expert on these stilts, and can not only walk on them, but even dance, and run so fast that it is hard for a stranger to keep up with them.

169. STABLE AND UNSTABLE EQUILIBRIUM.—The centre of gravity of every body tends to get to the lowest possible point.

carrying a weight on one side? Why do we find it easier to carry a pail of water in each hand than to carry but one? Why is an infant safer when creeping than when attempting to walk? Why does an intoxicated person reel? 167. When a person slips on one side, what does he do, and why? How do rope-dancers preserve their balance? 168. How have the shepherds of Landes turned the art of balancing to practical use? 169. What point does the centre of gravity tend to reach? Illustrate

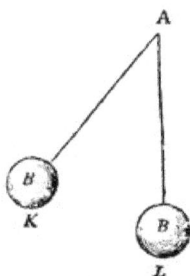

Fig. 74.

A ball suspended by a string, as in Fig. 74, and released from the hand at K, or any other point, will not come to rest till it reaches L, because there its centre of gravity, B, is at its lowest point. Hence, when a pendulum or plummet comes to rest, it always hangs vertically.

A hammer, no matter in what way it is thrown up, descends with its iron part first, because the centre of gravity, which is in that part, tends to get as low as possible. For the same reason, a shuttlecock or an arrow, when it has reached its highest point, turns and descends with its heaviest part foremost.

170. A solid body resting on a surface in such a way that its centre of gravity is lower than it would be in any other position, is said to be in Stable Equilibrium. If its centre of gravity could be brought lower by placing it differently, it is said to be in Unstable Equilibrium.

Fig. 75.

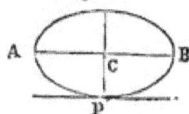

Thus, the oval body, A B, represented in Fig. 75, is in stable equilibrium, because its centre of gravity, C, is at its lowest possible point; and a force applied to either end will not cause it to fall over, but only to rock to and fro.

Fig. 76.

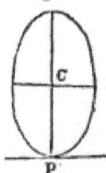

In the position shown in Fig. 76, it is in unstable equilibrium, because its centre of gravity might be brought lower; and a slight push will overturn it and bring it to the position shown in Fig. 75. It is hardly possible to balance an egg on either end; but placed on its side, it rests securely.

171. The stability of a sphere, or oval body like an egg, is increased by cutting it into two equal parts, as shown in Fig. 77. Bases of this shape are used in rocking toys, for supporting the figures of men and animals. Of this shape, also, are some of the huge Rocking Stones found in different parts of Europe, which are so nicely poised that the slightest push causes them to rock to and fro, while a dozen men can not overturn them.

Fig. 77.

172. Paradoxes.—The tendency of the centre of gravity to reach its lowest possible point sometimes produces wonderful effects, or Paradoxes, for which the unlearned are at a loss to account. Thus, we know that a ball will roll down a sloping surface; but a ball of light wood may be made to roll up a sloping surface by inserting a piece of lead in one side.

Fig. 78.

The ball A, for instance, loaded on one side with a plug of lead S, is placed on a sloping surface. The centre of gravity C, which is near S, at once tends to reach its lowest point; and owing to this tendency the ball rolls, till it reaches the position shown in B.

173. In like manner, a double cone, or body having the form of two sugar-loaves joined at their large ends, may be made to roll up an inclined plane. Fig. 79 represents two rails, joined at one

Fig. 79.

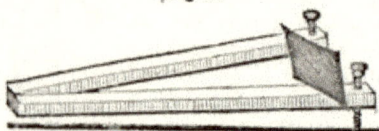

end, but apart and somewhat elevated at the other. Place the double cone at the middle of the rails just described, and instead of rolling down to the narrow end it will roll up to the wide end. This is because the centre of gravity, though apparently going up, is really going down; for, as the rails diverge, they let the double cone further down between them.

---

sphere or oval body be increased? For what are bases of this shape used? What stones are of this shape? 172. What are Paradoxes? How are they sometimes produced? How may a ball be made to roll up a sloping surface? Explain the principle involved, with Fig. 78. 173. Describe the experiment with the double cone, and explain the principle.

# CHAPTER VII.

## MECHANICS (CONTINUED).

THE MOTIVE POWER.—THE RESISTANCE.—THE MACHINE.—
STRENGTH OF MATERIALS.

174. In a previous chapter we have treated of the Laws of Motion; we now proceed to consider the following practical points:—

   I. The Motive Power, or Force by which motion is produced.

   II. The Resistance to be overcome, or *work* to be done, which is always opposed to the Power.

   III. The Machine, which is used by the Power in overcoming the Resistance, when it does not itself directly act.

   IV. The Strength of the Materials employed.

In the case of a steamboat, steam is the Power by which motion is produced; the weight of the boat is the Resistance, which constantly opposes the Power. Since steam can not be directly applied in such a way as to move the boat, a Machine is used to aid in overcoming the Resistance; and this Machine is the engine. On the strength of the materials employed depend the usefulness and safety of the whole.

### Motive Powers.

175. The chief powers used by man in producing motion are gravity, the elastic force of springs, his own strength, the strength of animals, wind, water, and steam.

176. *Gravity.—Springs.—*Gravity is applied by attaching weights to machinery, which they keep in motion by their constant downward tendency, as in certain kinds of

---

174. What four subjects connected with Mechanics are treated of in the present chapter? In the case of a steamboat, what is the power? What, the resistance? What, the machine? On what does the usefulness of the whole depend? 175. Name the chief powers employed by man in producing motion. 176. How is gravity ap-

4*

clocks. When the weight descends so far that it reaches a support, the machinery ceases to move, and is said to "run down". When there is no room to use weights, springs are often substituted for them, as in the works of watches. A spring is made of steel, or some other elastic substance; which, being bent, produces motion by a constant effort to unbend itself.

177. *Strength of Men and Animals.*—With his own strength man can produce a certain degree of motion, but not such as accomplishes the grandest results. From the strength of animals he derives important assistance. Even rude nations tame the animals around them, and turn their strength to account. The American Indians, when first discovered, had not learned to do this; and therefore, like other savages who rely entirely on their own strength, they had made no great advance in agriculture, manufactures, or any other branch of industry.

The horse is the animal whose strength is most widely and advantageously used. For continued labor, one horse is considered equal to five men. A horse of average strength can draw a load of a ton, on a good road, from 20 to 25 miles a day.

178. *Wind and Water.*—Still more powerful forces are found in wind and water, which are extensively used as moving powers by all civilized nations.

The wind is brought to bear, not only on the sails of vessels, but also in mills used for grinding grain, sawing wood, raising water, expressing oil from seeds, &c. Such machines are called Wind-mills; they were introduced into Europe from the East, about the time of the Crusades. The great objection to the wind as a moving power, is its irregularity, for in still weather the machines it moves are useless.

Water is a very powerful and useful agent. A little stream is often a

---

plied? When is the machinery said to run down? When there is no room to use weights, what are often substituted for them? How does a spring produce motion? 177. What is said of the strength of man as a source of motion? What, of the strength of animals? What animal is most widely used? To how many men is one horse considered equal? As regards drawing, what is a day's work for a horse of average strength? 178. What sources of motion are still more powerful? How is the wind brought to bear? What are machines moved by the wind called? Whence and when were wind-mills introduced into Europe? What is the great objection to

source of prosperity and wealth to an extensive region. Affording what is called "water-power", it moves huge machines, and thus affords the means of manufacturing easily and cheaply. Water was first used as a motive power by the Romans, in simple machines for grinding grain, about the commencement of the Christian era. It is now applied in various kinds of machines, for sawing, spinning, weaving, grinding, &c. Though a stream may run so high in spring and so low in summer as to be useless for a time, there is far less difficulty from these causes than from the irregularity of the wind.

179. *Steam.*—The greatest of all the powers employed by man is STEAM, or the vapor generated by submitting water to a high degree of heat. Steam being an elastic fluid, its properties and applications will be considered hereafter.

180. The uses of steam were unknown to the ancients; it was not till near the close of the seventeenth century that its importance began to be realized. Its application to machinery marks an era in the world's history, and has invested man with immense power over matter. Driving the boat and car, it bears him what was once a day's journey in an hour. Applied in countless varieties of machines, it is the means of supplying us with thousands of comforts unknown to our forefathers. The farmer is indebted to it for his spade, hoe, rake, scythe, ploughshare, and all his implements. It helps to make the shears with which he cuts the wool from his sheep, and then cards the wool, and weaves it into cloth. It separates his cotton from its seed, and turns it into muslin and calico. It aids the builder by making his tools, forging his nails and bolts, moulding his ornaments, polishing his marble, cutting his stone, and sawing his wood. It supplies our parlors with furniture, our kitchens with cooking utensils, our dining-rooms with glass and china, knives and forks. It knits, twists, washes, irons, dyes, gilds, grinds, digs, and prints; and hardly any work of art meets our eyes, in making which steam has not been directly or indirectly used. It does all this, moreover, with wonderful precision and rapidity. The pyramids of Egypt, we are informed, kept 100,000 men at work twenty years in their erection. It has been computed that one powerful steam-engine would have done as much work in the same time as 27,000 of these Egyptians.

### The Resistance.

181. Whatever opposes the Power is called the Resistance.

the wind as a moving power? What is said of water-power? By whom and when was it first used? For what purposes is it now employed? What are the disadvantages of water as a moving power? 179. What is the greatest of the powers employed by man? What is Steam? 180. When did its importance begin to be realized? What has been the result of its application to machinery? Enumerate the different articles which steam is constantly employed in producing. What interest-

182. The resistance is not always of the same character. It may be a weight to be lifted, as a pail of water from a well; or a body to be moved onward, as a train of cars; or a wheel to be turned, as in a mill; or particles to be compressed, as in packing cotton in bales; or cohesion to be overcome, as in splitting a log of wood. As the most usual form in which the resistance appears is that of a weight to be moved, the term Weight is often used instead of Resistance, with reference to work of any kind, or whatever opposes the moving power.

183. UNITS OF WORK.—The efficiency of a force is estimated by the resistance it can overcome, or the amount of work it can do. In order to compare different forces, we must have a uniform *unit of work.*

The unit of work adopted is the resistance encountered in raising one pound through the space of a foot. Hence, to raise a body any distance constitutes as many units of work as there are pounds in the body multiplied by the number of feet in the given distance. To raise 2 pounds of water from a well 6 feet deep, is equivalent to twice 6, or 12, units of work. To lift a load of 1,000 pounds 10 feet involves 10,000 units of work.

184. HORSE-POWERS.—In estimating large amounts of work, it is customary to use *horse-powers* as a measure. A horse can perform 33,000 units of work, that is, can raise 33,000 pounds a foot, in a minute. An engine, therefore, that can perform 33,000 units of work in a minute is said to be an engine of one horse-power; one that can do 66,000 units of work in a minute is an engine of 2 horse-powers; and so on. Hence the following

*Rule.*—To find the horse-power of an engine, divide the number of pounds it is capable of raising one foot in a minute by 33,000.

ing fact is stated with respect to the pyramids of Egypt? 181. What is the Resistance? 182. Mention some of the different forms in which the resistance appears, and give examples. What term is often used instead of *resistance,* and why? 183. How is the efficiency of a force estimated? To compare different forces, what is it necessary to have? What is the unit of work generally adopted? Give examples.

185. FRICTION.—The effect of the moving power is often diminished by Friction. ·

Friction is the resistance which a moving body meets with from the surface on which it moves.

If all surfaces were perfectly smooth, there would be no friction; but even those bodies that seem the smoothest are really covered with minute projections and depressions. These fit into each other, and a certain degree of force is required to raise the projections of the one surface over those of the other. With the naked eye we can not detect any unevenness on plate glass or polished steel; yet, if we view either through a microscope, we find that its surface is far from smooth, and hence there is some friction even when these substances are rubbed together.

186. Friction opposes motion in two ways:—

1. By increasing the resistance, as when a weight is dragged over the ground.

2. By diminishing the force before it is applied to the resistance; as in machinery, which sometimes loses as much as one-third of its power by the rubbing of its different parts against each other.

In estimating the working power of a machine for practical purposes, it is necessary to make allowance for the loss occasioned by friction; but, in merely investigating the principles of Mechanics and the construction of machines, we proceed as if the surfaces concerned were perfectly smooth, and no such thing as friction existed.

187. *Kinds of Friction.*—There are two kinds of friction:—

1. Sliding Friction, produced when a body slides on a surface, like the runners of a sleigh.

2. Rolling Friction, produced when a body rolls on a surface, like the wheels of a wagon.

188. Between any given surface and moving body, sliding friction is much greater than rolling friction. Hence we roll a barrel of flour over the ground instead of drag-

184. How are large amounts of work estimated? What is meant by a *horse-power?* Give an example. How may the horse-power of an engine be found? 185. By what is the effect of the moving power often diminished? What is Friction? How is it that friction is exhibited even between surfaces that appear smooth? Give an example. 186. In how many ways does friction oppose motion? Mention them. When is it necessary to make allowance for friction, and when not? 187. How many kinds of friction are there? Name them, and tell how each is produced. 188. Between any

ging it, and place a weight that is to be moved in a cart, or suspend it between wheels, instead of harnessing a horse directly to it.

On the same principle, we place rollers under a block of marble, and fasten castors, or small wheels, to the legs of heavy pieces of furniture. Rollers are

Fig. 80.

also used with advantage in pushing a ponderous packing-box up an inclined plane into a cart, as shown in Fig. 80. In all these cases, sliding friction is converted into rolling, and the resistance is thus diminished. The larger the wheels and rollers employed, up to a certain limit, the greater the gain; but even small ones materially lessen the friction.

Rolling friction, on the other hand, may be converted into sliding. This is done when the wheels of a heavily loaded stage or wagon descending a steep hill are *locked*, that is, prevented from turning by an apparatus provided for the purpose. The resistance is thus increased to such a degree that the load can descend in safety. On the same principle, *brakes* are applied to the wheels of cars, to stop them the sooner.

189. *Laws of Friction.*—Several important laws relating to friction have been settled by experiments. In making these, the apparatus represented in

Fig. 81.

Fig. 81 has been used. D E is a table, on which rests the block C. A string, passing over the pulley B, connects this block with a scale, A. By putting weights in the scale till the block moves, we are enabled to measure its friction; and, by making the block of different materials, varying its size and surface, and allowing it to remain a longer or shorter time on the table, the following laws have been established:—

1. The friction of a body is greater when it commences moving than after it has been moving for a time. Thus it

given surface and moving body, how does sliding friction compare with rolling friction? Mention some familiar cases in which we convert sliding into rolling friction, to lessen the resistance. What is said of the size of the wheels and rollers employed? In what cases is rolling friction converted into sliding? 189. How have the facts relating to friction been settled? Describe the apparatus employed for this purpose. When is the friction of a body greatest? Between what bodies and surfaces is fric-

takes a heavier weight to start the block C than it does afterwards to keep it in motion.

2. Friction is greater between soft bodies than hard bodies, and between rough surfaces than smooth ones. A sled that can hardly be moved over a newly ploughed field, is drawn without difficulty over a frozen pond.

3. In many cases, friction is increased by letting the surfaces remain in contact. At the end of five or six days, it has been found to be fourteen times as great as at first.

4. Between the same surfaces, friction is proportioned to the weight of the moving body. The friction of a block weighing 20 pounds is twice as great as that of a ten-pound block.

5. Within certain limits, friction is not increased by extent of plane surface. As long as the weight of a body remains the same, its friction will not vary, whether it rests on a larger or smaller base. In Fig. 81, the block C has its upper side hollowed out, so that, if turned over, it will rest merely on two ridges; yet the friction will be the same when it rests on that side as on the other.

190. *Modes of Lessening Friction.*—No means has yet been found of doing away with friction altogether; but it may be lessened in three ways:—

1. By smoothing and polishing the surfaces.

2. By putting grease or some other *lubricant,* as it is called, between the surfaces. This fills up their depressions. Finely powdered plumbago (the common black-lead used in pencils), dry for wooden surfaces and mixed with grease for metallic ones, is one of the best articles used for this purpose. The wood-sawyer greases his saw to make it move easily, and cartmen and carriage-drivers keep the

tion greatest? In many cases, how may friction be increased? Between the same surfaces, to what is friction proportioned? What effect is produced on the friction of a body by increasing its surface? Exemplify this with the figure. 190. Can friction be entirely removed? In how many ways may it be lessened? What is the first of these? What, the second? What article makes one of the best lubricants? By whom are lubricants used? How may the friction of a wheel be diminished? What

Fig. 82.

FRICTION WHEELS.

axles of their wheels well covered with some lubricating preparation.

3. The friction of a wheel may be diminished by making its axle, that is, the cylinder running through the centre, turn on the circumferences of two other wheels at each end, as shown in Fig. 82. Such wheels are called Friction Wheels. They are used in delicate machinery.

191. *Uses of Friction.*—Though friction occasions a great loss of power, it is not without its beneficial effects. A river is prevented from rushing madly through its channel by the friction of its waters on its banks and bed. A tempest gradually loses its force by the friction of the air against the projections on the earth's surface. It is friction that prevents the fibres of wool, hemp, and cotton, when twisted together, from slipping on each other and giving way. Without friction nails would be useless, for they would draw right out; the wheels of a carriage would turn on the ground without moving it forward; and neither man nor beast could walk. It is the friction of our feet on the ground that enables us to take steps: when the friction is lessened, as on smooth ice, we walk with difficulty; were there no friction, we should find it impossible to walk at all.

## Machines.

192. Machines are instruments used to aid the Power in overcoming the Resistance.

193. Simple machines used by the hand, are called Tools; as, the chisel, the saw.

194. Machines of great power are called Engines; as, the steam-engine, the fire-engine.

195. Machines merely aid the power in its action; *they can not create power.* This follows from the inertia of matter. The mightiest engine, therefore, remains at rest until acted on by some motive power; and, when thus acted on, it can not increase the power in the smallest degree, but on

the other hand diminishes it, more or less according to the friction of its parts.

If a man standing over a pit 100 feet deep can, in the space of a minute, just pull to the top a tub containing 100 pounds of coal, no machine can enable him to raise a single pound more in the same time. By using pulleys, he may, to be sure, raise 600, 800, or 1,000 pounds at a time, but it will take him 6, 8, or 10 times as long as before; and, therefore, in the same time he will do no more work than with his hands alone—but less, on account of the friction of the pulleys. So, a certain amount of steam, just capable of performing 50,000 units of work in a minute, can not by any machinery be made to perform a single additional unit of work in the same time. Hence the great universal law which follows:—

196. *What a machine gains in amount of work, it loses in time; and what it gains in time, it loses in amount of work.*

Let us apply this law. A quantity of steam capable of moving 50,000 pounds a foot in a second, may be made to move 100,000 pounds a foot, but it will be two seconds in doing it; or it may move the weight a foot in half a second, but in that case it will move no more than 25,000 pounds. Under no circumstances can there be a gain in units of work without a corresponding loss of time, or a gain in time without a corresponding loss of units of work.

197. PERPETUAL MOTION.—By Perpetual Motion is meant the motion of a machine, which, without the aid of any external force, on once being set in operation, would continue to move forever, or until it wore out.

Such a machine many have tried to invent, but without success. Friction and the resistance of the air are constantly opposing the action of machinery; and as matter, on account of its inertia, can generate no power that will compensate for this loss, every machine must in time come to rest, unless some external force, such as wind, water, or steam, keeps acting upon it. Hence Perpetual Motion is impossible.

198. ADVANTAGES OF USING MACHINERY.—If no additional power is generated by machinery, but there is an actual loss from the friction of its parts, why is it employed? —Because in other respects its use is attended with important advantages, among which are the following:—

1. Machinery enables us, with a certain amount of power, by taking a longer time, to do pieces of work that we could not otherwise do at all.

Fig. 83.

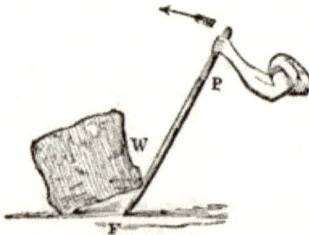

Thus, a farmer with a crow-bar, as shown in Fig. 83, can move a rock which with his hands alone he could not stir. With the aid of two other men, he could carry it or push it where he wanted, in one-third of the time that he could move it there alone with the crow-bar; but he may not have two others at hand to help him.

With machinery 10 men may do the work of 1,000. Of course it will take them 100 times as long; but this loss of time is of little consequence, compared with the difficulty of getting a thousand men together and placing them so as to work without interfering with each other. Some heavy pieces of work are of such a nature that but few laborers can get around them at a time; in these cases, unless the work can be divided, which is not always possible, it must remain undone without the aid of machinery.

2. Machinery enables us to use our power more conveniently.

The farmer removes a rock from his field with less difficulty and fatigue by means of a crow-bar than if he stooped over to lift it with his hands. The porter with his block and tackle hoists a box of goods to a loft with far greater ease than he could push or carry it up. The apparatus he uses enables him to hoist the load by pulling down upon a rope, and when pulling down his weight aids his strength.

Fig. 84.

3. Machinery enables us to use other motive powers besides our own strength.

A horse without machinery can not lift a weight; but he does it readily with the aid of the simple apparatus shown in Fig. 84. Steam, applied directly to a boat, can not move it forward; it is only with the help of machinery that it causes the wheel to revolve and thus produces motion. Here, as in all other cases, the power *is not created* by the machinery, but merely *transmitted* in a way to make it effective.

## Strength of Materials.

199. There is a limit to the power of all machinery; and this limit is the strength of the materials of which it is made. Machines that work well in small models sometimes utterly fail when made of full size, because, when the resistance is increased and their own weight is added, no material can be found strong enough to stand the strain.

Nature, also, recognizes this limit of size. Animals, after attaining a certain age, cease to grow. If they kept on growing, they would soon reach such a size and weight that they could not move. If there were an animal much larger than the elephant, it would stagger under its own weight, unless its bones and muscles were thicker and firmer than any with which we are now acquainted. Fish, on the contrary, being supported by the water, move freely, no matter how heavy they may be. Whales have been found over 50 feet long and weighing 70 tons—a monstrous size and weight, which no land animal could support.

200. To determine how great a strain given materials will bear, and how they may be put together with the greatest advantage, becomes an important question in Practical Mechanics. The relative strength of different substances has been treated of under the head of Tenacity, on page 23. The following general principles relating to rods, beams, &c., should be remembered.

1. Rods and beams of the same material and uniform size throughout, resist forces tending to break them in the direction of their length, with different degrees of strength, according to the areas of their ends.

Let there be two rods of equal length; if the areas of their ends are respectively 6 and 3 square inches, the one will bear twice as great a weight

sents itself? What is the second advantage of using machinery? How is this exemplified in the case of the farmer? How, in the case of the porter? What is the third advantage gained by using machinery? Illustrate this in the case of a horse. In the case of steam. In both of these cases, what does the machinery merely do? 199. What limit is there to the power of all machinery? Why do machines often fail, though small models of them work well? Show how nature recognizes a limit of size. How is it that fish can move, though much larger and heavier than land animals? 200. What important question is presented in Practical Mechanics? What is the first principle laid down respecting rods and beams? Give an example. When a rod is very long,

without breaking as the other. This law applies, no matter what the shape of the rods may be.

2. When a very long rod is suspended vertically, its upper part, having to support more of the weight of the rod than any other, is the most liable to break.

3. The strength of a horizontal beam supported at each end diminishes as the square of its length increases.

If two beams thus placed are respectively 6 feet and 3 feet long, the strength of the shorter will be to that of the longer as the square of 6 to the square of 3,—that is, as 36 to 9, or 4 to 1.

4. A horizontal beam supported at each end, is most easily broken by pressure or a suspended weight in the middle, and increases in strength as either end is approached. If, therefore, a beam of uniform strength is required, it should gradually taper from the middle towards the ends.

5. A given quantity of material has more strength when disposed in the form of a hollow cylinder than in any other form that can be given it. Nature constantly uses hollow cylinders in the animal creation, as in bones and the tubes of feathers; and the artisan, imitating nature, employs it in many cases where strength and lightness are to be combined.

### EXAMPLES FOR PRACTICE.

1. (*See* §§ 183, 184.) What is the horse-power of a steam-engine that can do 1,650,000 units of work in a minute?

2. What is the horse-power of an engine that can raise 2,376 pounds 1,000 feet in a minute?

3. What is the horse-power of an engine that can raise 1,000 pounds 2,376 feet in a minute?

4. A fire-engine can throw 220 pounds of water to a height of 75 feet every minute, what is its horse-power?

5. A cubic foot of water weighs 62½ pounds. How many horse-powers are required to raise 200 cubic feet of water every minute from a mine 132 feet deep?

what part of it is most likely to break? What law is given respecting the strength of a horizontal beam supported at each end? Give an example. In what part is a horizontal beam supported at each end most easily broken by pressure? What shape gives a beam uniform strength? In what form must a given quantity of material be disposed, to have the most strength?

6. A locomotive draws a train of cars, the resistance of which (caused by friction, &c.) is equivalent to raising 1,000 pounds, 15 miles an hour; what is its horse-power?

[*Find how many feet the locomotive draws the train in a minute, and then proceed as before.*]

7. How many pounds can an engine of 10 horse-powers raise in an hour from a mine 100 feet deep?

8. A certain man has strength equivalent to $\frac{1}{8}$ of one horse-power; how many pounds can he draw up in a minute from a pit 25 feet deep?

9. (*See* § 189, *Fourth Law.*) If the friction of a train of cars weighing 50 tons, on a level railroad, be equivalent to a weight of 500 pounds, what will be the friction of a train weighing 25 tons? of one weighing 100 tons? of one weighing 60 tons?

10. (*See* §§ 195, 196.) C can just draw 75 pounds of coal a minute out of a mine. With the aid of a system of pulleys, he can raise 225 pounds at a time; the friction being equivalent to 75 pounds, how many minutes will he be in raising the load?

[*In practical questions of this kind, the friction must be added to the resistance.*]

11. With a certain machine, one man can do as much as eight men without the machine. Allowing the friction of the machine to be equal to one-fourth of the resistance, how much longer will he be in doing a certain amount of work than they?

12. (*See* § 200.) [*The area of a rectangular surface is found by multiplying its length by its breadth; that of a triangle, by multiplying half its base by its perpendicular height.*] Which will support the greater weight without breaking, a joist whose section is 4 inches long by 5 broad, or one of the same kind of wood, 3 inches by 8?

13. Which, when suspended, will bear the greater weight without breaking, a square rod of iron whose end is 3 inches by 3, or a rod whose cross section is a triangle with a base of 6 inches and a perpendicular height of 2?

14. Two rods of copper, of equal length and uniform thickness, have ends respectively 4 inches by 2, and 17 inches by half an inch. Which, when suspended, will support the greater weight?

15. Two horizontal beams of the same material, breadth, and thickness, supported at both ends, are respectively 2 and 14 feet long. Which is the stronger of the two, and how many times?

# CHAPTER VIII.

## MECHANICS (CONTINUED).

### THE MECHANICAL POWERS.

NUMEROUS and varied as machines are, they are all combinations of six Simple Mechanical Powers, known as the Le'-ver, the Wheel and Axle, the Pulley, the Inclined Plane, the Wedge, and the Screw. These we shall consider in turn.

### The Lever.

201. A Lever is an inflexible bar, capable of being moved about a fixed point, called the Fulcrum.

The lever is the simplest of the mechanical powers. Its properties were known as far back as the time of Aristotle, 350 years B. C. Archimedes, a hundred years later, was the first to explain them fully.

202. KINDS OF LEVER.—In the lever three things are to be considered; the fulcrum, or point of support, the weight, and the power. Two of these are at the ends of the bar, while the other is at some point between them. According to their relative position, we have three kinds of levers:—

Fig. 85.

A Lever of the First Kind is one in which the fulcrum is between the power and the weight; as in Fig. 85, where F represents the fulcrum, P the power, and W the weight.

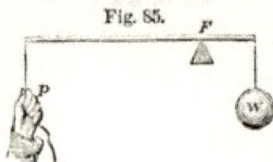

A Lever of the Second Kind is one in which the weight is between the power and the fulcrum; as in Fig. 86.

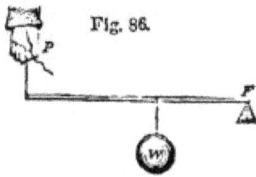
Fig. 86.

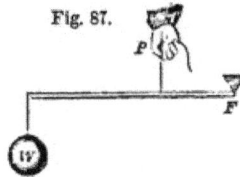
Fig. 87.

A Lever of the Third Kind is one in which the power is between the weight and the fulcrum; as in Fig. 87.

203. LEVERS OF THE FIRST KIND.—In levers of the first kind, the relative position of the three important points is

POWER FULCRUM WEIGHT OR WEIGHT FULCRUM POWER.

Fig. 88 shows one of the commonest forms in which this kind of lever appears,—the crow-bar. The power is applied at the handle. The weight is at the other end, and consists of something to be moved. The fulcrum is a stone on which the crow-bar rests. Using an instrument in this way is called *prying*.

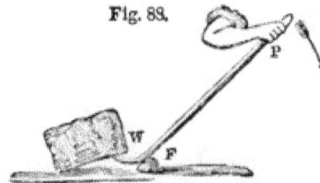
Fig. 88.

THE CROW-BAR.

204. The nearer the fulcrum is to the weight the greater the advantage gained, and consequently the greater the space that P will have to pass through in moving W a given distance. This principle is stated in the following

*Law.— With levers of the first kind, intensity of force is gained, and time is lost, in proportion as the distance between the power and the fulcrum exceeds the distance between the weight and the fulcrum.*

Thus, in Fig. 88, if the distance from P to F be five times as great as that from W to F, a pressure of 10 pounds at P will just counterbalance a weight of 50 pounds at W, and will therefore move anything under 50 pounds; while, for every inch that W is moved upward, P will have to move five inches downward.

The distance through which the power must pass, to move a weight vastly greater than itself, becomes an important matter in practical applications of the lever. When Archimedes saw the immense power that could be exerted with this instrument, he declared that with a place to stand on he could move the earth itself. He did not say how far he would have to travel to do this, in consequence of the great disproportion between his strength and the earth's bulk. Allowing that he had a place to stand on and a lever strong enough, and could pull its long arm with a force of 30 pounds through two miles every hour, it would have taken him, working ten hours a day, over one hundred thousand millions of years to move the earth a single inch!

205. *The Balance.*—When bodies of equal weight are supported by the arms of a lever, they will balance each other when placed at equal distances from the fulcrum, as in Fig. 89. They are then said to be *in equilibrium.*

Fig. 89.

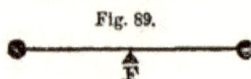

Fig. 90.

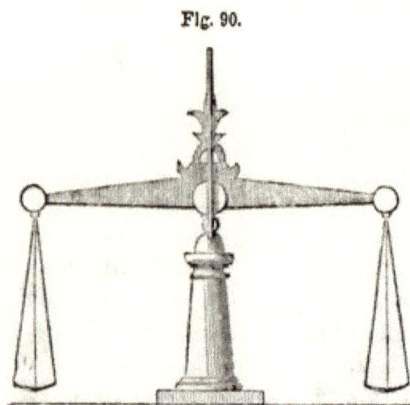

THE BALANCE.

On this principle the common Balance, represented in Fig. 90, is constructed. A beam is poised on the top of a pillar, so as to be exactly horizontal. From each end of the beam, at equal distances from the fulcrum, a pan is suspended by means of cords. The object to be weighed is placed in one of these pans, and the weights in the other.

When great accuracy is required, the beam is balanced on a steel knife-edge; the friction being thus lessened, it turns more easily. A balance capable of weighing ten pounds has been made so sensitive as to turn with the thousandth part of a grain.

206. The balance weighs correctly only when the arms of the beam are exactly equal. Hence dishonest tradesmen sometimes defraud those with whom they deal by throwing the fulcrum a little nearer one end of the beam than the other. When buying, they place the commodity to be weighed in the scale attached to the short arm; and, when selling, in the other, thus making double gains. To prove a balance, weigh an article first in one scale and then in the other; if there is any difference in the weight, the balance is not true.

tical applications of the lever? Show this in the supposed case of Archimedes. 205. When are two bodies of equal weight, supported by the arms of a lever, said to be in equilibrium? What is constructed on this principle? Describe the Balance. When great accuracy is required, how is the beam balanced? How sensitive has a balance been made? 206. When does the balance weigh correctly? How do dishon-

The true weight of a body may be determined, with a false balance, by placing it in either scale, balancing it with shot or sand, and then removing the body and replacing it with weights till equilibrium is established. This is called *double weighing*. It must give the true weight; for whatever error is made in the first weighing is corrected in the second.

207. *The Steelyard.*—When bodies of unequal weight are supported by the arms of a lever, they will balance each other whenever the weight of the one multiplied into its distance from the fulcrum, is equal to the weight of the other multiplied into its distance from the fulcrum.

In Fig. 91, let the distance W F be one inch and P F three inches. The weight of the one body, 30 pounds, multiplied into its distance from the fulcrum, 1, gives 30; the weight of the other, 10 pounds, multiplied into its distance from the fulcrum, 3, gives 30. These products being equal, the bodies will balance each other.

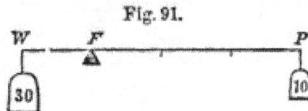

Fig. 91.

208. On this principle the Steelyard is constructed. The Steelyard is a kind of balance, which, though not so sensitive as the one described above, answers very well for heavy bodies, and is conveniently carried, as it requires but a single weight, and may be held in the hand or suspended anywhere.

Fig. 92 represents the steelyard. It is a lever of unequal arms; from the shorter of which the article to be weighed is suspended, either directly or in a scale-pan, while a constant weight is moved on the longer arm from notch to notch till equilibrium is established. The number at the notch on which the weight

Fig. 92.

THE STEELYARD.

then rests, shows the required weight in pounds. Thus, 15 pounds is the weight of the sugar-loaf in the Figure. The proper distances for the notches are found in the first place by experiments with known weights in the scale-pan.

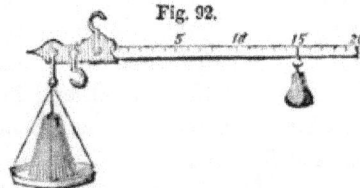

To enable the steelyard to weigh still heavier objects without increasing

est tradesmen sometimes defraud those with whom they deal? How may a balance be proved? How may the true weight of a body be determined with a false balance? What is this process called? 207. When will bodies of unequal weight supported by the arms of a lever be in equilibrium? Illustrate this with Fig. 91. 208. What is constructed on this principle? Describe the Steelyard, and the mode of weighing with it. How are the proper distances for the notches found in the first place? With

5

the length of its beam, it is often provided with an additional hook, hanging in an opposite direction from the other hook and nearer the point from which the article to be weighed is suspended. When the instrument is supported by this hook, a new fulcrum is formed, and the weight is shown by a new row of notches adapted to it. The greater the difference of length between the arms of a steelyard, the greater the number of pounds that it can weigh.

209. When more than two bodies are supported on the arms of a lever, if the weight of each be multiplied by its distance from the fulcrum, the lever will be in equilibrium (that is, the bodies will balance each other) when the sums of the products on the two sides of the fulcrum are equal.

Fig. 93.

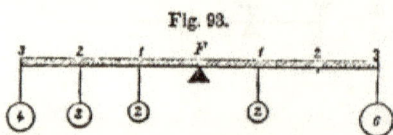

Thus, in Fig. 93 equilibrium is maintained, because the products of the weights on one side into their distances, added together, equal the sum of the products on the other :—

weights distances
2 × 1 = 2
3 × 2 = 6
4 × 3 = 12
Sum of products, 20

weights distances
2 × 1 = 2
6 × 3 = 18

Sum of products, 20

210. *Practical Applications.*—Familiar examples of levers of the first kind are found in the scissors and pincers; the rivet connecting the two parts being the fulcrum, the fingers the power, and the thing to be cut or grasped the weight. A poker introduced between the bars of a grate and allowed to rest on one of them, that *purchase* may be obtained for stirring the fire, is a lever of the first kind. So is the handle of a common pump.

Fig. 94.

When children teeter on a board balanced on a wooden horse, they use a lever of the first kind. According to the principles of the lever, if one is heavier than the other, to preserve the balance, he must sit nearer the fulcrum, as shown in Fig. 94.

what are some steelyards provided, and for what purpose? What steelyards weigh the greatest number of pounds? 209. If more than two bodies are supported on the arms of a lever, when will they balance each other? Apply this principle in Fig. 93. 210. Give some familiar examples of levers of the first kind. When children teeter on a board, what kind of lever do they use? If one is heavier than the other, where

211. *Bent Levers.*—Sometimes the arms of a lever are bent, instead of straight. In that case the same principles hold good, only that the arms of the lever are estimated, not by their actual length, but by the perpendicular distance from the fulcrum to the line of direction in which the power and weight respectively act.

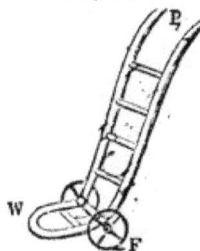

Fig. 95.

As an illustration of bent levers of the first kind, we may take the truck used for moving heavy articles, represented in Fig. 95. The axis on which the wheels turn represents the fulcrum; the weight is applied at W, and the power at P. The clawed side of a hammer, used in drawing out nails, is also a bent lever. The fixed point on which the head of the hammer rests is the fulcrum; the friction of the nail is the weight; and the power is applied at the extremity of the handle.

THE TRUCK.

212. *Compound Levers.*—Simple levers of the first kind may be combined into Compound Levers.

213. In compound levers, equilibrium is established when the power, multiplied by the first arms of all the levers, is equal to the weight multiplied by the last arms of all the levers.

Fig. 96.

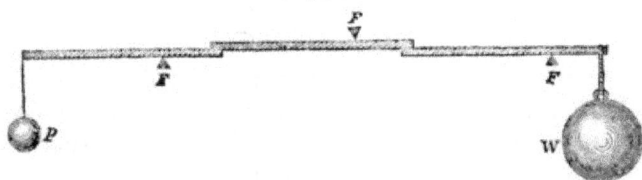

A COMPOUND LEVER.

Thus, in Fig. 96, which represents a compound lever formed of three simple ones, let the long arm of each lever be three times the length of its short arm; then 1 pound at P will balance 27 pounds at W, because

1 pound $\times 3 \times 3 \times 3 = 27$ pounds $\times 1 \times 1 \times 1$.

214. LEVERS OF THE SECOND KIND.—In levers of the second kind, the relative position is

POWER WEIGHT FULCRUM OR FULCRUM WEIGHT POWER.

Fig. 97.

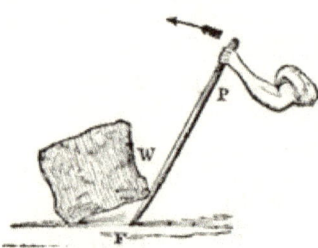

Fig. 97 shows how the crow-bar may be used as a lever of the second kind. The power is applied at the handle; the fulcrum is at the other end, and the weight to be moved is between them.

215. The nearer the weight is to the fulcrum the greater the advantage gained, and consequently the greater the space that P will have to pass through in moving W a given distance. This principle is stated in the following

*Law.— With levers of the second kind, intensity of force is gained, and time is lost, in proportion as the distance between the power and the fulcrum exceeds the distance between the weight and the fulcrum.*

Thus, in Fig. 97, if the distance P F be five times as great as W F, a pressure of 10 pounds at P will counterbalance a weight of 50 pounds at W, and move any thing under 50 pounds; while, for every inch that W is moved, P will have to move five inches in the same direction.

Fig. 98.

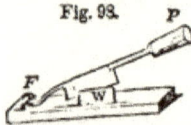

THE CHIPPING-KNIFE.

216. *Practical Applications.*—The common chipping-knife, used by apothecaries, and represented in Fig. 98, is a familiar illustration of levers of the second kind. The knife is fastened at one end, F, which thus becomes the fulcrum; the hand is applied, as the power, at the other end, P; and the substance to be cut is the resistance, or weight, between them. Nutcrackers and lemon-squeezers work on the same principle, and are levers of the second kind.

A door turned on its hinges, and an oar used in rowing, are also examples of this kind of lever. In the former case, the hinge is the fulcrum; the hand applied at the knob is the power; and the weight of the door, which may be re-

of the second kind, what is the relative position of the three important points? How may the crow-bar be used as a lever of the second kind? 215. What is the law of levers of the second kind? Apply this in Fig. 97. 216. What familiar articles will serve as illustrations of levers of the second kind? Show how a door turned on its hinges is

garded as concentrated in its centre of gravity somewhere between the two, is the resistance. In the latter case, the point at which the oar enters the water is the fulcrum; the rower's hand is the power; and the weight of the boat, acting at the row-lock, is the resistance. According to the law laid down in § 215, the further from the row-lock we grasp the oar, the more easily we overcome the resistance and produce motion.

217. Two persons carrying a weight suspended from a stick between them, use a double lever of the second kind. Power is applied at each end, and each end in turn becomes the fulcrum to the other, the weight resting on some intermediate point. The relation of the power at one end to the weight is governed by the same law as that of the power at the other end; and there-fore the weight, to be divided equally, must be suspended from the middle of the stick. If it is not so suspended, the man who is nearer the weight car-ries more than the other in proportion as he is nearer.

Thus, let a 12-pound weight, W, be suspended from a bar three feet long, at a distance of one foot from A and two feet from B. Then A will carry two-thirds of the weight, and B one-third. On

Fig. 99.

this principle, when it is desired that one of the horses harnessed to a car-riage should draw more than the other, it is necessary only to make the arm of the whiffle-tree to which he is attached proportionally shorter.

Fig. 100 shows how a weight may be equally distributed between three persons. B, being twice as far from E as D is, bears one-third of the weight, W; while A and C, at the extremities

Fig. 100.

of the equal-armed lever A D C, bear equal portions of the remaining two-thirds, or one-third each.

## 218. LEVERS OF THE THIRD KIND.—In le-vers of the third kind, the relative position is

Fig. 101.

FULCRUM POWER WEIGHT OR WEIGHT POWER FULCRUM.

The forceps, represented in Fig. 101, is a lever of the third kind. The two sides unite at one end to form the ful-crum; the article to be grasped is the weight; and the fin-gers, applied between the two, constitute the power.

## 219. Levers of the third kind, unlike those before described, involve a mechanical disad-

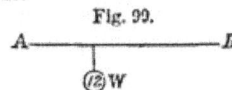

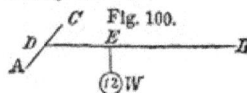

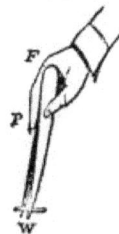

a lever of the second kind. Show how an oar acts as a lever of the second kind. 217. When two persons carry a weight suspended from a stick between them, what kind of a lever do they use? Where is the fulcrum? To be equally divided, where must the weight be suspended? If the weight does not hang from the middle of the

vantage; that is, to produce equilibrium, the power must always be greater than the weight.

*Law.* — *With levers of the third kind, intensity of force is lost, and time is gained, in proportion as the distance from the weight to the fulcrum exceeds the distance from the power to the fulcrum.*

Thus, in Fig. 101, if F W be three times as great as F P, it will require a power of three pounds at P to counterbalance a resistance of one pound at W. Levers of this class, therefore, are never used when great power is required, but only when a slight resistance is to be overcome with great rapidity.

220. *Practical Applications.*—The sugar-tongs, which resembles in shape the forceps above described, is a familiar example of the third kind of lever. So is the fire-tongs; and hence the difficulty of raising heavy pieces of coal with this instrument, particularly when the hand is applied near the rivet or fulcrum.

The sheep-shears is another lever of the third kind, admirably adapted to the work it performs; because the wool, being flexible, has to be cut rapidly, while it does not require any great degree of force.

A door becomes a lever of the third kind when one attempts to move it by pushing at the edge near the hinges. The mechanical disadvantage is shown by the great strength required to move it when the power is there applied. So, when a painter attempts to raise a ladder lying on the ground with its bottom against a wall, by lifting the top and walking under it grasping round after round in succession, he experiences great difficulty as he approaches the bottom, because the ladder, when he passes its centre of gravity, becomes a lever of the third kind.

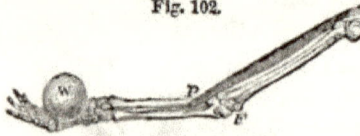

Fig. 102.

HUMAN ARM AND HAND.

Nature uses levers of the third kind in the bones of animals. The fore-arm of a man, represented in Fig. 102, will serve as an example.

stick, which man will carry the more? Illustrate this with Fig. 99. How may one of the horses harnessed to a carriage be made to draw more than the other? How may a weight be equally distributed between three persons? 218. In levers of the third kind, what is the position of the three important points? What instrument is an example of the third kind of levers? 219. To produce equilibrium in the third kind of levers, what is necessary? State the law for levers of the third kind. Illustrate this with Fig. 101. 220. What common articles are levers of the third kind? What is said of the sheep-shears? When does a door become a lever of the third kind?

The fulcrum, F, is at the elbow-joint; the biceps muscle, descending from the upper part of the arm and inserted near the elbow at P, operates as the power; while the weight, W, rests on the hand. If the distance F W be 15 times as great as F P, it will take a power of 15 pounds at P to counterbalance one pound at W; and when the arm is extended, the disadvantage is still greater, in consequence of the muscle's not acting perpendicularly to the bone, but obliquely.

This accounts for the difficulty of holding out a heavy weight at arm's length. In proportion as power is lost, however, quickness of motion is gained; a very slight contraction of the muscle moves the hand through a comparatively large space with great rapidity. Here, as in all the works of creation, the wisdom of Providence is shown in exactly adapting the part to the purpose for which it is designed. With so many external agents at his command, man does not need any great strength of his own; quickness of motion is much more necessary to him, and this the structure of his arm ensures.

### The Wheel and Axle.

221. The Wheel and Axle is the second of the simple mechanical powers. It consists of a Wheel attached to a cylinder, or Axle, in such a way that when set in motion they revolve around the same axis.

222. In the simplest form of the wheel and axle, the power is applied to a rope passing round the wheel, while the weight is attached to another rope passing round the axle.

Fig. 103.

This form of the machine is shown in Fig. 103. C D is a frame; B is the wheel; A is the axle, attached to the frame at its extremities E and F by gudgeons, or iron pins, on which it turns. P is the power, and W is the weight.

223. The wheel and axle is simply a revolving lever of the first kind. One application of the lever can not move a body any great distance; but, by means of the wheel and axle, the action of the lever is continued unin-

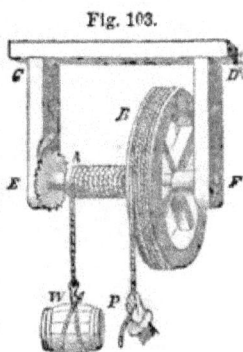

THE WHEEL AND AXLE.

Under what circumstances does a ladder become a lever of the third kind? In what does Nature use levers of the third kind? Show, by Fig. 102, how the fore-arm is a lever, and point out the relation between power and weight. How is the wisdom of Providence shown, in making the arm such a lever? 221. What is the second simple mechanical power? Of what does the Wheel and Axle consist? 222. In the simplest form of this machine, how is the power applied, and how the weight? Illus-

terruptedly.   This machine has therefore been called *the perpetual* or *endless lever*.

224. The wheel and axle must turn round their common axis in the same time.   In each revolution, a length of rope equal to the wheel's circumference is pulled down from the wheel, while only as much rope is wound round the axle as is equal to the axle's circumference.   There is, therefore, a loss of time, greater or less according as the circumference of the wheel exceeds that of the axle; but, by the law of Mechanics already stated, there must be a corresponding gain of power.

Viewing the wheel and axle as a lever of the first kind, we have the circumference of the wheel for the long arm, and that of the axle for the short arm.   If the diameters of the wheel and the axle are given instead of their circumferences, they may be taken for the two arms; and so with the radii, if they are given.   In practice, an allowance of 10 per cent. of the weight must be made for the stiffness of the ropes and the friction of the gudgeons.   —From these principles is deduced the following law :—

225. LAW OF THE WHEEL AND AXLE.— *With the wheel and axle, intensity of force is gained, and time is lost, in proportion as the circumference of the wheel exceeds that of the axle.*

Thus, in Fig. 103, if the circumference of the wheel B is five feet and that of the axle A one foot, a power of 40 pounds at P will counterbalance a weight of 200 pounds at W, and of course lift any thing under 200 pounds.

226. DIFFERENT FORMS.—The wheel and axle is extensively used, and assumes a variety of forms.

Fig. 104.

Instead of having a rope attached to it, the wheel is often provided with projecting pins, as shown in Fig. 104, to which the hand is directly applied.   This form of the machine is used in the pilot-houses of steamboats for moving the rudder.   In calculating the advantage in this case, instead of the circumference of the wheel we must take the circumference of the circle described by the point to which the hand is applied.

A still more common form, much used in drawing water from wells and loaded buckets from mines, is shown in Fig. 105.   Instead of a wheel, we

trate this with Fig. 103.   223. What has the wheel and axle been called, and why? 224. Explain the operation of the wheel and axle, and show how great the loss of time and gain of power will be.   Viewing the wheel and axle as a lever, what is the long arm?   What is the short arm?   What, besides the circumference, may be taken as the arms of the lever?   What allowance must be made in practice?   225. State the law of the wheel and axle.   Illustrate this law with Fig. 103.   226. Describe the form of

have here a *Winch*, or handle, attached to the axle. In this case, to calculate the advantage gained, we must compare the circle described by the extremity of the handle (shown in the Figure by a dotted line) with the circumference of the axle.

Fig. 105.

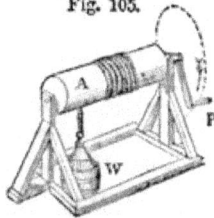

Fig. 106.

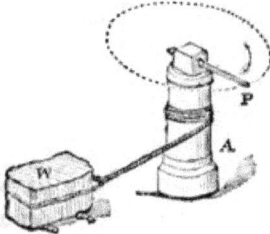

Fig. 106 shows a third form of the wheel and axle. Here the axle A is vertical, instead of horizontal. A bar inserted in its head, at the extremity of which the hand is applied, takes the place of the wheel. If the circumference of A is 3 feet and the circle described by P is 12 feet, a power of 1 pound at P will counterbalance a weight of 4 pounds at W.

227. *The Capstan.*—The Capstan (see Fig. 107) is a familiar example of this form of the wheel and axle. It is used by sailors for warping vessels up to a dock, raising anchors, &c.; and consists of a massive piece of timber, round which a rope passes. This is surmounted by a circular head, perforated with holes, into which, when the instrument is to be used, strong bars, called

Fig. 107.

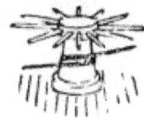

THE CAPSTAN.

*handspikes,* are inserted. Several men may work at each handspike, pushing it before them as they walk round the capstan. The handspikes act on the principle of the lever. The longer they are, therefore, the more easily the men overcome the resistance, but the further they have to walk in doing it.

228. *The Windlass.*—This is a similar form of the wheel and axle, used on shipboard for various purposes.

The windlass is not vertical, like the capstan, but horizontal or parallel to the deck. It is a round piece of timber, supported at each end, and perforated with rows of holes. Pushing against handspikes inserted in these

the wheel and axle used in the pilot-houses of steamboats. In calculating the advantage in this case, what must we substitute for the circumference of the wheel? Describe the form of the machine used in drawing water from wells. How is the advantage ascertained in this case? Describe a third form of the wheel and axle, exhibited in Fig. 106. 227. What machine is a familiar example of this third form? For what is the Capstan used? Of what does it consist? How is it worked? How do the handspikes act? 228. What similar instrument is often substituted for the cap-

boles, the boatmen turn the barrel of the windlass halfway over. It is held there by a suitable apparatus, till the handspikes are removed and put in a new row of holes, when the process is repeated. The windlass acts on the same principle as the capstan, but is less convenient, on account of the manner in which the force is applied, and the necessity of removing the handspikes to new holes from time to time.

229. Wheels enter largely into machinery. The modes of connecting them will be considered hereafter.

### The Pulley.

230. The Pulley is the third of the simple mechanical powers. It consists of a wheel with a grooved circumference, over which a rope passes, and an axis or pin, round which the wheel may be made to turn. The ends of the axis are fixed in a frame called a *block*.

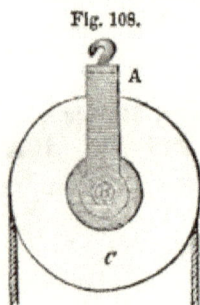

Fig. 108.

THE PULLEY.

Fig. 108 gives a view of the pulley. A represents the block, B the axis, and C the wheel. Round the groove in the wheel passes a rope, at one end of which the power acts, while the weight is attached to the other.

231. KINDS OF PULLEY.—Pulleys are of two kinds,—Fixed and Movable.

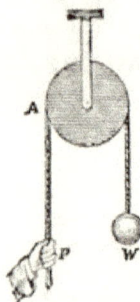

Fig. 109.

FIXED PULLEY.

232. *Fixed Pulleys.*—A Fixed Pulley is one that has a fixed block.

Fig. 109 represents a fixed pulley. The block is attached to a projecting beam. P is the power, and W the weight. For every inch that P descends, W ascends the same distance. There is, therefore, no loss of time, and no gain in intensity of force. One pound at P will just counterbalance one pound at W.

233. In this rule, as well as all the others pertaining to the Mechanical Powers, it must be remembered that friction is not taken into account. In the case of the pulley, in consequence of the stiffness of the rope and the friction of the pin, an allowance of 20 per cent. of the weight, and often more, must be made in practice.

234. Though no power is gained with the fixed pulley, it is frequently used to change the direction of motion. The sailor, instead of climbing the mast to hoist his sails, stands on deck, and by pulling on a rope attached to a pulley raises them with far less difficulty. With equal advantage the builder uses a fixed pulley in raising huge blocks of stone or marble, and the porter in hoisting heavy boxes to the lofts of a warehouse.

Fig. 110.

235. With two fixed pulleys, horizontal motion may be changed into vertical; horses are thus enabled to hoist weights, as shown in Fig. 84.

236. Fig. 110 shows how a person may raise himself from the ground, or let himself down from a height, by means of a fixed pulley. In lofty buildings an apparatus of this kind is sometimes rigged near a window, to furnish means of escape in case of fire.

FIRE ESCAPE.

237. *Movable Pulleys.*—A Movable Pulley is one that has a movable block.

Fig. 111.

Fig. 111 represents a movable pulley. A is the wheel. One end of the rope is fastened to a support at D, while the power is applied to the other at P.

238. To raise the weight a given distance with the movable pulley, the hand must be raised twice that distance. Time, therefore, being lost in the proportion of 2 to 1, the intensity of the force is doubled. A power of one pound at P will counterbalance two pounds at W, and raise anything under two pounds.

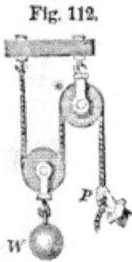

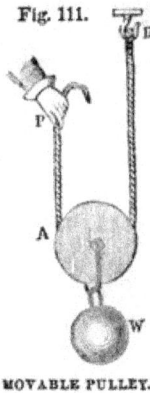

Fig. 112.

239. A movable pulley is seldom used alone. It is generally combined with a fixed pulley, as shown in Fig. 112. No additional power is thus

MOVABLE PULLEY.

what does the Pulley consist? What is the Block? Point out the parts of the pulley in Fig. 103. 231. How many kinds of pulleys are there? 232. What is a Fixed Pulley? Point out the parts in the Figure. What is the gain with this pulley? 233. What allowance must be made for friction in the case of the pulley? 234. If no power is gained by the use of the fixed pulley, why is it used? Give examples. 235. How may horizontal motion be changed into vertical? 236. What does Fig. 110

Fig. 113.

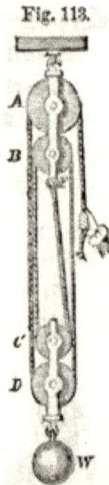

gained; on the contrary, there is a loss, the friction of two pulleys being double that of one. But this loss is more than counterbalanced by the greater convenience of pulling downward.

240. When a high degree of force is required, several movable and fixed pulleys may be combined, as represented in Fig. 113. A and B are fixed pulleys; C and D are movable ones, from the block of which the weight W is suspended. One end of the rope is attached to the lower extremity of the fixed block, F; to the other end the power is applied, after the rope has passed in succession over each of the four pulleys.

To move W an inch with this combination, each length of rope must be shortened an inch, and therefore P must move as many inches as there are lengths of rope. Since there are two lengths of rope for each movable pulley, we may lay down the following law :—

241. *Law of Movable Pulleys.*— *With movable pulleys, a power will balance a weight as many times greater than itself as twice the number of movable pulleys employed.*

In Fig. 113, a power of 1 pound will balance a weight of 4 pounds. If three movable pulleys were used, 1 pound at P would balance 6 pounds at W; if four were used, 8 pounds, &c. Friction, however, nullifies much of this gain.

242. *White's Pulley.*—To lessen the friction, when a number of pulleys are required, the wheels are made to turn on the same axis. This is effected by having but one block for all the upper pulleys, and one for the lower; grooves being cut in each, to take the place of separate wheels. The friction in each block is thus reduced to that of a single wheel. This system is called, from its inventor, White's Pulley.

Fig. 114 gives a front and a side view of White's Pulley. A is the fixed

show ? For what is an apparatus of this kind sometimes used ? 237. What is a Movable Pulley? Describe it with Fig. 111. 238. To move a weight a given distance with a movable pulley, how far must the power travel? What, then, is the law of this machine? 239. With what is a movable pulley generally combined? What is gained by this combination? 240. Describe the combination of movable and fixed pulleys represented in Fig. 113. 241. What is the law of movable pulleys? Apply this law in the case of the pulley represented in Fig. 113. By what is much of this gain nullified? 242. When a number of pulleys are required, how is the friction lessened?

block, with grooves of different sizes representing the separate wheels. B is the movable block, similarly prepared. A single rope is used, which is fastened at one end to the smallest fixed pulley, and acted on by the power at the other. Here again, if friction is left out of account, the power will counterbalance a weight as many times greater than itself as twice the number of movable pulleys. In Fig. 114 there are six movable pulleys; consequently, with a pressure of 1 pound at P, equilibrium will be established when W is twice six, or 12, pounds.

Fig. 114.

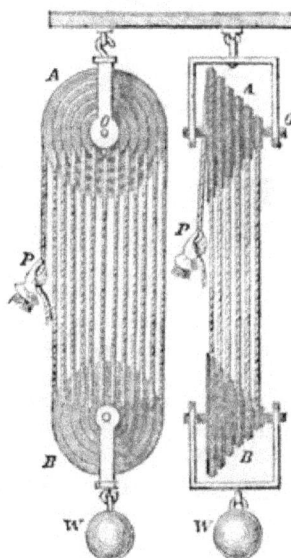

WHITE'S PULLEY.

243. Fig. 115 shows another system of movable pulleys, each of which has a separate rope of its own attached at one end to a fixed support.

Fig. 115.

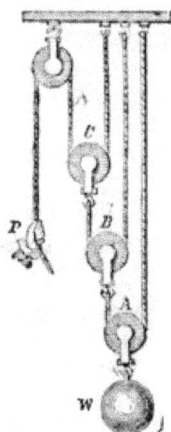

To raise the lowest pulley, A, and the weight suspended from it one inch, two inches of its rope must be pulled up. This is done by pulling up twice 2, or 4, inches of B's rope; and this, in turn, by pulling up twice 4, or 8, inches of C's rope. P, therefore, must descend 8 inches, to raise W one inch. If there were four movable pulleys, P would have to descend 16 inches to raise W one inch; if 5, 32 inches, and so on,—P's distance doubling for each new pulley added. Hence, with this combination, *the power balances a weight as many times greater than itself as 2 raised to the power denoted by the number of movable pulleys.*

244. The pulley is so cheap and convenient that it is much used in its simple forms. In complicated systems, more than half the advantage is lost by friction and the stiffness of the ropes; and consequently such systems are used only when immense weights are to be raised.

## The Inclined Plane.

245. The Inclined Plane is the fourth of the simple mechanical powers. It is a plane surface, inclined to the horizon at any angle. Every road not perfectly level is an inclined plane.

Fig. 116.

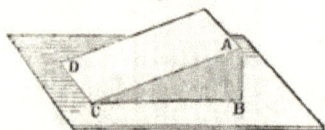

THE INCLINED PLANE.

A D, in Fig. 116, is an inclined plane, of which A C is the length, A B the height, and B C the base. In theory, an inclined plane is perfectly smooth and hard. No such surface, however, exists; and, therefore, in estimating the advantage of this machine for practical purposes, allowance must be made for friction, according to the irregularity or softness of the surface.

246. When a body is moved over a horizontal surface, its weight is supported, and the resistance of the air and friction are all that have to be overcome. When a body is lifted perpendicularly, there is no friction, but we must overcome the whole weight and the resistance of the air. When a body is drawn up an inclined plane, the resistance of the air, friction, and a portion of the weight must be overcome,—more or less of the weight being supported, according to the inclination of the plane. It is, therefore, harder to move a body up an inclined plane than over a level surface, as we know by dragging a wagon up hill; but it is easier than to lift it to the same height.

247. *Law.* — *With an inclined plane, intensity of force is gained, and time is lost, in proportion as its length exceeds its height.*

Fig. 117.

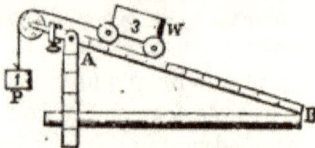

Thus, in Fig. 117, let the length of the plane A B be 12 feet, and its height 4 feet; then 1 pound at P will counterbalance 3 pounds at W.

With a given height, the longer the plane the easier it is to raise an object upon it. Hence, on steep mountains, the road is not carried from the bottom

said of the pulley in its simple forms? What is said of complicated systems of pulleys? 245. What is the fourth simple mechanical power? What is the Inclined Plane? For what must allowance be made, in estimating the advantage of the inclined plane, and why? 246. As regards the resistance to be overcome, show the difference between moving a body over a horizontal surface, lifting it, and drawing it up an inclined plane. 247. What is the law of the inclined plane? Illustrate this law with Fig. 117. How is the road up a steep mountain frequently made, and why? How

directly to the top, but winds round the sides. Instinct teaches a horse this principle; for, if left to himself in ascending a hill, he does not go straight up, but moves in a zigzag course from one side of the road to the other, thus taking more time, but making the ascent easier.

248. *Practical Applications.*—When hogsheads or heavy boxes are to be raised into carts or pulled up a pair of stairs, the work is facilitated by laying long planks, or *skids*, in such a way as to form an inclined plane. A piece of board is similarly placed, if a carriage or wheelbarrow has to be raised over a high curb-stone.—The marine railway, on which ships of immense weight are drawn out of the water, to be repaired, is one of the most useful applications of this machine.

249. The inclined plane was known to the ancients. It is supposed that the Egyptians used it in raising the huge blocks of stone employed in the construction of their pyramids.

250. *Law of Bodies rolling down an Inclined Plane.*—When bodies are allowed to roll down an inclined plane, they have a uniformly accelerated motion, and attain the same velocity by the time they reach the bottom that they would have if dropped perpendicularly from the starting point.

A ball dropped from a height of $64\frac{1}{2}$ feet, when it strikes the ground, has a velocity of $64\frac{1}{2}$ feet in a second. If it were allowed to roll from the same height, down an inclined surface a mile long, perfectly smooth and hard, it would have the same velocity on reaching the bottom. The shorter the plane, the less time it would take for its descent and the sooner it would acquire the velocity in question.

251. When the perpendicular height is considerable, objects rolling or sliding down an inclined plane acquire, near the bottom, a prodigious velocity. A remarkable instance of this was exhibited at a slide near Lake Lucerne, Switzerland, down which fir-trees were allowed to descend, from the top of a mountain. The slide was about eight miles long; and, though the descent was but 800 feet to a mile and the road was often circuitous, the trees went tearing along with frightful speed, performing the whole distance in six minutes.

---

does a horse ascending a hill display his instinct? 248. In what familiar cases is the inclined plane used? What is one of the most useful applications of this machine? 249. By whom is the inclined plane thought to have been used in ancient times? 250. What is the law of bodies rolling down an inclined plane? Illustrate this. 251. When do bodies sliding down an inclined plane acquire a prodigious velocity?

112        MECHANICS.

## The Wedge.

252. The Wedge is the fifth of the simple mechanical powers. It appears in two forms, according to the use for which it is designed.

253. FIRST KIND OF WEDGE.—In its first form, the wedge is simply a solid and movable inclined plane. It is used for raising great weights a short distance, and follows the law of the inclined plane; that is, *the power counterbalances a weight as many times greater than itself as the height of the wedge is contained in its length.*

Fig. 118.

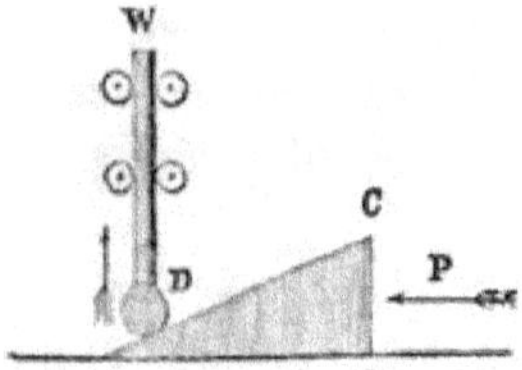

Fig. 119.

Fig. 118 shows how the wedge may be used for raising weights. W D is a pillar, so fixed that it can not move, except perpendicularly upward. A B is a wedge resting on its base. The sharp edge being brought near the extremity of the pillar, power is applied to the side B C. W must rise, as it can not move in any other direction. By driving the wedge under to C, the pillar is raised the distance B C.

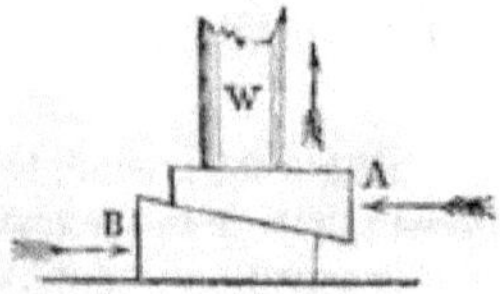

A more common mode of raising bodies with this machine is shown in Fig. 119. A and B are similar wedges. Simultaneous blows are given them at A and B in opposite directions with heavy mallets, and the weight W is slowly raised. The same power must be applied to each as if it acted alone. Twice as much power, therefore, is required as when but one wedge is used, but the weight is raised twice as high in a given time.

254. Thus applied, the wedge is an efficient and useful machine. It raises immense weights, though to no great distance. With its aid, ships are brought up on the dry dock, and houses thrown out of line by the sinking of their foundations are restored to the perpendicular. Wedges are also used in extracting oil from seeds. The seeds are placed between immovable timbers, in bags that allow the liquid, as it is pressed out, to ooze through. Between the bags are then inserted wedges, which are gradually driven in. So intense is their pressure that every particle of oil is extracted, and the seeds, when taken out, are found mashed together, into a dense solid mass.

---

What instance of this is mentioned? 252. What is the fifth mechanical power? In how many forms does the wedge appear? 253. Describe the first kind of wedge. For what is it used? What law does it follow? Describe the operation of this sort of wedge with Fig. 118. What is the more common mode of raising bodies with this

255. *Familiar Applications.*—Chisels and other tools sloped, or *chamfered*, as it is called, on only one side, are familiar examples of this sort of wedge. The longer the chamfered part in proportion to its thickness, the more easily the chisel overcomes the resistance of the wood into which it is driven.

256. SECOND KIND OF WEDGE.—The second kind of wedge (see Fig. 120) has the shape of two inclined planes united at their bases. It is used for splitting timber and rending rocks in quarries.

Fig. 120.

THE WEDGE.

The resistance overcome by the wedge, when thus used, is the cohesion of the substance to be split. As long as the wedge is merely *pressed* against this substance, little or nothing is effected; but, when *driven in with blows*, it becomes a highly useful instrument. When once forced in, it is prevented from receding by the friction of the wood against its sides. Thus every blow begins to act where the preceding blow left off acting.

257. *Advantage gained.*—The exact advantage gained by this sort of wedge when driven in by blows, can not be computed. The percussion gives such a shock to the particles that they open a little in advance of the wedge, as shown in Fig. 120, and readily allow it to enter.

The only law we can lay down is this: — *With a given thickness, the longer a wedge is, the more easily it penetrates.*

258. *Familiar Applications.*—Knife and razor blades, the heads of axes and hatchets, nails, and all cutting instruments chamfered on both sides, are examples of this kind of wedge. Pins and needles may be looked upon as wedges with an infinite number of sides. In all these, the longer the instrument in proportion to its thickness, the greater the advantage gained.

259. In seeking to increase the advantage of the wedge by lengthening it, care must be taken not to make it too long. A slender tool will answer for

machine? What is said of the power in this case? 254. For what is the first kind of wedge used? Describe the mode of extracting juices from seeds. 255. What tools are examples of this kind of wedge? On what does the ease with which they overcome the resistance depend? 256. Describe the second kind of wedge. For what is it used? What sort of power must be applied to the wedge, when thus used? What prevents the wedge from receding? 257. What is said of the advantage gained by the wedge? What is the only law that can be laid down for this machine? 258. Mention some familiar examples of the second kind of wedge. 259. In the case of the

soft substances, but not for hard. A carpenter's chisel, for instance, whose chamfered edges make an angle of 30 degrees, would soon break if used on iron. When this metal is to be cut, the edges should make an angle of 60 degrees, and for copper at least 80.

## The Screw.

260. The screw is the sixth and last of the simple mechanical powers. It consists of a cylinder with a spiral
Fig. 121. ridge and groove winding alternately round it in parallel curves. The portions of the ridge passing successively from one side of the cylinder to the other are called the Threads of the screw.

Fig. 121 represents a screw. If we could unwind the threads from the cylinder, commencing at the end A, we should have a continuous wedge. The back of this wedge is applied to the cylinder; and on its thickness depends the distance between the threads of the screw.

261. KINDS OF SCREW.—Screws are of two kinds:—

1. The Exterior or Convex Screw, represented in Fig. 121, in which the ridge and groove are on the outside of the cylinder.

2. The Interior or Concave Screw, in which the ridge and groove are on what may be regarded as the inside surface of a cylinder.

These two forms are used together, and are generally called the Screw and the Nut. Every screw must have a nut grooved in such a way as to receive its ridge.

262. *Advantage gained.*—The power is applied at the head of the screw. The resistance is to be overcome by pressure produced at its other end. Every time the screw is turned once round in the nut, it advances as far as the distance between two of its threads, and compresses to that

wedge, what must be avoided? What difference is there between a carpenter's chisel and one suitable for iron and copper? 260. What is the sixth mechanical power? Of what does the Screw consist? What is meant by the Threads of the screw? If we could unwind the threads from the cylinder, what would they form? 261. How many kinds of screws are there? Name and describe each. How are these two forms of the screw used, and what are they generally called? 262. With the screw, how

extent any fixed object against which it is directed. *With the screw, therefore, the power produces a pressure as many times greater than itself, as the circumference of the head is greater than the distance between the threads.*

Here, again, however, friction lessens the effect; and, to gain greater power, a lever is generally combined with the screw. The mode of doing this is shown in Fig. 122, in which S is the screw and L the lever.

In calculating the advantage in this case, instead of the circumference of the head take the circle described by the point of the lever at which the hand is applied. In Fig. 122, let the distance between the threads be 1 inch and the dotted circle 100 inches; then (friction being left out of account) a power of 1 pound at the extremity of the lever will produce a pressure of 100 pounds at the lower end of the screw.

Fig. 122.

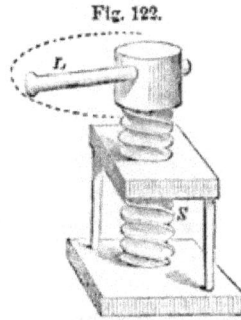

THE SCREW AND NUT.

### 263. BOOK-BINDER'S PRESS.

—The Book-binder's Press, represented in Fig. 123, exhibits one of the most useful and convenient modes of applying the screw.

Fig. 123.

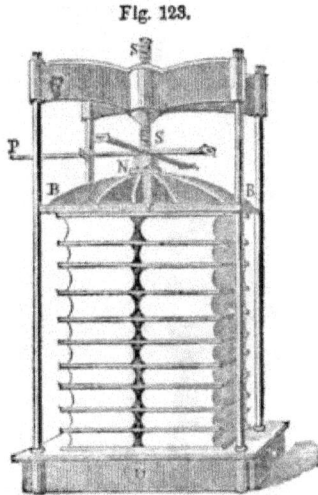

BOOK-BINDER'S PRESS.

S S is a screw, playing in a stationary nut in the head of the press. Attached to the screw near its bottom are two bars at right angles to each other, at the extremities of which the hand is applied when the press is to be worked. Still greater leverage is obtained by applying the power at the end of a bar, P, introduced successively into holes in the extremities of the cross-pieces, as in working the windlass. A fall or platen, B B, is attached to the screw, in such a way that it does not turn as the screw revolves, but must rise or descend with it. Between this fall and the bed of the press, D, the books to be pressed are

great a pressure does the power produce? In practice, what lessens the effect? How is greater power obtained? When a lever is combined with the screw, how may we find the advantage gained? Illustrate this with Fig. 122. 263. What machine exhibits a useful application of the screw? Describe the book-binder's press. How is

placed. Here again, to obtain the advantage, divide the circumference of the circle described by P, by the distance between the threads.

264. Screws, applied in this or some similar way, are extensively used when a great and continued pressure is required within a small space. Cotton is compressed into bales, juices are extracted from fruit, coins are stamped, and houses are raised from their foundations, with the aid of the screw.

265. HUNTER'S SCREW.—When intense pressure is required, the threads of the screw have to be so close together that they are necessarily thin and liable to break. To prevent this, an ingenious contrivance, called after its inventor Hunter's Screw, is used.

Hunter's Screw consists of two screws, working one within the other, in such a way that as the larger descends the smaller ascends, though not quite so far. The difference between the respective distances of the threads in the two screws determines how far on the whole the screw advances. With Hunter's Screw, therefore, the power produces a pressure as many times greater than itself, as the difference between the respective distances of the threads in the two screws is contained in the circle described by the power.

Fig. 124.

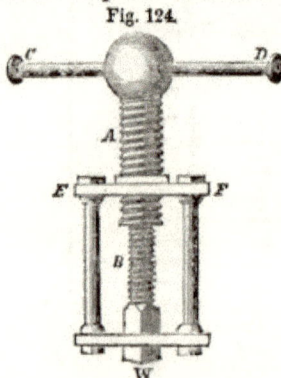

HUNTER'S SCREW.

A is the larger screw, B W the smaller one. C D is the lever by which it is worked, and E F the stationary nut. The pressure is produced at W. If the threads of the larger screw are 1 inch apart, and those of the smaller $^3/_4$ of an inch, the difference is $^1/_4$ of an inch. Then, if the extremities of the lever describe a circle of 100 inches, the advantage will be equal to 100 divided by $^1/_4$, or 400; that is, a power of 1 pound applied at either end of the lever will produce a pressure of 400 pounds at W.

the advantage gained by this machine to be calculated? 264. For what purposes are screws used? 265. When great pressure is required, what difficulty attends the use of the screw? To remedy this, what ingenious contrivance is used? Describe Hunter's Screw. With this screw, how great a pressure does a given power produce?

By making the threads of the two screws nearly the same distance apart, an immense power is obtained without diminishing the size and strength of the threads. The action of the screw is of course proportionally slow, time being always lost as power is gained.

266. THE ENDLESS SCREW.—Instead of working in a nut, a screw is sometimes made to act on teeth cut in the circumference of a wheel. In this case, the only motion of the screw is round its axis. The winch being turned, the threads of the screw catch the teeth of the wheel and move it forward. As fast as one tooth passes out of reach, another is caught; and, the motion being thus continuous, the machine is called the Endless Screw. Its operation will be understood from Fig. 125, where it is combined with a wheel and axle for the purpose of lifting a weight.

Fig. 125.

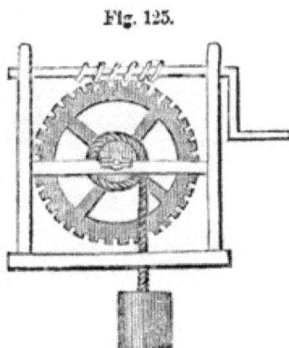

THE ENDLESS SCREW.

### EXAMPLES FOR PRACTICE.

1. (See § 204.) A lever of the first kind is 20 inches in length: the long arm is 15 inches; the short arm, 5. How great a power will balance a weight of 112 pounds? With the same lever, how great a weight will a power of 50 pounds balance?

2. A farmer, in forcing a stump from the ground, uses a crow-bar 6 feet long, which he rests on a stone five feet from the end where his hand is applied. The resistance of the stump is equal to a weight of 500 pounds; how great a pressure must he exert, to move it?

3. A man weighing 180 pounds, and a boy of 60 pounds, are teetering on a board 12 feet long. That they may balance each other, how near must the man sit to the horse on which the board rests?

4. A man whose strength enables him to use a pressure of 120 pounds, wishes to move a rock weighing 600 pounds with a lever of the first kind. What must be the comparative length of the arms of the lever?

   If with his unaided strength he could move 120 pounds thirty feet in one minute, how long will it take him to move the rock with the lever the same distance?

---

Illustrate this with Fig. 124. How may an immense power be gained with Hunter's Screw? 266. Describe the Endless Screw and its mode of operation. With what is it combined for lifting weights?

5. (*See* § 207.) The short arm of a steelyard is 2 inches long; at its end a 10-pound weight is suspended. How great a weight must be attached to the other end to balance it, the length of the steelyard being one foot?

6. (*See* § 213.) There is a compound lever formed of two simple ones, the first arms of which are 10 inches each, and the short arms 2 inches each. How great a weight at the extremity of the last short arm will be supported by a power of 1 pound at the other end?

7. (*See* § 215.) A lever of the second kind is 20 inches long; the weight is 5 inches from the fulcrum. How great a power must be applied, to balance a weight of 112 pounds?

8. With the same lever as in the last sum, how great a weight will a power of 50 pounds balance?

9. A is rowing with an oar 9 feet long, and has his row-lock 2 feet from his hand; B rows with an eight-foot oar, and his row-lock is 1 foot from his hand. If they strike the water with an equal length of oar, which exerts the greater power on the boat?

10. (*See* § 217.) A man and a boy, at opposite ends of a bar 5 feet long, are carrying a 150-pound weight suspended between them. The boy can carry but 30 pounds; how far from his end must the weight hang, to give him that portion of it, and the man the rest?

11. Three men are bearing a weight suspended from a bar in the manner shown in Fig. 100. The single man at one end is twice as strong as each of the two at the other end. How must the weight be placed (the bar being 4 feet long), that each may bear a part proportioned to his strength?

12. (*See* § 219.) A lever of the third kind is 20 inches long; the power is 5 inches from the fulcrum. How great must it be, to balance a weight of 112 pounds?

13. A pair of pincers is 6 inches long. How great a force must be applied, 2 inches from the top, to overcome a resistance of 3 ounces?

14. The distance of a man's hand from his elbow is 16 inches. The biceps muscle is inserted in his fore-arm 2 inches from the elbow. With how great power must the muscle act to sustain a weight of 56 pounds in the extended hand?

15. (*See* § 225.) The circumference of a wheel is 8 feet; that of its axle, 16 inches. The weight, including friction, is 60 pounds; how great a power will be required to raise it?

16. The pilot-wheel of a boat is 3 feet in diameter; the axle is 4 inches. The resistance of the rudder is 180 pounds, to which one-tenth of itself must be added for friction, &c. How great a power must be applied to the wheel, to move the rudder?

17. An axle one foot in circumference, fitted with a winch that describes a circle of 6 feet, is used for drawing water from a well. How great a power will it take to move 60 pounds of water, allowing one-tenth for friction?

18. Four men are drawing in an anchor that weighs 1,000 pounds, with a capstan. The barrel of the capstan has a radius of 6 inches. The circle described by the handspikes has a radius of 5 feet. How great a pressure must each of the four men exert, to move the anchor?

19. (*See* § 232.) With a fixed pulley, how great a power will it take to hoist a weight of 50 pounds, 20 per cent., or one-fifth, being added for friction?

20. (*See* § 238.) With a movable pulley, how great a power will it take to hoist a weight of 50 pounds, twenty per cent. being allowed for friction?

21. (*See* § 239.) With a fixed and a movable pulley, how great a power will it take to hoist a weight of 50 pounds, 40 per cent., or two-fifths, being allowed for friction?

22. (*See* § 241.) With two fixed and two movable pulleys, how great a power will it take to hoist a weight of 50 pounds, 60 per cent., or three-fifths, being allowed for friction?

23. (*See* § 242.) How great a power will it take to hoist a weight of 100 pounds with one of White's Pulleys having five grooves in each block, 35 per cent., or seven-twentieths, being allowed for friction?

24. (*See* § 243.) With a system of six movable pulleys, having each its own rope, and arranged as shown in Fig. 115, how great a weight (including friction) will a power of 20 pounds raise?

25. With a similar system of five movable pulleys, how great a power will it take to balance a weight of 64 pounds, to which the friction of the pulleys adds 50 per cent., or one-half of itself?—*Ans.* 3 pounds.

$$[64 + 32 = 96 \qquad 2^5 = 32 \qquad 96 \div 32 = 3 \; Answer.]$$

26. (*See* § 247.) How great a power will be required to balance a weight of 40 pounds (friction included), on an inclined plane, whose length is 8 times its height?

27. (*See* § 253.) A weight of 1,500 pounds is to be raised with a wedge 60 inches long and 12 inches high at its head. How great must the power be?

28. A builder desires to raise a weight of 900 pounds with two similar wedges, as shown in Fig. 122. Each wedge is 3 feet long and 9 inches through at the head. How great a power must be applied to each?

29. A weight of 1,020 pounds is to be lifted 1½ feet. The greatest power that can be applied is 255 pounds. Give the dimensions of the wedge.

30. (*See* § 257.) Of two wedges 4 inches thick at the head and respectively 6 and 8 inches long, which can be driven into a log the more easily? Which will break the sooner, both being made of the same material?

31. (*See* § 262.) How great a pressure (including friction) will be exerted by a power of 15 pounds applied to a screw whose head is 1 inch in circumference, and whose threads are one-eighth of an inch apart?

32. A book-binder has a press, with a screw whose threads are one-third of an inch apart, and a nut worked by a lever which describes a circle of 8 feet. How great a pressure will a power of 5 pounds applied at the end of the lever produce, the loss by friction being equivalent to 240 pounds?

33. (*See* § 265.) How great a pressure is produced by a power of 1 pound with one of Hunter's Screws, worked by a lever which describes a circle of 75 inches; the threads of the larger screw being half an inch apart and those of the smaller one-third of an inch, 33⅓ per cent., or one-third, of the pressure being deducted for friction?

# CHAPTER IX.

## MECHANICS (CONTINUED).

267. ALL machines, however complicated, are combinations of the six simple mechanical powers described in the last chapter. The chief objects in combining them are to gain a sufficient degree of power, and to give such a direction to the motion as will make the machinery do the work required.

### Wheelwork.

268. The wheel enters more largely into machinery than any other of the Mechanical Powers.

269. Several wheels combined in one machine are called a Train.

270. In a train of two wheels, the one that imparts the motion is called the Driver; the one that receives it, the Follower.

271. MODES OF CONNECTION.—There are three ways in which motion may be transmitted from one wheel to another:—1. By the friction of their circumferences. 2. By a band. 3. By teeth on their outer rim.

272. *Friction of the Circumferences.*—One wheel may move another by rubbing on its circumference, or outer rim. The wheels are so placed that their rims touch, and one of them is set in motion. The circumference of each

having been previously roughened, friction prevents the moving wheel from slipping over the one at rest, and motion is imparted to the latter. Wheels thus connected work regularly and with little noise, but will not answer when a great resistance is to be overcome, and hence are not much used.

273. *Bands.*—One wheel may be made to move another by means of a band passed round both circumferences. Such a band is known as a Wrapping Connector. It is also called an Endless Band, because, its ends being joined, we never seem to reach them, though the motion is continuous in the same direction. The band must be stretched so tight that its friction on the wheels may be greater than the resistance to be overcome.

Fig. 126 shows how wheels are connected by an endless band. If the follower is to turn in the same direction as the driver, the band is passed over it without crossing, as in A; if in the opposite direction, the band is crossed, as in B.

Fig. 126.

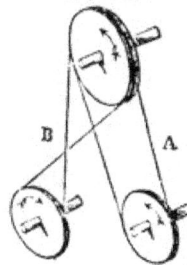

274. The bands used for this purpose are generally made of leather, or gutta percha [*pert'-sha*]. The wheels may be far apart, if necessary; and on this account, as well as because a great amount of power may thus be transmitted, the wrapping connector is much used. The motion imparted is exceedingly regular, any little inequalities being corrected by the stretching of the band.

275. Fig. 127 shows the different forms given to the circumferences of wheels, in order that the band may not slip off. A's circumference is concave, or hollows towards the centre, with a rim on each side. B's is the same, with a row of pins down the centre. C's circumference is even across, with a rim on each side. D has no rim, but bulges out in the centre, so that when the band tends to approach one side it is pulled back by the tightening on the other.

Fig. 127.

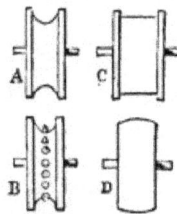

vantage, and what the disadvantage, of this mode of connection? 273. What is a Wrapping Connector? What other name is given to it, and why? How tight must the band be? In passing from the driver to the follower, when is the band crossed, and when not? 274. Of what are endless bands usually made? By what advantages is their use attended? What renders the motion imparted by wrapping connectors exceedingly regular? 275. Describe the different forms given to the circumferences of wheels on which a wrapping connector is to act. 276. What is the third way in

6

Fig. 128.

276. *Teeth.*—One wheel may be made to move another by means of teeth on the circumference of each. A toothed wheel is shown in Fig. 128.

277. Small toothed wheels combined with large ones are called Pinions, and their teeth Leaves.

278. Two or more wheels connected by teeth are called Gearing. When so arranged that the teeth work in each other, they are said to be *in gear;* and when not, *out of gear.*

Fig. 129.

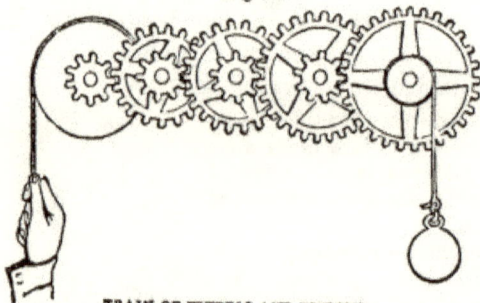

TRAIN OF WHEELS AND PINIONS.

Figure 129 shows a train of wheels and pinions in gear. To find how great a weight will be balanced by a given power with such a train, multiply the power successively by the number of teeth on the wheels, and divide by the product of the number of teeth on the pinions. For instance, in Fig. 129, let the first large wheel have 18 teeth, the second 18, the third 27, and the fourth 27; and let each pinion have 9 teeth. Then (leaving friction out of account) a power of 2 pounds will balance a weight of 72 pounds. For

$$2 \times 18 \times 18 \times 27 \times 27 = 472392$$
$$9 \times 9 \times 9 \times 9 = 6561$$
$$472392 \text{ divided by } 6561 = 72$$

279. KINDS OF TOOTHED WHEELS. — There are three kinds of toothed wheels; viz., Spur-wheels, Crown-wheels, and Bevel-wheels.

280. *Spur-wheels.*—Spur-wheels have their teeth perpendicular to their axes, as shown in Fig. 129.

The teeth are either made in one piece with the rim, or

consist of separate pieces set into the rim. In the latter case, they are called Cogs.

In mills, Cog-wheels are generally used with Trundles, or Lanterns, as represented in Fig. 130.

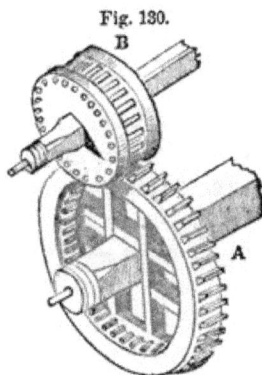

Fig. 130.

A is a large cog-wheel. B is a trundle, consisting of two parallel discs and an intervening space traversed by round pins called Staves, so arranged as to receive the cogs of the other wheel.

Mill-wheels are generally made of cast-iron; but they are found to work most smoothly when one of them has wooden instead of iron teeth. Wooden teeth are therefore often set in the larger one, which is then called a Mortice-wheel.

COG-WHEEL AND TRUNDLE.

281. *Crown-wheels.* — Crown-wheels have their teeth parallel to their axes.

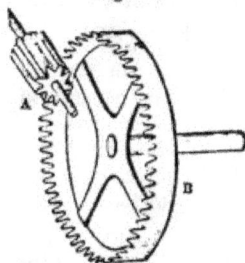

Fig. 131.

CROWN-WHEEL AND PINION.

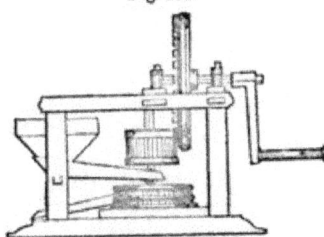

Fig. 132.

HAND-MILL.

Fig. 131 represents the contrate-wheel and pinion of a watch. B, whose teeth run the same way as its axis, is a crown-wheel. A, whose teeth are at right angles to its axis, is a spur-wheel.

Fig. 132 shows how a crown-wheel worked by a winch is combined with a trundle in a hand-mill used in Germany and Northern Europe. The crown-wheel moves vertically, but it communicates a horizontal motion to the trundle, which in turn imparts it to the mill-stone.

282. *Bevel-wheels.* — Bevel-wheels are wheels whose teeth

form any other angle with their axes than a right angle.

A pair of bevel-wheels in gear are shown in Fig. 133.

Fig. 133.

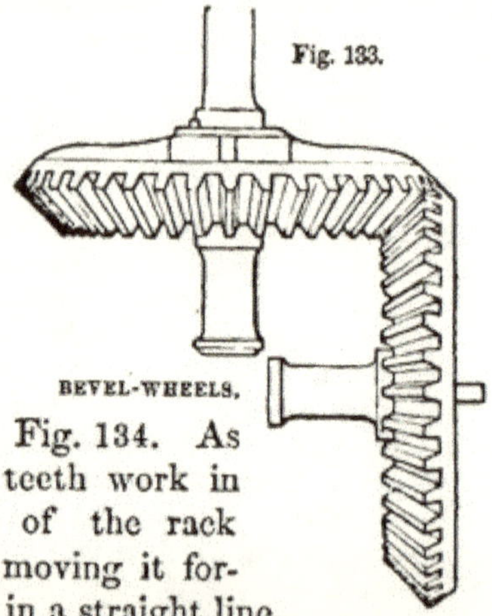

BEVEL-WHEELS.

283. Rack and Pinion.—Circular motion is converted into rectilinear (that is, motion in a straight line) by means of the rack and pinion, represented in Fig. 134. As the pinion A revolves, its teeth work in those of the rack B C, moving it forward in a straight line.

Fig. 134.

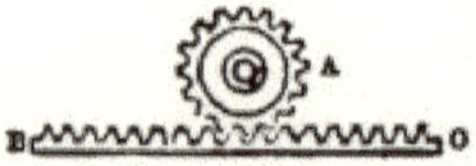


RACK AND PINION.

284. Forge-Hammer.—A toothed wheel may produce an alternate up-and-down motion, as in the case of the Forge-hammer, represented in Fig. 135.

Fig. 135.

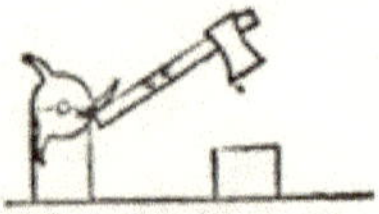

THE FORGE-HAMMER.

The wheel is so placed that its teeth successively come in contact with the handle of the hammer, which turns on a pivot. As the wheel revolves, a long tooth carries the lower end of the handle down and raises its head. As soon as the tooth releases the handle, the head of the hammer falls on the anvil by its own weight. A new tooth then comes into play, and the operation is repeated.

285. Cranks.—The Crank is much used in machinery for converting circular motion into rectilinear, or rectilinear into circular. It has different forms, but is generally made by bending the axle in the way represented in Fig. 136. As the wheel to which it is attached turns, the crank A also revolves, and causes the rod B, with which it is connected, to move alternately up and down.

Fig. 136.

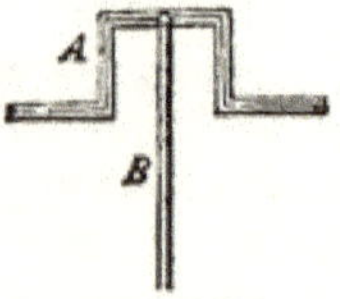


THE CRANK.

---

does Fig. 133 represent? 283. How may circular motion be converted into rectilinear? Describe the working of the Rack and Pinion. 284. What kind of motion does a toothed wheel produce in the case of the forge-hammer? Explain the working of the forge-hammer. 285. For what is the Crank used? Describe its usual form, and

The point at which the rod stands at right angles to the axle (as in the Figure) is called the Dead-point. Two dead-points occur in each revolution. When at either, the crank loses its power for the instant; but the impetus carries it along, and as soon as the dead-point is passed it again begins to act.

286. Another form of the crank is exhibited in Fig. 137, which shows how a wheel is moved by a treadle-board worked by the foot. A is the treadle; B C is a cord passed round the pulley D, and attached to the crank E, which is connected with the axle of the wheel F. When the foot bears the treadle-board down, the end of the crank is raised to its highest point. Here it would remain if the foot were kept on the board; but, the foot

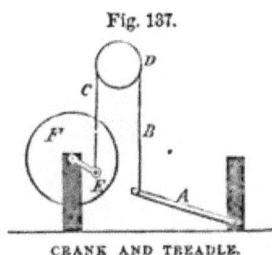

Fig. 137.

CRANK AND TREADLE.

being removed, the impetus of the wheel carries the crank round again to its lowest point, raising in turn the end of the treadle-board. The foot is now applied again with the same effect as before, and continuous motion is thus imparted to the wheel.

287. FLY-WHEELS.—The motion of machinery must be even and regular. Both power and resistance must therefore act uniformly; if either increases too rapidly, the sudden strain is apt to break some part of the works. To prevent this, the fly-wheel is used.

The fly-wheel appears in various forms, but generally consists of a heavy iron hoop with bars meeting in the centre. It is set in motion by the machinery, and by reason of its weight acquires so great a momentum that irregularities either in power or resistance, unless long continued, have but little effect. If, for instance, the power ceases to act for a moment, or the resistance suddenly increases or diminishes, the great momentum of the fly prevents the motion of the machinery from varying to any great extent.

288. The fly-wheel also accumulates power, and thus enables a machine to overcome a greater resistance than it could otherwise do. The power,

explain its operation. What is meant by the Dead-point of the crank? What is said of the crank at its dead-point? 286. What does Fig. 137 represent? Explain the operation of the crank and treadle. 287. For what is the Fly-wheel used? Of what does it generally consist? Explain how the fly-wheel prevents irregularities of motion. 288. For what other purpose is the fly-wheel used? How does the fly-wheel

allowed to act on the fly alone for a short time, gives it an immense momentum; and this momentum directly aids the power, when the machine is applied to the required work.

## Clock and Watch Work.

289. One of the commonest and most ingenious applications of wheelwork is exhibited in clocks and watches.

290. HISTORY.—The advantages of combining wheels and pinions were partially known as far back as the time of Archimedes; yet they were comparatively little used in machinery, and not at all for the measurement of time.

Instead of clocks and watches, consisting of trains of wheels, the ancients used the sun-dial, and clep'-sy-dra or water-clock. The former indicated the hour by the position of the shadow cast by a style, or pin, on a metallic plate; the latter, by the flow of water from a vessel with a small hole in the bottom. The dial was of course useless at night; and neither it nor the clepsydra, however carefully regulated, could measure time with any great degree of accuracy.

Even Alfred the Great, 985 years after Christ, had no suitable instrument for measuring time. To tell the passing hours, he used wax candles twelve inches long and of uniform thickness, six of which lasted about a day. Marks on the surface at equal intervals denoted hours and their subdivisions, each inch of candle that burned showing that about twenty minutes had passed. To prevent currents of air from making his candles burn irregularly, he enclosed them in cases of thin, transparent horn,—and hence the origin of the lantern.

291. Clocks moved by weights were known to the Saracens as early as the eleventh century. The first made in England (about 1288 A. D.) was considered so great a work that a high dignitary was appointed to take care of it, and paid for so doing from the public treasury. The usefulness of clocks was greatly increased by the application of the pendulum, which was made about the middle of the seventeenth century.

Watches seem to have been first made in the six-

aid the power? 289. In what do we find one of the most ingenious applications of wheel-work? 290. What is said of the knowledge of wheel-work possessed by the ancients? What did the ancients use for the measurement of time? How did the sun-dial indicate the hour? How, the clepsydra? What is said of the accuracy of these instruments? How did Alfred the Great measure time? What was the origin of the lantern? 291. When were clocks moved by weights first made by the Saracens? When was the first made in England? How was this clock regarded? What

teenth century, though it is not known who was their inventor. For a time they were quite imperfect, requiring to be wound twice a day, and having neither second nor minute hand. The addition of the hair-spring to the balance, by Dr. Hooke, in 1658, was the first great improvement. Others have since been devised; and chronometers (as the best watches, manufactured for astronomers and navigators, are called) are now made so perfect as not to deviate a minute in six months, even when exposed to great variations of temperature.

292. CLOCK-WORK.—In clocks, except such as are moved by springs similarly to watches, the moving power is a weight; to which, when wound up, gravity gives a constant downward tendency. In its effort to descend, it sets in motion a train of wheels and pinions; and they move the hands which indicate the hours and minutes on the face.

The motion of the wheels, though caused by the weight, is regulated by the pendulum and an apparatus called the Escapement, shown in Fig. 138. The *crutch* A B C moves with the pendulum. As the latter vibrates, the *pallets* B, C, are alternately raised far enough to let one tooth of the *scape-wheel* pass, its motion at other times being checked by the entrance of one of the pallets between the teeth. Hence, though the weight is wound up, the clock does not go till the pendulum is set in motion. If the pendulum and escapement are removed, the weight runs down unchecked, turning the various wheels

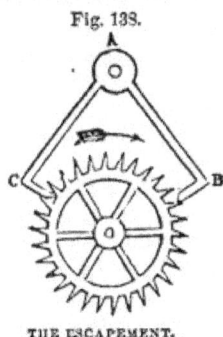

Fig. 138.

THE ESCAPEMENT.

with great rapidity. The motion of the wheels is thus made uniform by the pendulum; and by shortening or lengthening it we can make the clock go faster or slower.

293. WATCH-WORK.—In a watch, there is no room for a weight or pendulum; hence a spring, called the *main-*

greatly increased the usefulness of clocks? When were watches first made? What was the character of those first constructed? What was the first great improvement? What is said of the chronometers made at the present day? 292. What is the moving power in clocks? How does the weight set the clock in motion? How is the motion of the wheels regulated? Explain, with Fig. 138, how the Escapement regulates the motion. If the pendulum and escapement are removed, what is the consequence? How is the clock made to go faster or slower? 293. In a watch, what

*spring*, is substituted for the former as a moving power, while the *balance* and *hair-spring* take the place of the latter as a regulator.

The main-spring is either fixed to an axle capable of revolving, as shown at O P in Fig. 140, or is contained within a hollow barrel, connected by a chain with a conical axle, called the *fusee*, represented in Fig. 139. A is the barrel,

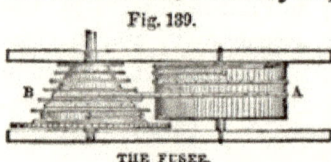
Fig. 139.

THE FUSEE.

within which and out of sight is the main-spring, having one end attached to the inner surface of the barrel, and the other fastened to a fixed axle passing through the barrel. B is the fusee.

The watch is wound up with a key, applied to the square projecting from the fusee. By turning the square the chain is drawn off from the barrel and wound round the fusee. The barrel is thus turned till the spring in the inside is tightly coiled. This spring, by reason of its elasticity, tends to uncoil, and in so doing moves the barrel round, drawing off the chain from the fusee, and winding it again around the barrel. The fusee is thus turned, and carries with it the first wheel of the train, which imparts motion to all the rest. When the spring has uncoiled itself, the chain, being entirely wound round the barrel, ceases to move the fusee, and all the wheels come to rest. The watch is then said *to run down*.

The reason of the peculiar shape of the fusee is this. The power of the spring is proportioned to the tightness with which it is coiled, and hence is greatest when the watch is first wound. The chain is consequently then made to act on the smallest part of the fusee ; because, the nearer to the axis the force is applied, the less its power of producing motion. As the spring gradually uncoils, its power is weakened and it is made to act on a larger part of the fusee. By thus adjusting the size of the fusee to the varying power of the spring, a uniform effect is secured.

294. An escapement similar to that used in clocks connects the moving power with the balance. To the latter, also, a very fine spiral spring is attached, which is fastened at its other end to a fixed support. The watch is regulated by shortening or lengthening this spring, the balance being made to vibrate faster or slower accordingly.

295. The works of an ordinary watch are shown in Fig. 140. For convenience of inspection, they are arranged in a line, and the distance between the two plates, and also between the upper plate and the face, is increased.

---

takes the place of the weight, and what of the pendulum? What two ways are there of fixing the main-spring? Explain Fig. 139. How is the watch wound up? Explain the working of the fusee. When does the watch run down, and why does motion then cease? What is the reason of the peculiar shape of the fusee? 294. What connects the moving power with the balance? What is attached to the balance? How

Fig. 140.

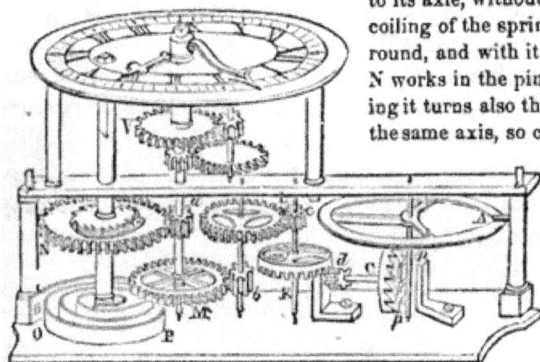

WORKS OF A WATCH.

O P is the *main-spring*, attached to its axle, without a fusee. The uncoiling of the spring carries the axle round, and with it the *great wheel* N. N works in the pinion *a*, and by turning it turns also the *centre-wheel* M on the same axis, so called from being in the centre of the watch. M turns the pinion *b* and the *third wheel* L, which in turn works in the pinion *c* and causes the *second* or *contrate-wheel* R, on the same axis,

to revolve. R works in the pinion *d* and carries round the *balance* or *crown wheel* C, which is on the same axis with it.

The saw-like teeth of the balance-wheel are checked (as in the case of the escapement of a clock) by the *pallets p, p*, which are projecting pins on the *verge* of the *balance* A. The *hair-spring*, fastened at one end to a fixed support, and at the other to the balance, may be shortened by the *curb* or *regulator*, if the watch goes too slow, or lengthened if it goes too fast, thus controlling the motion of the balance and consequently that of the other wheels.

296. The force of the main-spring is so adjusted as to make the great wheel N revolve once in four hours. The spring generally turns it seven or eight times round before it is uncoiled, so that with one winding the watch runs twenty-eight or thirty-two hours. The great wheel N has forty-eight teeth, the pinion *a* but twelve; so that *a* and the centre-wheel M revolve once every hour, and their axle, carried through to the face, bears the minute-hand.

Between the face and the upper plate is a train of pinions and wheels connected with the axle of the centre-wheel. They are so adjusted that the wheel V revolves once in twelve hours. V carries the hour-hand. It is attached to a hollow axle, through which the axle of the centre-wheel passes to carry the minute-hand.

297. Thus we see that the works of a watch are nothing more than an ingenious combination of wheels, moved by a spring and regulated by a balance. The arrangement of the

Is the watch regulated? 295. What does Fig. 140 represent? With the aid of Fig. 140, describe the works of a watch and their mode of operation. How is the watch regulated? 296. How great a force is generally given to the main-spring? How long does the watch run with one winding? Explain the arrangement of the minute-hand. Explain that of the hour-hand. 297. Of what, as we have seen, do the works

6*

wheels and pinions is such, that there is a constant increase of velocity and a corresponding loss of power. The great wheel, which begins the train, revolves once in four hours; the balance, which closes it, revolves in one-fifth of a second; but the force of the spring becomes so attenuated by the time it reaches the balance, that the slightest additional resistance there, a particle of dust or even a thickening of the oil used to prevent friction, deranges, and may stop, the action of the whole.

# CHAPTER X.

## MECHANICS (CONTINUED).

### HYDROSTATICS.

298. HYDROSTATICS and Hydraulics are branches of Mechanics that treat of liquids.

Hydrostatics is the science that treats of liquids at rest.

Hydraulics is the science that treats of liquids in motion, and the machines in which they are applied.

299. The principles of Hydrostatics and Hydraulics are equally true of all liquids; but it is in water, which is the commonest liquid, that we most frequently see them exhibited.

Water abounds on the earth's surface. It covers more than two-thirds of the globe, and constitutes three-fourths of the substance of plants and animals.

300. NATURE OF LIQUIDS.—Liquids differ from solids in having but little cohesion.

---

of a watch consist? What is said of the arrangement of the wheels and pinions? What is the comparative velocity of the great wheel and the balance? What is said of the force of the spring by the time it reaches the balance?

298. What sciences treat of liquids? What is Hydrostatics? What is Hydraulics? 299. What is said of the principles of hydrostatics and hydraulics? How much of the globe is covered with water? How much of the substance of plants and animals consists of water? 300. In what respect do liquids differ from solids? What shows

Cohesion is not entirely wanting in liquids, as is proved by their particles' forming in drops; but it is so weak as to be easily overcome. Thick and sticky liquids, like oil and molasses, have a greater degree of cohesion than thin ones, like water and alcohol.

301. Liquids were long thought to be incompressible, but experiment has proved the reverse. Submitted to a pressure of 15,000 pounds to the square inch, a liquid loses one-twenty-fourth of its bulk. Were the ocean at any point a hundred miles deep, the pressure of the water above on that at the bottom would reduce it to less than half its proper volume.

302. To distinguish them from the gases, liquids are often called non-elastic fluids; yet they are not devoid of elasticity.

To prove this, after compressing a body of water, remove the pressure, and it will resume its former bulk. Again, if a knife-blade be brought in contact with a drop of water hanging from a surface, the drop may be elongated by slowly drawing away the blade; but it immediately returns to its original shape, if the blade is entirely removed without detaching the drop from the surface.

## Law of Hydrostatics.

### 303. *Water at rest always finds its level.*

No matter what the size or shape of a body of water may be, its surface has the same level throughout; that is, it is equally distant at every point from the earth's centre. Accordingly, the surface of the ocean is spherical; and this we know to be the case from always seeing the mast of a vessel approaching in the distance before we see the hull. In small masses of liquids, no convexity is perceptible; and we may consider their surfaces as perfectly flat.

304. The tea-pot affords us a familiar illustration of this law. The tea always rises as high in the spout as in the body of the pot; and, if the body is higher than the spout, it will pour out from the latter when the pot is filled.

So, let there be a number of vessels having communication at their bases, as shown in Fig. 141. If water be poured into any of them, it will rise to

that cohesion is not entirely wanting in liquids? What liquids have the most cohesion? 301. What is said respecting the compressibility of liquids? If the ocean were a hundred miles deep, what would be the consequence of the pressure? 302. What are liquids often called, to distinguish them from gases? Is the name strictly correct? Prove that liquids are elastic? 303. What is the great law of Hydrostatics? What do we mean, when we say that a body of water has the same level throughout? What sort of a surface must the ocean have? What evidence is there of this? How may we regard the surfaces of small bodies of liquids? 304. Show how the tea-pot illus-

Fig. 141.

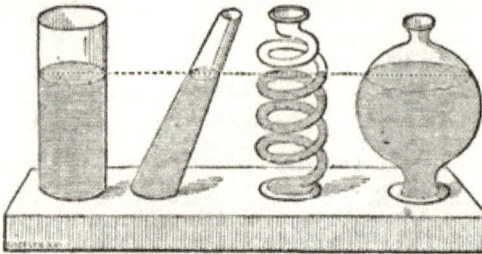

the same level in all, no matter how they may differ in shape or size. In like manner, if there be subterraneous connection between a river affected by the tide and pools near its banks, the water in the pools will rise and fall simultaneously with that in the river.

305. We take advantage of this law in supplying cities with water from elevated ponds or streams. The water may be conveyed in pipes any distance, may be carried beneath deep ravines or the beds of rivers, and when released from the pipe at any point will rise to the level from which it started.

Fig. 142.

Thus, in Fig. 142, the pond A is made to supply the house D with water by means of pipes carried down into the valley, under the stream B and over the bridge C. In the house it will reach the level of the pond from which it was taken, shown by the dotted line.

Fountains formed by tapping the pipe at any point, rise, theoretically, to the same level, as seen in the plate, but are prevented from quite reaching it by the resistance of the air and the check which the ascending stream receives from the falling drops.

306. The ancient Romans appear to have known that water conducted in pipes will find its level; yet so difficult did they find it to make water-

tight joints, that, instead of employing pipes, they conveyed their water through vast level aqueducts, bridging at an immense expense such ravines and valleys as lay in their course. In modern times, iron pipes laid beneath the surface, however much it may be depressed, accomplish the same object with much less cost, the water always rising to its original level when allowed to do so. The lower the pipes are sunk, the stronger they should be ; for the upward pressure of the water, tending to resume its level, increases in proportion to the depth.

307. *Artesian Wells.*—It is on this principle, also, that Artesian Wells are made. They are so called from the province of Artois [*ahr-twah'*], in France, the first district of Europe where they were extensively introduced, though known to the Chinese for centuries.

The outer crust of the earth consists of different strata, or layers; some of which (rock and clay, for instance) are impervious to water, and others not (such as gravel and chalk). If a stratum which allows water to flow through it is enclosed, after leaving the surface, between two impervious layers, and thus descends to a lower level, the water received by this stratum at the surface, unable to pass out above or below, collects in it throughout its whole length. Let an opening then be made at any point into this reservoir through the impervious stratum above, and the water will at once rise to find its level.

Such openings are Artesian wells. They have been carried in some cases a third of a mile below the surface ; and so abundant is their supply of water that a single well of this kind at Paris has been computed to yield 14,000,000 gallons daily. The elevated end may be several hundred miles distant ; it matters not how far. It is thought that the deserts of Arabia and Africa might be supplied with water, and thus rendered habitable, by means of Artesian wells.

308. *Springs.*—Springs have a similar origin. The rain drunk up by the earth's surface gradually sinks, till it reaches an impervious stratum. Along this it runs, receiving additions as it goes, till it finds vent in some natural opening.

In ordinary wells, the water does not rise to the earth's surface, because it does not come from an elevated stratum.

---

Why did they not employ pipes? What precaution must be taken, in consequence of the upward pressure of the water? 307. What wells are made on this principle? Why are Artesian Wells so called? Explain their working. How low have they been carried? How much water does the well at Paris supply? How far off may the elevated end of the stratum be? What is thought respecting the deserts of Arabia and Africa? 308. Explain the origin of springs. Why does not the water rise in

309. *Locks.*—We are enabled to run canals through uneven tracts by taking advantage of the fact that water always finds its level. If the bottom of the canal were not of
a uniform grade, the water would run towards the lower end
and inundate the surrounding country. When, therefore,
the ground is uneven, the canal is built in sections, each level
in itself, but of a different grade from the one next to it,
with which it is connected by a compartment called a Lock.

Fig. 143.

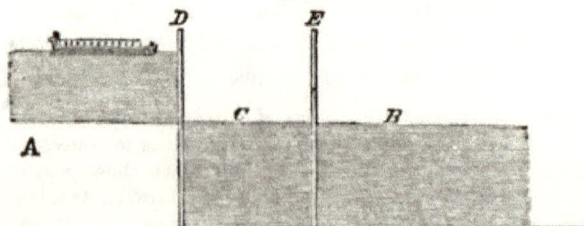

Let A B represent a canal, the upper section of which, A, is fifteen feet
higher than the lower section B. A boat is passed from one to the other by
means of the lock C, which communicates with either section, as may be desired, by opening sliding valves in the lock-gates D, E. When a boat is going
down, the gate E is closed and D is opened till the water in the lock assumes
the same level as in A. The boat is then brought into the lock; the gate D is
closed and E is opened. The water, gradually sinking in the lock, bears the
boat along with it till it reaches the same level as in B. In going up, the operation is reversed. The boat having passed from B into the lock, E is closed
and D opened. The water rushes in to find its level, and the boat is raised
till it stands at the same height as the water in A.

310. *The Spirit Level.*—The Spirit Level, an instrument much used by surveyors, masons, and others, operates

Fig. 144

THE SPIRIT LEVEL.

on this same principle. It consists
of a glass tube (see Fig. 144) nearly filled with colored alcohol, just
enough air being allowed to remain
in it to form a bubble. The tube is then closed, and fixed
in a wooden or metallic case.

On being applied to a surface, if the latter is perfectly level, the air-bub-
ble will rest midway of the tube, in its highest point which has been found

ordinary wells? 309. How are we enabled to run canals through uneven tracts of
country? With the aid of Fig. 143, show the workings of a Lock. 310. What is the

by previous experiment and marked. If the bubble rests in any other place, it shows that one end of the tube is higher than the other, and consequently that the surface on which it rests is not level.

The tube is sometimes made of a different form, and nearly filled with water instead of alcohol; the instrument is then known as the Water Level.

## Pressure of Liquids.

**311. FIRST LAW.**—*Liquids, subjected to pressure, transmit it undiminished in all directions.*

Solids transmit pressure only in the line in which it is exerted; liquids transmit it in every direction. This is proved by experiment.

In Fig. 145, A represents a glass vessel of water, to the neck of which a piston, B, is tightly fitted. Tubes are inserted at intervals through orifices in the sides. As the piston is driven down, the pressure is felt alike at all points of the vessel, as is shown by the flow of the water from the tubes.

Fig. 145.

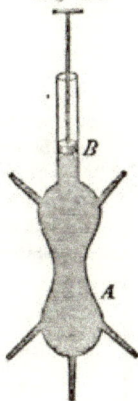

**312. SECOND LAW.**—*Liquids, influenced by gravity alone, press in all directions.*

Bore a hole in the bottom of a pail filled with water; the water rushes out—this proves its downward pressure.

Bore a hole in the side of the same pail: the water rushes out—this proves its lateral pressure.

Bore a hole in the bottom of a boat; the water rushes in—this shows its upward pressure.

**313. THIRD LAW.**—*The pressure of liquids in every direction is proportioned to their depth.*

The downward pressure of liquids increases with their depth. To prove this, take four tubes of equal diameter, and over one end of each tie a piece of very thin india rubber. Fill them with water to different heights, say 5, 10, 20, and 30 inches. The india rubber will be distended the most in the one containing the greatest depth of liquid.

The lateral pressure of liquids increases with their depth. Hence dams

Spirit Level? Of what does it consist? How is the spirit level used? What is the Water Level? 311. What is the first law relating to the pressure of liquids? What is the difference between solids and liquids in this respect? Illustrate this law with Fig. 145. 312. What is the second law relating to the pressure of liquids? Prove the downward pressure of liquids. Prove their lateral pressure. Prove their upward pressure. 313. What is the third law relating to the pressure of liquids? What experiment proves that the downward pressure of liquids is proportioned to their depth?

and sea-walls should increase in strength towards their bases. On the same principle, barrels holding liquids should be more securely hooped at bottom than at top.

Fig. 146.

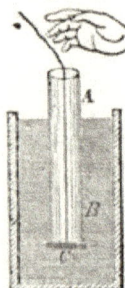

The upward pressure of liquids increases with their depth. This is shown by the experiment represented in Fig. 146. A B is an open tube, ground perfectly smooth on the lower end. C is a plate of lead attached to a string. Pass the string through the tube, and with it keep the lead plate close against the ground end; then introduce the whole into a deep vessel of water. When it has descended an inch or two, let go the string, and the lead will sink. Let it go near the bottom of the vessel, and, as shown in the Figure, the lead will be supported by the water. The upward pressure has therefore increased with the depth.

314. At great depths the pressure of water becomes immense; neither divers nor fish can endure it. Strong glass bottles, empty and tightly corked, are often let down with cords at sea, and the pressure is generally sufficient to break them at a depth of 60 feet. If the bottle does not break, either the cork is driven in or water enters through its pores. The hardest wood, sunk to a great depth, has its pores so thoroughly filled with water as to become incapable of rising. Hence, when a ship goes down at sea, her timbers are never seen again.

Fig. 147.

315. This law leads to wonderful results. Effects almost incredible may be produced by an insignificant body of liquid so disposed as to have considerable depth.

We may, for example, burst a stout cask with a few ounces of water. Having filled the cask with water and inserted in its top a long tube communicating with the inside, we may force the staves asunder, however tightly hooped, by simply pouring water into the tube.

316. Similar effects are often produced in nature. Let D (see Fig. 148) be a mass of rock through which runs a long crevice, A B, communicating with C, a large cavity below, full of water, and having no outlet. When a shower fills the crevice, so great a pressure may be generated as to rend the rock in fragments. It is in this way that many of the great convulsions of nature are produced.

What should be the strongest part of dams, sea-walls, and barrels,—and why? Describe the experiment which proves that the upward pressure of liquids increases with their depth. 314. What is said of the pressure of water at great depths? What experiment is often made with strong glass bottles? What is the effect of this pressure on wood sunk to a great depth? 315. How may wonderful effects be produced by an insignificant body of liquid? How, for example, may a cask be burst? 316. What similar effect is produced in nature? 317. What is meant by the Hydro-

**317.** *Hydrostatic Paradox.*—Pressure being proportioned to depth alone, a very small quantity of liquid may balance any quantity, however great. This principle is called the Hydrostatic Paradox.

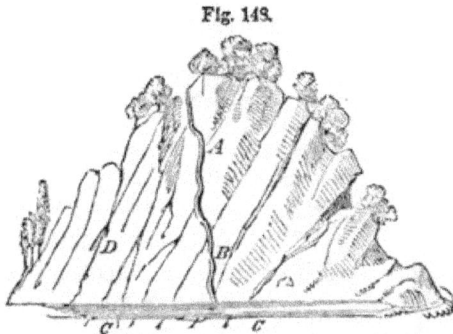

Fig. 148.

Improbable as it appears at first, its truth is proved in various ways.

In Fig. 149, let A be a vessel holding 50 gallons, and B a tube of the same height, communicating with A, and having a capacity of one gallon. Water poured in either rises to the same height in both. When both are full, the pressure of the one gallon in the tube must be as great as that of the 50 gallons in the vessel; otherwise, the latter would force its way into the tube and cause the water there to overflow.

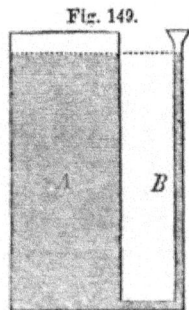

Fig. 149.

**318.** *Rule for finding the Pressure on the Bottom of a Vessel.*—To find the pressure of a body of liquid on the bottom of the vessel containing it, multiply its height into the area of the vessel's bottom.

According to this rule, different quantities of liquid may produce equal pressure. In Fig. 150, let A, B, and C be three vessels having equal bases, and containing the same depth, though different quantities, of liquid; then the pressure on their bottoms will be equal.

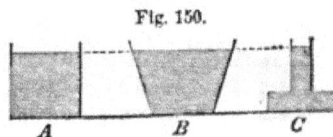

Fig. 150.

**319.** *Hydrostatic Bellows.*—Interesting experiments may be performed with the Hydrostatic Bellows, represented in Fig. 151.

---

Fig. 151.

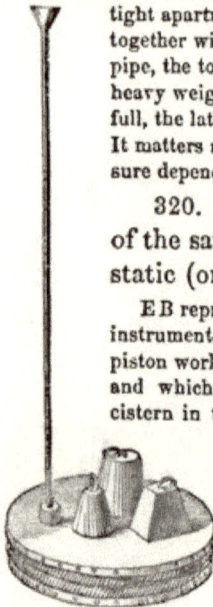

HYDROSTATIC BELLOWS.

A metallic pipe, about four feet long, is screwed into a water-tight apartment, formed of two circular pieces of board fastened together with a broad leather band. As water is poured into the pipe, the top of the bellows rises, and with such force as to lift heavy weights placed upon it. When both pipe and bellows are full, the latter will support from three to four hundred pounds. It matters not how small the bore of the pipe may be; the pressure depends solely on its height.

320. *Hydrostatic Press.*—A useful application of the same principle is made in Bramah's Hydrostatic (or Hydraulic) Press, exhibited in Fig. 152.

E B represents a forcing-pump worked by the lever A. This instrument, which is fully described on page 188, consists of a piston working within a small tube to which it is tightly fitted, and which descends, as shown by the dotted lines, into a cistern in the bottom of the frame of the press. F G is a tube connecting E B with the large cylinder C, to which is fitted a smaller wrought-iron cylinder D, free to move up and down within it. D has a platen, H H, attached to it, between which and the top of the frame, the cotton, hay, cloth, or other substance to be pressed, is placed.

Fig. 152.

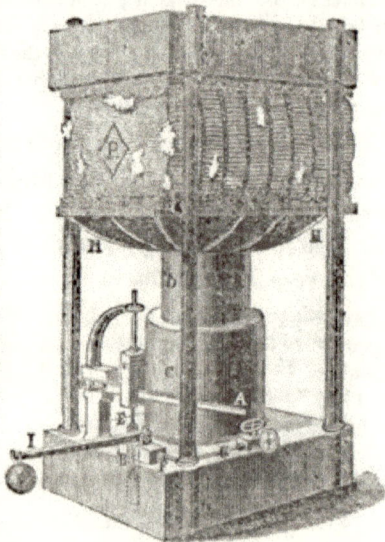

HYDROSTATIC PRESS.

To work the press, raise the long arm of the lever A. Water is by this means drawn up from the cistern into the tube E B; and, when A is lowered and the piston thus made to descend, being prevented from returning to the cistern by a valve which closes, it is forced through the tube F G into the lower part of the cylinder C. D being thus driven up and with it the platen, whatever is confined between the latter and the top of the frame is

performed with it. How great a weight will it support? 320. Describe the Hydrostatic Press, with the Figure. How is it worked? How great pressure may be obtained

subjected to pressure, greater or less according to the quantity of water forced into C.

With the Hydrostatic Press any degree of pressure may be obtained that is not too great for the strength of the materials employed. The machine is extensively used, not only for pressing, but also for extracting stumps, testing cables, and raising vessels out of water.

## Specific Gravity.

321. If we weigh a cubic inch of water, and then the same bulk, or volume, of silver, and of cork, we find the silver heavier than the water, and the cork lighter. If we proceed to compare the weights of various other substances, taking a cubic inch of each, we shall find that they all differ more or less. To express the comparative weight of different substances, the term Specific Gravity is used.

322. The Specific Gravity of a substance is the weight of a given bulk of it compared with the weight of an equal bulk of some other substance taken as a standard. The standard employed is distilled water at the temperature of 60 degrees.

A standard of this kind must be invariable. Hence the temperature of the water is fixed; for at a higher degree of heat it would become rarer,—and at a lower degree, denser. Distilled water is taken, because it is pure; the intermixture of vegetable and mineral matter in spring and river water affects their density, and makes them unfit for a standard.

A cubic inch of silver weighs $10\frac{1}{2}$ times as much as a cubic inch of water; accordingly, the specific gravity of water being 1, that of silver is $10\frac{1}{2}$. A cubic inch of cork weighs $\frac{24}{100}$ as much as the same bulk of water; the specific gravity of cork, therefore, is set down at $\frac{24}{100}$ or .24.

323. Fluids that do not mix, when brought together, arrange themselves in the order of their specific gravities, the heaviest at the bottom. Thus, if mercury, water, and oil be thrown into a tumbler, the mercury will settle at the

with the hydrostatic press? For what is this machine used? 321. If we weigh equal bulks of different substances, what do we find? What term is used to express the comparative weight of different substances? 322. What is Specific Gravity? What is taken as a standard? Why is the temperature of the water fixed? Why is distilled water taken? What is the specific gravity of silver, and why? What is the specific gravity of cork, and why? 323. How do fluids that do not mix, when brought

bottom, because its specific gravity is greatest; next will
come the water; and on top, the oil, which is the lightest
of the three.

Cream rises on milk, because its specific gravity is less than that of milk.
For the same reason, the oily particles of soup float on the top.

The negroes in the West Indies take advantage of this law of specific
gravity. When they want to steal rum out of a cask, they introduce through
the hole in its top the neck of a bottle filled with water. The water descends
on account of its greater weight, and rum takes its place in the bottle.

324. Gases, like liquids, differ in their specific gravity. Smoke rises, because it is lighter than air. Hydrogen is so much inferior to air in specific
gravity, that it not only rises itself, but also carries up a loaded balloon.
Carbonic acid gas, on the other hand, is somewhat heavier than air; it is
therefore found at the bottom of wells and mines, where its poisonous properties sometimes prove fatal to those who descend.

325. If a solid floats on a liquid, like cork on water, its
specific gravity is less than that of the liquid; if it sinks,
like lead, its specific gravity is greater. If solid and liquid
have the same specific gravity, the solid will remain stationary at any depth at which it is placed, without rising
or sinking.

That a solid may float, it is not essential that, in a compact mass, it weigh
less than a like bulk of the liquid. A solid may therefore float or sink in
the same liquid, according to the form it is made to assume. A cubic inch
of iron weighs $7\frac{1}{4}$ times as much as a like bulk of water, and will therefore
sink in the latter; but, if beaten out into a vessel containing more than $7\frac{1}{4}$
cubic inches, this same iron will float, because then it is lighter than an equal
bulk of water. It is on this principle that iron ships float.

Fig. 153.

326. A floating solid displaces its own
weight of liquid.

To prove this, fill the vessel A with water up to the
opening B. Drop in a ball of wood. As it becomes
partially immersed, it raises the water and causes it to
flow through B. Catch the water thus displaced, and
it will be found to weigh exactly the same as the ball.

327. A body immersed in water is

together, arrange themselves? Give an example. Why does cream rise on milk?
What use do the negroes in the West Indies make of this principle? 324. What is
said of the specific gravity of gases? Why does smoke rise? How does hydrogen
compare with air in specific gravity? Carbonic acid? 325. When will a solid float
on a liquid, when sink, and when remain stationary without rising or sinking? How
may a solid which in a compact mass is heavier than water, be made to float?

buoyed up, and loses as much weight as the water it displaces weighs.

A boy can bring up from the bottom of a pond a heavy stone which he could not lift on land. In raising a bucket from a well, we find it become heavier the moment it leaves the water. In each case, the weight of the object, while in the water, is diminished by its upward pressure.

That the weight thus lost equals that of the water displaced, is shown with the apparatus represented in Fig. 154. From one side of a balance suspend a solid cylinder B, and on the same scale place a hollow cylinder A, which just contains the other. Balance the whole with a weight C in the opposite scale. If, now, we immerse B, still suspended, in a vessel of water, C will be found to outweigh A B, but the difference is exactly made up by filling A with water; and as A just holds B, it is evident that it holds as much water as B displaces.

Fig. 154.

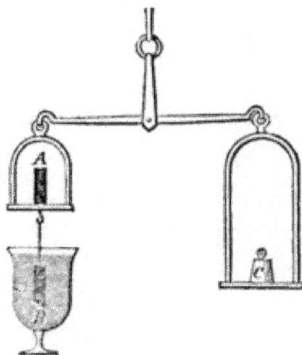

328. SPECIFIC GRAVITY OF LIQUIDS.—The specific gravity of a body is simply its weight compared with that of a like bulk of water. Hence the specific gravity of a liquid may be easily obtained in the following way : Fill a glass vessel, whose weight is known, with water up to a certain mark, and weigh it ; subtract the weight of the vessel, and you have the weight of the water alone. Then fill the vessel to the same height with the liquid in question, weigh it again, and subtract the weight of the vessel as before. To find the specific gravity of the liquid, divide its weight by that of the water.

A flask that will hold 1,000 grains of water, called the Thousand Grain Bottle, is often used for this purpose. A glass stopper, with a narrow opening running lengthwise through it, is fitted to the neck. The flask being filled, this stopper is inserted ; as it descends, it forces out the excess of liquid through its opening, and thus always ensures the same volume of liquid

Give an example. 326. How much liquid does a floating solid displace? Prove this with Fig. 153. 327. How much weight does a body immersed in water lose? Give some familiar examples of this loss of weight. Prove, with the apparatus represented in Fig. 154, that the weight lost equals that of the water displaced. 328. How may the specific gravity of a liquid be obtained? What is the Thousand Grain Bottle?

inside. A flask containing 1,000 grains of water will hold 13,568 grains of mercury and 792 grains of alcohol; dividing according to the rule, we find the specific gravity of mercury to be 13.568, and that of alcohol .792.

329. *The Hydrometer.*—The specific gravity of liquids may also be determined by the Hydrometer. This instru-ment consists of a hollow ball, C, from which rises a graduated scale, A; while to its lower side is attached a solid ball, B, of sufficient weight to keep the instrument in a vertical po-sition.

Fig. 155.

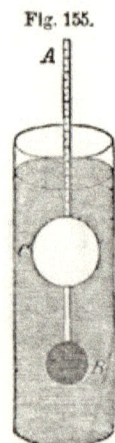

THE HYDROM-ETER.

To find the specific gravity of any liquid, place the hydrom-eter in it. The rarer the liquid, the farther it descends; and the figure on the scale at the point where it meets the surface, is noted. A table accompanies the instrument, which tells the specific gravity of a liquid when the height to which it rises on the scale is known.

The hydrometer is used by dealers in spirits, oils, and chemicals, to test their strength. The height to which the pure article rises on the scale being known, any different re-sult when a liquid is tested, indicates adulteration.

330. SPECIFIC GRAVITY OF SOLIDS. — The simplest way of finding the specific gravity of a solid would be to take a certain bulk of it (say a cubic inch or cubic foot), ascertain its weight, and divide it by the weight of a like bulk of water. It is so difficult, however, to obtain any given bulk exactly, that other methods have to be resorted to.

331. If the solid sinks in water, weigh it first in air, and then in water by means of a balance prepared for the pur-pose. Divide its weight in air by the weight it loses in water, and the quotient will be its specific gravity.

This is the same thing as dividing the weight of the solid by that of an equal bulk of water, for we have already seen that a solid weighed in a liquid loses as much weight as the liquid it displaces weighs.

---

How many grains of mercury will such a flask hold? Of alcohol? What, then, is the specific gravity of mercury and alcohol? 329. What instrument is used for ob-taining the specific gravity of liquids? Describe the Hydrometer. How is the spe-cific gravity obtained with this instrument? By whom is the hydrometer chiefly used? How does it indicate adulteration? 330. What would be the simplest mode of finding the specific gravity of a solid? What difficulty stands in the way? 331. How may we find the specific gravity of a solid that sinks in water? Give an

A piece of platinum weighs 22 grains in air, and 21 in water. Dividing 22, the weight in air, by 1, the loss of weight in water, we get 22 for the specific gravity of platinum.

332. To find the specific gravity of a solid that floats on water, attach to it some body heavy enough to sink it. Weigh the two, thus attached, in air and in water; and by subtraction find their loss of weight in water. In the same way, find how much weight the heavy body alone loses in water. Subtract this from the loss sustained by the two, and you get the weight of a volume of water equal to the body under examination. Divide the body's weight in air by this remainder, and you have its specific gravity.

*Example.* Required the specific gravity of a piece of elm wood weighing 2 ounces. Attach to it 4 ounces of lead.

The combined solids weigh in air 2 + 4 = 6 ounces.
In water we find them to weigh ......... 3.15 ounces.

Loss of the combined solids in water, 2.85 ounces.

The lead alone weighs in air............ 4 ounces.
The lead alone weighs in water.......... 3.65 ounces.

Loss of the lead in water, ........... .35 ounce.

Weight of a volume of water equal to the wood, 2.85 — .35 = 2.50
Specific gravity of elm wood, 2 ÷ 2.50 = .8

333. SPECIFIC GRAVITY OF GASES.—The specific gravity of gases is found by a process similar to that employed for liquids. Air is taken for the standard. A glass flask furnished with a stop-cock is weighed when full of air, and again when exhausted by means of an air-pump; the difference between these weights is the weight of a flask-full of air. The flask is then filled with the gas in question, and again weighed; this weight, less that of the exhausted flask, is the weight of a flask-full of the gas. Divide the weight of the gas by that of the air, and the quotient is the specific gravity required.

334. TABLES OF SPECIFIC GRAVITIES.—The following

tables give the specific gravity of some of the most important substances :—

### SPECIFIC GRAVITY OF SOLIDS AND LIQUIDS.—*Standard, Distilled Water*, 1.

| | | | | | |
|---|---|---|---|---|---|
| Iridium .... | 23.000 | Iron, cast ...... | 7.207 | Ice ......... | .930 |
| Platinum... | 22.069 | The earth ...... | 5.210 | Living men.. | .891 |
| Gold ...... | 19.358 | Diamond....... | 3.536 | Cork........ | .240 |
| Mercury.... | 13.568 | Parian Marble .. | 2.838 | Human blood | 1.045 |
| Lead....... | 11.445 | Anthracite coal.. | 1.800 | Milk ........ | 1.030 |
| Silver...... | 10.474 | Bituminous coal. | 1.250 | Sea water ... | 1.026 |
| Copper, cast | 8.788 | Lignum vitæ.... | 1.333 | Olive oil .... | .915 |
| Tin ........ | 7.285 | Oak............ | .970 | Alcohol ..... | .792 |

### SPECIFIC GRAVITY OF GASES.—*Standard, Air*, 1.

| | | | |
|---|---|---|---|
| Hydriodic Acid ........ | 4.300 | Air .................. | 1.000 |
| Carbonic Acid ........ | 1.524 | Nitrogen .............. | 0.972 |
| Oxygen .............. | 1.111 | Hydrogen ............ | 0.069 |

335. By examining the above tables, it will be found that solids generally have a greater specific gravity than liquids, and liquids than gases. Among solids, the metals are the heaviest.

The heaviest known substance is the metal iridium, which, bulk for bulk, weighs 23 times as much as water. The lightest substance is hydrogen gas. It would take about 12,000 cubic feet of hydrogen to weigh as much as one cubic foot of water.

Sea-water, being impregnated with salts, is somewhat heavier than fresh water. It is therefore more buoyant; and this every swimmer that has tried it knows. A vessel passing from fresh water to the sea, draws less water in the latter, that is, does not sink to so great a depth.

336. Water is 828 times heavier than air; that is, it would take 828 cubic inches of air to weigh as much as 1 cubic inch of water. Hence, by confining air in tight chambers in different parts of life-boats, they are made so buoyant that they can not sink even when filled with water. Life-preservers act on the same principle. The air confined in them, being 828 times lighter

cific gravity of gases be found? 334. [*Questions on the Tables.*—Which is the densest of the metals? Which is the densest of liquids? Will the wood called *lignum vitæ* float in water? What liquid will it float in? Which weighs more, a cubic foot of water or the same bulk of ice? In which would a boat sink deepest, olive oil, alcohol, or sea-water? Could a man swim in alcohol? Would a balloon rise most easily in hydrogen, carbonic acid, or air? Would a balloon filled with oxygen rise in air?] 335. How do solids, as a general thing, compare with liquids in specific gravity? How do gases compare with liquids? Among solids, what class of bodies are heaviest? What is the heaviest known substance? How does its weight compare with that of water? What is the lightest substance? How many cubic feet of hydrogen would it take to weigh as much as one cubic foot of water? How does sea-water compare with fresh water in specific gravity? In which is it easier to swim? In which does a vessel draw less water? 336. How does air compare with water in specific gravity?

than the same bulk of water, helps to keep up the bodies to which they may be attached. Many species of fish are provided with bladders, which they can fill with air or exhaust at pleasure; they are thus able to increase or diminish their specific gravity instantaneously, and to rise or sink accordingly.

337. The specific gravity of living men is set down at .891, or less than ⁹⁄₁₀ of that of water. The human body, therefore, will float; and, if the head is thrown back so as to bring the mouth uppermost, there is no danger of drowning, even in the case of those who can not swim. If the air is expelled from the lungs, and water takes its place, the specific gravity is increased; consequently the bodies of drowned persons sink. After remaining under water for a time, they again float; this is owing to the generation of light gases within them, by which their specific gravity is lessened.

338. If we know the specific gravity of a body, we can easily find how much any given bulk of it weighs. A cubic foot of water is found to weigh 1,000 ounces, or $62\frac{1}{2}$ pounds avoirdupois; the weight of a cubic foot of any given substance will, therefore, be equal to $62\frac{1}{2}$ pounds multiplied by its specific gravity.

*Example.* Required the weight of a cubic foot of gold. The table makes the specific gravity of gold 19.358. Multiplying this into 62.5, we get 1209.875 pounds for the weight required.

339. Two solids of equal bulk will displace equal quantities of a liquid in which they are immersed; but two solids of equal weight will not do so, unless their specific gravity is the same. This principle has been applied in testing the purity of the precious metals.

If, for instance, we wish to find whether a piece of silver is pure, we put it in a vessel even full of water, and catch what overflows: we do the same with an equal weight of what is known to be pure silver. If equal quantities of water are displaced, the article tested is pure, for it has the same specific gravity as pure silver; but if not, it is adulterated.

340. The fact above stated was discovered and first applied by Archimedes. Hiero, king of Syracuse, having purchased a golden crown and suspecting the purity of the metal, asked the philosopher to test it, without injury to its costly workmanship. In vain Archimedes tried to solve the prob-

On what principle are life-boats and life-preservers constructed? How are fish enabled to rise or sink at pleasure? 337. How does the body of a living man compare with water in specific gravity? What follows, as regards danger of drowning? Why do the bodies of drowned persons at first sink, and afterwards rise? 338. If we know the specific gravity of a body, how may we find the weight of any given bulk of it? Give an example. 339. When will two solids immersed in a liquid displace equal quantities? To what has this principle been applied? How, for example, may we find whether a piece of silver is pure? 340. By whom was this principle discovered?

7

lem; till one day, when bathing, he observed, that, as more and more of his body became submerged, the water rose proportionally higher and higher in the vessel. It at once occurred to him that any body of equal weight and exactly the same density, *but no other*, would cause an equal rise of the liquid; and here was a clue to the solution of the problem that had troubled him. Naked as he was, he rushed home from the bath, shouting "*Heureka!*" *I have found it!* He immediately procured a quantity of pure gold equal in weight to the crown, and a like weight of pure silver. Then successively plunging the gold, the silver, and the crown, in a vessel brim-full of water, he caught and weighed the liquid displaced in each case. Finding that the crown displaced more than the gold and less than the silver, he inferred that it was neither pure gold nor pure silver, but a mixture of the two. Archimedes afterwards investigated the subject further, and discovered the leading principles connected with specific gravity.

## Capillary Attraction.

341. If one end of a fine glass tube be placed in a vessel of water, the other end being left open, the water will rise in the tube above its level. The force that causes the water to rise is known as Capillary Attraction. It is so called from the Latin word *capillus*, a hair, because it is most strikingly exhibited in tubes as fine as a hair.

A liquid will not rise by capillary attraction in tubes that exceed one-fifteenth of an inch in diameter.

342. CAUSE.—The rise of liquids in capillary tubes is owing, it is thought, to the attraction of the inner surface of the solid. In proof of this, we find that the surface of the liquid in the tube is concave, being raised where it comes in contact with the sides of the tube.

Fig. 156.

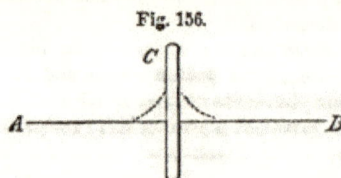

The same thing is seen when a glass plate, C, is placed perpendicularly in water, A B: the surface, instead of maintaining the same level throughout, rises near the glass on both sides, as represented by the dotted lines.

The above experiment seems to show that the attraction of glass for water is sufficiently great to overcome

Relate the circumstances. 341. What is Capillary Attraction? Why is it so called? What is the limit of size for capillary tubes? 342. To what is the rise of a liquid in capillary tubes owing? What proof is there of this? When a glass plate is placed perpendicularly in water, what may be observed? What does this experiment show?

the gravity of the latter. It is, also, greater than the cohesion subsisting between the particles of water; for, if the glass be removed, some of the liquid will be found adhering to its surface,—that is, it will be *wet*.

343. This attraction, however, does not exist between all solids and liquids; on the contrary, we sometimes find as decided a repulsion.

Let the glass plate, for instance, in the last experiment, be greased, and the water, now acted on by a repelling force, instead of being elevated near the sides, will be depressed, as shown by the dotted lines in Fig. 157. A similar appearance is presented when a glass plate is plunged into a dish of mercury. When this repulsion exists, the liquid does not wet the solid; when the glass plate is drawn out of the mercury, not a particle of the liquid adheres to it.

Fig. 157.

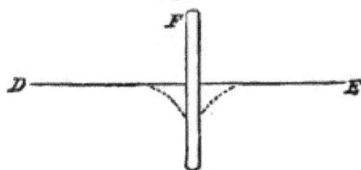

The repulsion just mentioned may be so great as to prevent a solid from sinking in a liquid lighter than itself. A fine needle smeared with grease, if carefully laid in a horizontal position on the surface of still water, will remain floating there. It is thus that insects are able to walk on water; the repulsion between their feet and the liquid prevents them from sinking or even becoming wet.

344. FAMILIAR EXAMPLES.—Examples of capillary attraction meet us on all sides.

If one end of a towel be left in a basin of water, the part outside soon becomes wet, the liquid being drawn up through its minute fibres. The same thing happens if a piece of sponge, of bread, or of sugar, remains in contact with a liquid, the pores of the substance acting like capillary tubes. Blotting paper drinks up ink on the same principle.

The common lamp affords a good illustration of capillary attraction. The oil or burning-fluid is drawn up through the fibres of the wick fast enough to support the flame. There is a limit, however, beyond which capillary attraction does not act; and, therefore, if the oil gets low, the lamp grows dim and finally goes out. To allow a free passage to the oil, the little tubes

---

843. What sometimes takes the place of this attraction between solid and liquid surfaces? Give an example. When a glass plate is plunged into a dish of mercury, what phenomenon is presented? What is sometimes the consequence of this repulsion? Give an example. How is it that insects walk on water? 844. How may capillary attraction be illustrated with a towel and a piece of bread or sugar? How is the flame

must be kept clear; and, as impurities gather in them from the ascending liquid, the wick must be changed from time to time.

Capillary attraction is strikingly exhibited in wood. Water is drawn up into its pores, distending them, and causing a perceptible increase of size. This expansion is turned to practical account in the south of France. A large cylinder of free-stone, several feet in length, has circular grooves made at intervals in its surface. Into these grooves are driven wedges of dry wood, which are then kept wet with water. As the wood absorbs the liquid, it gradually expands, till it rends the solid cylinder into rough mill-stones, which require but little labor to fit them for market.

It is capillary attraction that renders the banks of streams so productive; the water drawn in through the pores of the earth, fertilizes the adjacent parts. On the same principle, a potted plant may be supplied with the necessary moisture by filling the saucer in which it stands with water. Houses are rendered damp by the absorption of external moisture, the pores of the brick or stone, of which the walls are built, acting as capillary tubes.

345. LAWS OF CAPILLARY ATTRACTION.—*Different liquids rise to different heights in tubes of the same size.* Ether, for example, rises about one-half, and sulphuric acid only one-third, as high as water.

*The same liquid always rises to the same height in a tube of given size; and this height is proportioned to the fineness of the bore.* In a tube $\frac{1}{100}$ of an inch in diameter, water rises $5\frac{3}{10}$ inches.

Fig. 158.                                              Fig. 159.

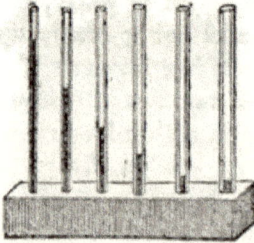

346. Fig. 158 represents six tubes of different bore, communicating at the bottom with a vessel containing colored water. The water rises according to the fineness of the bore, standing highest in the smallest tube.

of a lamp supplied with fuel? How is capillary attraction exhibited in wood? What use is made of this principle in France? What is the effect of capillary attraction on the banks of streams? How may a potted plant be supplied with moisture? How are houses made damp? 345. What is the law of capillary attraction, as regards dif-

347. The same principle is illustrated with two glass plates (see Fig. 159), joined at one end and slightly diverging so as to form an angle of about two degrees. Let the plates rest in colored water to the depth of an inch, and the liquid will rise between them, reaching the greatest height where the surfaces are nearest together, and thus forming the curve called the hy-per'-bo-la.

348. INTERESTING FACTS.—If a capillary tube capable of raising water four inches be broken off at three, there will be no overflow, as might be expected. The water will rise three inches to the top of the tube, and there stop. But it will be supplied as fast as evaporation takes place. Hence, to prevent waste in a spirit lamp, an extinguisher is put over the wick when it is not burning.

It is a remarkable fact that no evaporation takes place unless the liquid reaches *the top* of the capillary tube. Tubes containing as much water as they could hold under the influence of capillary attraction, have been hung in the sun for months, without losing any part of their contents by evaporation.

349. FLOATING BODIES.—Motion is produced in bodies floating near each other, by a force resembling capillary attraction. This may be shown with two balls, as represented in Figs. 160, 161, 162.

A and B are cork balls, capable of being wet with water. When they are brought close together, the attraction of their surfaces raises the water around them; the column that separates them becomes thinner and thinner, till at last they touch.

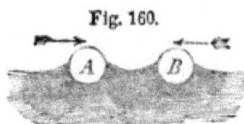

Fig. 160.

C and D are similar balls, greased so that they can not be wet. In this case, the surface of the surrounding water is repelled, forming little hollows in which they rest. Since there is not enough liquid between them to balance the pressure from without, the balls again approach each other.

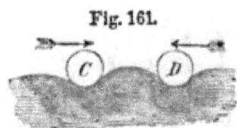

Fig. 161.

ferent liquids? Give an example. What is the law for the same liquid in a tube of given size? How high does water rise in a tube $1/100$ of an inch in diameter? 346. What does Fig. 158 represent? 347. Describe the experiment with two glass plates. 348. What fact is stated respecting a capillary tube broken off at the top? Why is it necessary to put an extinguisher on a spirit-lamp? What fact is stated respecting evaporation from capillary tubes? 349. How are floating bodies affected by a force resembling capillary attraction? What, for example, is the effect on cork balls capable of being wet? On balls greased so that they can not be wet? On balls, one of

Fig. 162.

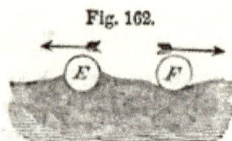

E and F are a pair of similar balls, one of which, E, can be wet, while the other, F, can not. The water, attracted by E, rises around it, whereas around F it is depressed. If these balls are placed near together, F, being repelled from the wall of water around E, will recede from it.

350. ENDOSMOSE AND EXOSMOSE.—Two peculiar results of capillary attraction, known as Endosmose and Exosmose, remain to be mentioned.

Endosmose is the inward motion of a fluid, through a membranous or porous substance, into a vessel containing a different fluid. Exosmose is the outward motion of the contained fluid through the same substance.

Fill a vessel with alcohol, tie over the top a bladder that has been soaked in water, and immerse the whole in water. In a few hours it will be found that water has passed into the vessel through the bladder, and that alcohol has passed out into the water. The former movement is called Endosmose; the latter, Exosmose. The inward current is stronger than the outward one. Water passes in faster than alcohol escapes; and consequently the bladder soon becomes puffed out. All membranous and porous substances, such as india rubber, plaster of paris, wood, &c., permit the passage of these currents, which are owing to capillary attraction.

351. Endosmose and exosmose are exhibited in the case of gases, as well as liquids.

If a phial full of air, with a piece of thin bladder tied over its mouth, be placed in a jar of carbonic acid gas, the latter will force its way into the phial while air will pass out. Here, again, the inward current is the stronger; the bladder is puffed out, and finally bursts.

The facility with which gases thus pass in and out through porous substances is proportioned to their rarity. Hydrogen, the rarest of known bodies, exhibits these movements in their greatest perfection. This is the reason why the rose balloons, recently so popular as toys, lose their buoyancy in a few days. They are made of thin india rubber, and filled with hydrogen. When allowed to remain in the air, endosmose and exosmose take place. Hydrogen passes out through the pores of the rubber, and air takes its place. The balloon gradually becomes less buoyant, ceases to rise, and at last, as it loses more of its hydrogen, is carried to the ground by the weight of the india rubber.

which can be wet and the other not? 850. What is Endosmose? What is Exosmose? Show how endosmose and exosmose operate. Through what sort of substances do they take place? 851. What, besides liquids, are affected by these movements? Give an example. What gases most readily pass in and out through porous substances? What gas exhibits endosmose and exosmose most distinctly? What is the

352. The skin being porous, a liquid with which it remains in contact will find its way through by endosmose and be absorbed by the body. If a drop of the powerful poison called prussic acid be placed on the arm, a sufficient quantity to cause death will thus be taken into the system.

353. Endosmose and exosmose enter largely into the operations of nature. They cause the ascent and descent of sap in trees and vines. The inside of living plants consists of minute cells, containing fluids of different densities. These fluids are constantly passing in and out through the porous walls which separate them, under the influence of exosmose and endosmose, modified by the vital action at the same time going on.

## EXAMPLES FOR PRACTICE.

1. (*See* § 328.) A phial weighing 4 ounces when empty, weighs 6 ounces when filled with water, and 7 when filled with nitric acid. Required, the specific gravity of the acid.—*Ans.* 1.5.

2. A vessel filled with ether weighs 13.575 ounces; filled with water, 15 ounces; when empty, 10 ounces. What is the specific gravity of ether?

3. An empty jar weighs 7.5 pounds; filled with sulphuric acid, it weighs 12.1125 pounds; and filled with water, 10 pounds. Find the specific gravity of sulphuric acid.

4. A Thousand Grain Bottle is found to hold 870 grains of oil of turpentine, and 1,036 grains of oil of cloves. What is the specific gravity of these oils?

   In which would a cork ball sink the deeper?

5. (*See* § 331.) A piece of crown-glass weighs 5 ounces in the air, and 3 in water. What is its specific gravity?—*Ans.* 2.5.

6. A beef-bone weighs 2.6 ounces in water, and 6.6 ounces in air. What is its specific gravity?

7. What is the specific gravity of a piece of ivory, which weighs 16 ounces in air, and loses 8³/₄ ounces when weighed in water?

8. (*To solve the next two sums, see* § 332 *and Example. In each case, we may suppose a pound (16 ounces) of lead, weighing 14.6 ounces in water, to be used for sinking the solid.*)

   A piece of wax weighs 8 ounces; when it is fastened to a pound of lead, the whole weighs in water 13.712 ounces. What is the specific gravity of the wax?—*Ans.* .9.

9. Fastening a piece of ash to a pound of lead, I find their weight in water to be 12.76 ounces. The ash alone weighs 10 ounces in the air. What is its specific gravity?

10. (*See* § 333.) A glass flask, with the air exhausted, weighs 4 ounces; filled with air, it weighs 4.25 ounces; and filled with cy-an'-o-gen, 4.45125 oz. What is the specific gravity of cyanogen?—*Ans.* 1.805.

11. A flask full of chlorine weighs 11.222 ounces. Filled with air, it weighs 10.5 oz., and when the air is drawn out, 10 oz. Required, the specific gravity of chlorine.

12. According to the answers of the last two sums, in which would a balloon rise most easily, air, cyanogen, or chlorine?

13. (*See* § 336.) How many cubic feet of air would it take to weigh as much as 4 cubic feet of water?

14. (*See* § 338, *and Table.*) How much would a cubic foot of gold weigh? How much, the same bulk of silver?

15. What would be the weight of 4 cubic feet of Parian marble?

16. What is the weight of a block of anthracite coal, 6 feet long, 4 feet wide, and 3 feet high? (*To find the number of cubic feet in the block, multiply the length, breadth, and thickness together.*)

17. Suppose a room 10 feet high, long, and wide, to be filled with gold, what would the gold weigh?

---

# CHAPTER XI.

## MECHANICS (CONTINUED).

### HYDRAULICS.

354. HYDRAULICS treats of liquids in motion, whether issuing from orifices or running in pipes and the beds of streams. It shows how water is applied as a moving power, and describes the machines used for raising liquids.

355. FLOW OF LIQUIDS THROUGH ORIFICES.—If an orifice be made in the side or bottom of a vessel containing a liquid, the latter will escape through it. The particles of liquid near the orifice are forced out by the pressure of those above.

356. *Velocity.*—The velocity of a stream flowing through
an orifice depends on the distance of the latter below the
surface of the liquid, being equal to the velocity which a
body would acquire in falling that distance.

If, for instance, in a reservoir full of water, three orifices be made at
depths of $16^1/_{12}$, $64^1/_3$, and $144^3/_4$ feet, the liquid (leaving friction, &c., out of
account) will issue from them with velocities of $32^1/_6$, $64^1/_3$, and $96^1/_2$ feet per
second, because such, as we have found, would be the velocity of a body falling through the different distances first named.

The distances above mentioned are to each other as 1, 4, 9 ; the velocities
are to each other as the square roots of these numbers, 1, 2, 3.   Consequently, *the velocities of streams issuing from different orifices in the same vessel
are to each other as the square roots of their respective distances below the surface of the liquid.*   Friction, however, and other causes, produce more or
less deviation from this rule.

357. As long as the liquid is kept at the same height in
the vessel, it issues from a given orifice with the same velocity; but, if the vessel is not replenished, as the liquid
gets lower, the pressure diminishes, and the velocity of the
stream diminishes with it.   It takes twice as long to empty
an unreplenished vessel through a given orifice, as it would
for the same quantity of water to escape if the liquid were
kept at its original level.

358. *The Clepsydra.*—Among the ancients, time was
measured by the flow of water through an orifice, in an instrument called the Clepsydra, or Water-clock.   It consisted of a transparent vessel with a hole in the bottom that
would empty it in a certain time.   A scale on the side of
the vessel indicated, by figures at different levels, the number of hours which it took the liquid to reach them successively in its descent.   As the discharge was most rapid
when the vessel was full, the divisions were of course longest
at the top of the scale.

The clepsydra was necessarily inaccurate, inasmuch as the flow of the

---

354. Of what does Hydraulics treat?   355. What causes a liquid to flow through
an orifice in the vessel containing it ?   356. On what does the velocity with which a
stream issues from an orifice depend?   Give an example.   What is the law for the
velocities of streams issuing from different orifices in the same vessel ?   357. What
difference does it make, as regards the velocity of a stream through an orifice, whether
the vessel is kept replenished or not ?   858. What did the ancients use for measuring

---

water varied in rapidity according to its temperature and the density of the atmosphere. Yet it answered for general purposes; indeed, it was the only instrument used for measuring small intervals of time in astronomical observations.

359. *Course of Streams flowing from Orifices.*—A liquid issuing from an orifice descends in the same line as a projectile (see § 127). The curve described is called a parabola. In a given vessel, a stream will spout to the greatest horizontal distance, from an orifice midway between the surface and the bottom of the liquid. Streams flowing through orifices equally removed from this central one, will spout to the same distance.

Fig. 163.

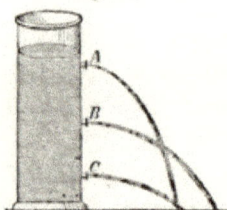

In Fig. 163, if the orifice B be midway between the surface and the bottom of the liquid, the stream passing through it will spout to the greatest distance; and if A and C be equi-distant from B, the streams passing through them will reach the same point.

360. *Volume discharged.*—To find the volume of liquid discharged in a given time from an orifice in a vessel that is kept replenished, multiply the area of the orifice by the velocity of the stream per second, and this product by the number of seconds.

No allowance is here made for friction; in practice, therefore, the discharge is less than would appear from this rule.

*Example.* How much water will be discharged from an orifice of 2 square inches in 5 seconds, the velocity of the stream being 10 inches in a second, and the vessel being kept replenished?—*Ans.* 2 × 10 × 5 = 100 cubic inches.

361. The quantity discharged through a given orifice in a given time differs in the case of different liquids. Alcohol, for instance, flows more slowly than water, and mer-

time? Describe the clepsydra. What rendered the clepsydra inaccurate? 359. What curve does a stream issuing from an orifice describe? At what part of a vessel will a stream from an orifice spout to the greatest distance? What is said of streams equally removed from the central one? Exemplify these principles with Fig. 163. 360. What is the rule for finding the volume of liquid discharged from an orifice in a given time? What causes deviations from this rule in practice? Give an example. 361. What is said of the quantity discharged in the case of different liquids? Give an

cury more rapidly; the discharge of alcohol will therefore
be less, and that of mercury greater, than the discharge
of water.

362. A circular orifice of given area discharges more
liquid in a given time than one of any other shape. This is
because a circle is the smallest line that can enclose a given
space; in passing through a circular orifice, therefore, the
liquid comes in contact with a less extent of solid surface,
and is less retarded by friction.

363. The volume discharged through an orifice in a given time may be
increased by heating the liquid. Heat lessens its cohesion, and enables it to
flow more rapidly.

364. The discharge may also be increased by fitting a short tube, or Adjutage, to the orifice. The minute currents of the particles are thus prevented from obstructing each other in the act of passing out. The best shape for
such a tube is that of a bell with the large
end out, as shown at A in Fig. 164. When
such a tube is used, the discharge in a given
time is increased one-half; and there is a
still greater gain if the bottom of the vessel is rounded outward to meet the tube,
as at B in Fig. 165.

Fig. 164.    Fig. 165.    Fig. 166.

If, however, the tube extends into the
vessel, as at C in Fig. 166, instead of increasing the discharge, it obstructs
and diminishes it.

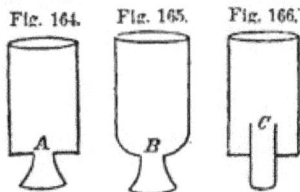

365. FLOW OF LIQUIDS IN PIPES AND THE BEDS OF
STREAMS.—The friction of water against the sides of pipes
in which it is conveyed, retards its velocity and diminishes
the quantity discharged.

When the distance is great, or there are sudden turn-
ings, allowance must be made for friction by increasing the
size of the pipes, or the quantity discharged will fall far
below what is required. If, for instance, leaving friction
out of account, pipes 6 inches in diameter would yield the
desired supply, nine-inch pipes would be none too large to
use.

example. 362. With a given area, what shape must an orifice have, to discharge the
most liquid? Why is this? 363. How may the volume discharged be increased?
364. What other mode of increasing the discharge is there? Describe the kinds of
adjutage mentioned in the text, and state the effect of each. 365. What is the effect
of friction on the flow of liquids? How great an allowance should be made for fric-

366. *Rivers.*—The friction of a stream against its banks and bottom materially retards its motion. Hence the velocity of a river is always less near its banks than towards the centre, and near the bottom than at the surface.

The windings of a stream also lessen its velocity. Were it not for their numerous bends, many large rivers would flow so rapidly that they could not be navigated.

367. The velocity of a stream depends much on the slope of its bed. A river with but few bends, and a fall of three inches to the mile, moves at the rate of about three miles an hour. As the slope increases, the velocity rapidly increases also; and a fall of three feet in a mile gives the impetuosity of a torrent.

Sometimes the bed of a river has a considerable fall at first, and then becomes comparatively level. In such cases, the impetus of the water keeps it in motion at a rate proportioned to its volume. The fall of the Amazon, in the last 700 miles of its course, is only 12 feet.

368. The quantity of water discharged by a stream depends on its size and velocity. In large rivers, it is almost incredible. The discharge of the Mississippi is estimated at twelve billions of cubic feet every minute, and that of the Amazon is nearly four times as great.

369. *Waves.*—Waves are caused by the action of wind on a liquid surface. As the particles of a liquid move freely among each other, the undulations produced directly by the wind extend along the surface to a great distance, farther than the wind itself.

The wind is enabled to take hold, as it were, of the water, and produce waves, by the friction at the surface. This friction may be diminished, just as in the case of machinery, by covering the surface with oil. The wind then slips over it, and the water becomes comparatively calm. It is said that boats have been enabled to get through a dangerous surf in safety, by emptying barrels of oil upon it.

370. Waves appear to move forward, but in deep water they only move

---

tion? 366. Where has the water of a river the least velocity, and why? What effect have the windings of a stream on its velocity? 367. On what does the velocity of a stream chiefly depend? How great a velocity does a fall of three inches in a mile produce? How great a fall produces the velocity of a torrent? How great a fall has the bed of the Amazon near its mouth? What keeps its waters in motion? 368. On what does the quantity of water discharged by a stream depend? How great is the discharge of the Mississippi? Of the Amazon? 369. By what are waves caused? What enables the wind to produce waves? How may a rough sea be calmed?

up and down. A floating body, after rising and falling with successive waves, when the sea becomes calm is found in the same spot as before. If, however, shoals or rocks interfere with the undulations, an onward motion is produced, and breakers are formed. Waves are always found breaking on a rocky shore, whatever way the wind may blow.

371. Waves do not generally exceed 20 feet in height, —that is, do not rise more than 10 feet above, and sink more than 10 feet below, the usual level of the sea. They sometimes, however, attain a height of 40 feet. Vast and mighty as they are, their effects are confined to the surface, never extending to the great body of the ocean. The severest gales are not felt at a depth of 200 feet.

372. *Tides.*—In the ocean, and the bays, rivers, &c., communicating with it, there is an alternate rise and fall of water, each lasting about six hours. These movements are called Tides. When rising, the tide is said *to flow;* when falling, *to ebb.*

373. Tides are caused chiefly by the attraction of the moon. This body, when opposite any given part of the earth, attracts the water at that part most strongly towards itself, and causes high tide. At the same time it is high tide at the opposite point of the globe, because the moon, attracting the mass of the earth more strongly than the more distant water on its surface, draws the former, as it were, away from the latter. These elevations are accompanied with corresponding depressions, or low tides, at other points.

The sun, also, attracts the water on the earth's surface; but not so strongly as the moon, in consequence of its vast distance. When sun and moon act in the same direction, which happens at every new and full moon, the tides are highest, and are called Spring-tides. When sun and moon act in opposite directions, the tides are lowest, and are called Neap-tides.

374. The height of the tide is affected by prevailing winds, the shape of adjacent coasts, and other circumstances; accordingly, it is different in different places. At St. Helena, the rise of water is only 3 feet; in parts of the British Channel, it is 60. The highest tides known are in the Bay of Fundy, where they attain a height of 70 feet.

370. How do waves appear to move? How do they really move? What proof is there of this? What is the effect of shoals or rocks? 371. What is the height of waves? How far below the surface do they extend? 372. What are Tides? 373. By what are tides caused? What, besides the moon, attracts the water? What are Spring-tides, and how are they caused? What are Neap-tides, and when do they occur? 374. By what circumstances is the height of tides affected? How great is

158 HYDRAULICS.

This makes the average rise one foot every five minutes,—so rapid a flow that animals feeding on the shore are sometimes overtaken and drowned.

375. WATER-WHEELS.—Running water is exceedingly useful as a moving power. Made to act on wheels, it causes them to revolve by its momentum, turns the shafts or axles connected with them, and thus sets machinery of various kinds in motion.

The wheels moved by water-power are of four kinds; the Undershot, the Overshot, the Breast-wheel, and the Turbine.

Fig. 167.

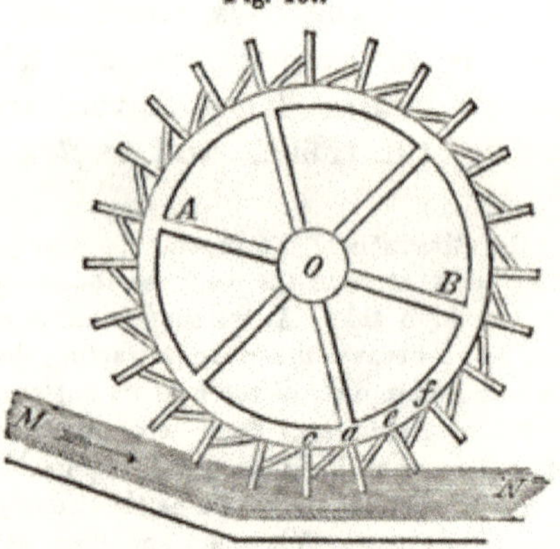

THE UNDERSHOT WHEEL.

376. THE UNDERSHOT WHEEL is represented in Fig. 167. A wheel, A B, attached to an axle, O, has a number of *float-boards*, *c, d, e, f*, fitted into its rim, at right angles, and at equal distances from each other. The whole is hung in such a way that the lowest float-board, *c*, is immersed in running water, M N. The current, striking against several float-boards, which are more or less submerged, carries the wheel around.

The stream is often conducted to the wheel through a narrow passage called a *Race;* and its power is sometimes increased by giving the race a slight inclination (see Figure). In other cases, the water is made to strike the wheel immediately after issuing from the bottom of a dam, with a high degree of velocity produced by the pressure of a large body of water. Yet, under the most favorable circumstances, as the *weight* of the water does not act on the wheel, but only the force of the current, no more than one-fourth of the moving power can be made available.

377. THE OVERSHOT WHEEL is represented in Fig. 168. It consists of a wheel, A B, attached to an axle, O, and having a number of *buckets, c, d, e, f*, on its rim, at equal distances. A stream is conducted through a race, G H,

the rise at St. Helena? In the British Channel? In the Bay of Fundy? 375. How is running water turned to account? Name the four kinds of water-wheels. 376. Describe the Undershot Wheel. How is the stream often conducted to the wheel? How is its power increased? In other cases, how is a high degree of velocity produced? How much of the moving power can be made available with this wheel?

and made to fall on the wheel from above. The weight of the water and the force with which it descends cause the wheel to revolve. Another bucket is brought under the stream, which in its turn is filled, and a new one is presented.

As the wheel turns, the descending buckets gradually lose their water, so that by the time they commence rising they are entirely empty. As the de-

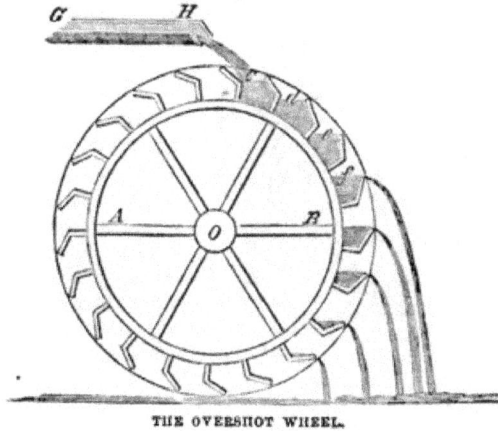

Fig. 168.

THE OVERSHOT WHEEL.

scending buckets contain more or less water and the ascending ones contain none, the wheel is kept revolving; and the weight of the stream, as well as its velocity, being turned to account, three-fourths of the moving power is saved.

378. In THE BREAST-WHEEL, shown in Fig. 169, there is a similar arrangement of apartments on the rim. The water is received half way up, or still higher in the High Breast-wheel commonly used in this country; and its weight is thus made available. This wheel ranks between the Overshot and the Undershot in efficiency, saving three-fifths of the moving power.

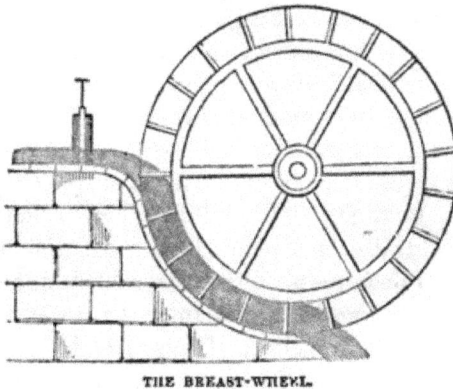

Fig. 169.

THE BREAST-WHEEL.

379. THE TURBINE, a section of which is represented in Fig. 170, instead of being vertical, like the wheels just described, is horizontal. It consists of a wheel, A B, divided into a number of apartments, $c, d, e, f$, by curved partitions. To the inner rim of the wheel is fitted a fixed cylinder, G H, divided into apartments corresponding with those of the wheel, but running in the opposite direction.

377. Describe the Overshot Wheel. Explain its operation. How much of the moving power does it utilize? 378. In the Breast-wheel, how is the water received? How much of the moving power is utilized? 379. Describe the Turbine. Explain its

Fig. 170.

THE TURBINE.

This fixed cylinder is connected with the base of an upright tube, J K, through the middle of which runs another tube, I.

The water which is to set the machinery in motion enters J K, runs through the apartments of G H, is discharged by them into the corresponding apartments of the wheel, and passes out into a course provided for its escape. It strikes the partitions nearly at right angles, and with great force in consequence of the pressure of the liquid in the tube. The wheel is thus made to revolve; and a shaft connected with it from below and passing through the inner tube I, communicates the motion to machinery above. Wherever there is a fall of water, turbines are found very useful. They have been known to utilize, or turn to account, four-fifths of the motive power,—more than is saved by any other wheel.

380. Propulsion of Boats.—The wheels of steamboats are not turned by running water, like those described above, but by machinery worked by steam. As they strike the water, the latter reacts on them; and the boats are forced forward or backward, according to the direction in which their wheels turn. Paddles on their rim enable the wheels to strike the water more forcibly.

As the paddles descend and ascend, they have to overcome a considerable resistance in a vertical direction, which retards their motion; it is only when they are vertical in the water that their full effect is felt. The rolling of the boat, also, often interferes with their action, burying them too deep or raising them entirely out of water. These disadvantages have led some to prefer a submerged screw to the paddle-wheel. The screw is placed in the stern; and vessels moved by its means are called Screw Propellers.

381. The resistance which a vessel encounters in passing through water depends on its shape. The narrower the vessel and the sharper its prow, the more readily it pene-

trates the water, on the principle of the wedge. Too great narrowness, on the other hand, is dangerous in boats that navigate stormy waters, and does not allow sufficient room for freight. To determine the shape that best combines speed, safety, and capacity, is the work of the ship-builder. It is a difficult problem, and one that is perhaps not yet solved, though great advances have been made of late years in naval architecture.

382. BARKER'S MILL.—An ingenious hydraulic machine, called Barker's Mill, and represented in Fig. 171, remains to be described.

Fig. 171.

A is an upright hollow cylinder, turning freely on a vertical axis. Through its lower end runs a horizontal tube, B C, communicating internally with the cylinder. On opposite sides of this tube, at its extremities, are two small openings. A continuous stream is introduced, through the pipe D E, into the funnel at the top of the cylinder A. It runs down into the cross-tube B C; and, if there were no opportunity of escape, it would there rest, pressing equally in every direction. The moment, however, that the two holes in the ends are opened, the water runs through; and the pressure at the holes being thus removed, while that on the opposite sides remains undiminished, the tube is forced round in the direction of the pressure, that is, in an opposite direction to the jets of water. The cylinder A turns with the tube, and thus motion is communicated to the mill-stone S. H is a hopper, which feeds the mill with grain.

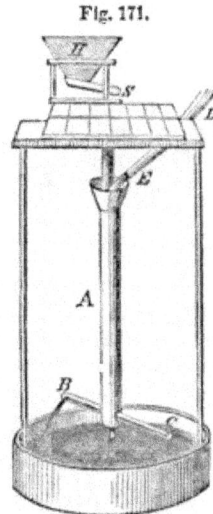

BARKER'S MILL.

383. MACHINES FOR RAISING WATER. —It is often desirable to raise water from a lower to a higher level. Well-sweeps, acting on the principle of the lever, are used for this purpose, as is also the wheel and axle in a variety of forms. But, when a large supply is required, other machines, worked with less expense of time and labor, are employed. Some of these involve the principles of Pneumatics, and will be treated under that head.

Those that belong exclusively to Hydraulics are described below.

384. *Archimedes' Screw.*—The Screw of Archimedes, called after the philosopher that invented it, is one of the simplest machines for raising water. It consists of a tube wound spirally round a solid cylinder, as represented in Fig. 172.

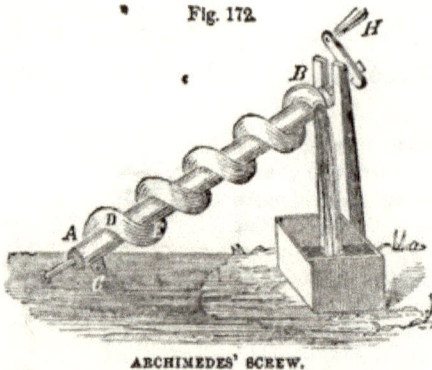

Fig. 172.

ARCHIMEDES' SCREW.

To work the machine, let one end of the tube, C, rest just below the surface of the water. The cylinder, A B, must be inclined at an angle of about 35 degrees, and be fastened at the lower end in such a way as to revolve freely when turned by the handle, H. When the cylinder is turned, the open end of the tube, C, scoops up some of the water. When it has got half way round, the point D is lower than the end C, and the water descends to D by the force of gravity. Another half-revolution brings the point E lower than D, and again the water descends. This is continued till the water is discharged at the upper end. As new water is constantly scooped up, there will be a continuous flow as long as the handle is turned.—Archimedes' Screw operates only at short distances.

385. *The Chain Pump.*—The Chain Pump is much used for raising water. The principle it involves is also applied in dredging-machines, for cleaning out the channels of rivers.

This machine (see Fig. 173) consists of a continuous chain, to which circular plates, c, d, e, f, &c., are attached at equal distances. The plates are of such a size as exactly to fit the cylinder G H, the lower end of which rests in the water. The chain passes over the two wheels, I, J; to the upper one of which, I, a handle is attached. When the handle is turned, the chain is set in motion. The plates, ascending through G H, carry up water before them, which has no opportunity of escaping till it reaches the opening K.

machines for raising water? Of what does Archimedes' Screw consist? Describe its mode of operation. At what distances does Archimedes' screw operate? 385. What machine is much used for raising water? What other application is made of the principle it involves? Describe the Chain Pump, and its mode of operating.

There it is discharged, as long as the handle is turned.

386. *The Hydraulic Ram.*—The Hydraulic Ram was invented in France, in 1796. It raises water by successive impulses, which have been compared to the butting of a ram, and hence its name. The requisite power is gained by momentarily stopping a stream in its course, and causing its momentum to act in an upward direction.

Fig. 174 represents a simple form of the Hydraulic Ram. To a stream or reservoir at A, is adapted an inclined pipe, B, through which the water that works the ram is conveyed. Near the lower end of the pipe B rises an air-chamber, D, with which an upright pipe, F, is connected. The passage connecting B with the air-chamber is commanded by a valve opening upward. At the extremity of the pipe B is another valve, E, opening downward, and made just heavy enough to fall when the water in B is at rest.

Fig. 173.

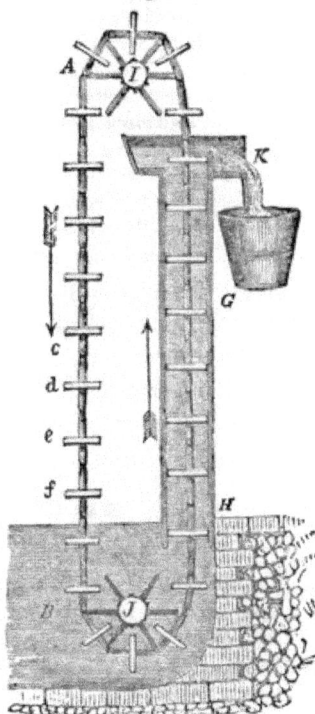

THE CHAIN PUMP.

Fig. 174.

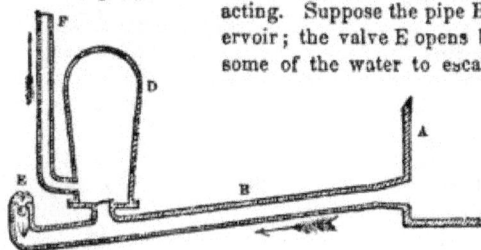

THE HYDRAULIC RAM.

The play of the valve E makes the machine self-acting. Suppose the pipe B to be filled from the reservoir; the valve E opens by its weight, and allows some of the water to escape. Soon, however, the water acquires momentum enough to raise the valve and close the opening. The stream is thus suddenly stopped, and the pipe would be in danger of bursting from the shock were it not for the valve in the air-chamber D, which is at once forced upward, and allows

386. When and where was the Hydraulic Ram invented? Why is it so called? How is the requisite power gained in the ram? Describe the hydraulic ram, and

some of the water to enter. The air in D is at first condensed by the pressure of the water thus admitted ; but, immediately expanding by reason of its elasticity, it drives the water into F, for the closing of the valve prevents it from returning to B. By this time the water in B is again at rest, the valve E opens, and the whole process is repeated.

By successive impulses the water may be raised in F to a great height. A descent of four or five feet from the reservoir is sufficient. Care must be taken to have the valve E just heavy enough to fall when B is at rest, and not so heavy as to prevent it from readily rising as the momentum of the stream increases. The pipe B must also be of such length that the water, when arrested in its course, may not be thrown back on the reservoir.

387. Hydraulic Rams afford a cheap and convenient means of raising water in small quantities to great heights, wherever there is a spring or brook having a slight elevation. They are used for a variety of purposes, and particularly when a supply of water is needed for agricultural operations.

### EXAMPLES FOR PRACTICE.

☞ *Friction is left out of account in these examples.*

1. (*See* § 356, *rule in italics.*) Two streams issue from different orifices in the same vessel with velocities that are to each other as 1 to 6. How many times farther from the surface is the one than the other?

2. The stream A runs from an orifice in a vessel three times as fast as the stream B. How do their distances below the surface of the liquid compare?

3. In a vat full of beer there are two orifices of equal size; one 9 inches below the surface, and the other 25. How does the velocity of the latter compare with that of the former?

4. There are three apertures in a reservoir of water, 1, 4, and 16 feet below the surface. With what comparative velocity will their streams flow?

5. A stream flows from an aperture in a vessel at the rate of 4 feet in a second. I wish to have another stream from the same vessel with a velocity of 16 feet per second. How much farther below the surface than the first must it be?

6. (*See* § 359.) A vat full of ale, 3 feet high, has four apertures in it, 3, 12, 18, and 24 inches respectively from the top. Through which will the liquid spout to the greatest horizontal distance? Which next? Which next?

7. (*See* § 360.) How much water will be discharged every minute from an orifice of 3 square inches, the stream flowing at the rate of 5 feet in a second, and the vessel being kept replenished?

---

its mode of operating. How great a descent is required? What precautions are necessary? 387. In what case may hydraulic rams be used with advantage?

How much will be discharged every minute from another orifice in the same vessel, equally large, but situated four times as far below the surface of the liquid?

8. A stream flows from a hole in the bottom of a vessel with a velocity of 6 feet in a second. The hole has an area of 5 square inches, and the vessel is emptied in 15 seconds. How much water does the vessel hold?

9. (*See* § 376.) A stream having a momentum equivalent to 100 units of work is applied to an Undershot Wheel; how many units of work will it perform?—*Ans.* 25.

(*See* § 377.) How many units of work will it perform, if applied to an Overshot Wheel?

(*See* § 378.) How many, if applied to a Breast-wheel?

(*See* § 379.) How many, if applied to a Turbine?

# CHAPTER XII.

## PNEUMATICS.

388. PNEUMATICS is the science that treats of air and the other elastic fluids, their properties, and the machines in which they are applied.

389. DIVISION OF ELASTIC FLUIDS.—The elastic fluids are divided into two classes:—

I. GASES, or such as retain their elastic form under ordinary circumstances. Some of the gases, under a high degree of pressure, assume a liquid form; as, carbonic acid and chlorine; others, such as oxygen and nitrogen, can not be converted into liquids by any known process.

II. VAPORS, or elastic fluids produced by heat from liquids and solids. When cooled down, they resume the liquid or solid form. Steam, the vapor of water, is an example.

390. All gases and vapors have the same properties.

388. What is Pneumatics? 389. Into what two classes are elastic fluids divided? What are gases? What difference is there in the gases? What are vapors? 390. In

The principles of Pneumatics, therefore, relate to all alike, though they are most frequently exhibited and applied in the case of air, with which we have far more to do than with any other elastic fluid.

## Air.

391. Air is the elastic fluid that we breathe. It surrounds the earth to a distance of about fifty miles from its surface, and forms what is called the Atmosphere. It exists in every substance, entering the minutest pores.

392. VACUUMS.—Air may be removed from a vessel with an instrument called the Air-pump. A Vacuum is then said to be produced. Vacuums sometimes result from natural causes; but they last only for an instant, as the surrounding air at once rushes in to fill them. Hence the old philosophers used to say, *Nature abhors a vacuum.*

393. PROPERTIES OF AIR.—Air can not be seen, but it can be felt by moving the hand rapidly through it. It is therefore material, and has all the essential properties of matter.

394. Air is impenetrable.

395. *The Diving-bell.*—The impenetrability of air is shown by the Diving-bell, represented in Fig. 175. A C is a large iron vessel, shaped somewhat like an inverted tumbler, and attached to a chain, by which it is let down in the water. As the vessel descends, the air in it is condensed by the upward pressure of the liquid, and water enters. The lower it gets, the more the air is compressed, and the greater the amount of water admitted. The impenetrability of the air, however, keeps the greater part of the bell

Fig. 175.

THE DIVING-BELL.

what are the principles of Pneumatics most frequently exhibited, and why? 391. What is Air? How far does it extend from the earth's surface? What does it constitute? 392. What is a Vacuum? What did the old philosophers say, and why? 393. What proves the air to be material? 394. What apparatus shows the impenetrability of air? 395. Describe the Diving-bell. Explain how descents are made with

clear of water, so that several persons may descend in it to the bottom of the sea.

As fast as the air is vitiated by the breath, it is let off by a stop-cock, while fresh air is supplied from above by a condensing syringe, through the pipe B. Air may be thus forced down in sufficient quantities to expel the water altogether from the bell, so that the divers can move about without difficulty on the bottom of the sea. If air were not impenetrable, the bell would be filled with water, and the divers drowned.

When the diving-bell was invented, is not known. History makes no mention of it before the sixteenth century. At that time, we are told, two Greeks, in the presence of the emperor Charles V. and several thousand spectators, let themselves down under water, at Toledo in Spain, in a large inverted kettle, and rose again without being wet. In 1665, a kind of bell was employed off the Hebrides, for the purpose of recovering the treasure lost in several ships belonging to the Invincible Armada. From that time to the present, various improvements have been made in the diving-bell; and it is now extensively used for clearing out harbors, laying the foundation of submarine walls, and recovering articles lost by shipwreck.

**396. Air is compressible.**

This is proved with the diving-bell. If the air were not compressible, no water would enter the bell as it descended.

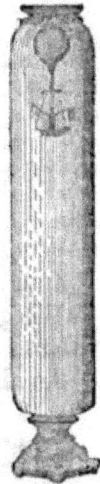

Fig. 176.

**397. Air is elastic.**

This also may be shown with the diving-bell. When, on its descent, water has entered, on account of the air's being compressed, let the bell be raised, and the air will resume its original bulk, expelling the water.

*Bottle Imps.*—The compressibility and elasticity of air may be exhibited in an amusing way with the apparatus represented in Fig. 176. In a vessel nearly full of water are placed several small balloons, or hollow figures of men, &c., made of colored glass, and called Bottle Imps. Each figure has a little hole in the bottom, and is of such specific gravity that it will just float in water. A piece of thin india rubber is tied over the mouth of the vessel, so as to cut off communication with the external air. Now press on the india rubber cover. The

BOTTLE IMPS.

water at once transmits the pressure to the air in the hollow figures. This air is condensed, water enters, the specific gravity of the figures is increased,

---

it. What is the first mention made of the diving-bell in history? In 1665, for what purpose was it used? For what is it now extensively used? 396. How does the diving-bell prove air to be compressible? 397. How does it prove air to be elastic? What properties in air do the Bottle Imps illustrate? Describe the bottle imps, and

and they descend. On removing the fingers from the cover, the air, by reason of its elasticity, resumes its original bulk, and the figures rise. By thus
playing on the india rubber, the figures may be made to dance up and down.

398. *Mariotte's Law.*—The elastic fluids are the most
easily compressed of all substances. *The greater the pressure to which they are subjected, the less space they occupy,
and the greater their density.* A body of air which under
a certain pressure occupies a cubic foot, under twice that
pressure will be condensed into half a cubic foot; under
three times that pressure, into one-third of a cubic foot, &c.
This principle, variously stated, is called, from its discoverer, Mariotte's Law.

The more the elastic fluids are compressed, the greater
is their resistance to the pressure. Hence, their elastic force
increases with their density.

399. *The Air-gun.*—By subjecting a body of air to a great pressure, we
may increase its elastic force sufficiently to produce wonderful effects. The
Air-gun is an example. It consists of a strong metallic vessel, into which
air is forced till it is in a state of high condensation. The vessel is then attached to a barrel like that of an ordinary gun, to the bottom of which a bullet is fitted. Pulling a trigger opens a valve, the condensed air rushes forth,
and drives the bullet out with great force.

One supply of condensed air is sufficient for several discharges, though
each is weaker than the preceding one. The labor required for condensing
the air prevents this instrument from being much used; but as it makes less
noise, when discharged, than the ordinary gun, it is sometimes employed by
assassins.

400. Air has weight.

Weigh a flask full of air, and then weigh the same flask
with the air exhausted. The difference indicates the weight
of the air contained.

401. Experiments show the weight of 100 cubic inches of air to be about
$30\frac{1}{2}$ grains. This makes it 828 times lighter than water. It has been computed that the weight of the whole atmosphere surrounding the earth is equal
to that of a globe of lead 60 miles in diameter.

explain the principle on which they dance up and down. 398. What substances are
the most easily compressed? What is Mariotte's Law? To what is the elastic force
of gases and vapors proportioned? 399. How may a body of air be made to produce
wonderful effects? What instrument proves this? Describe the Air-gun, and its
operation. Why is not the air-gun used more? By whom is it sometimes employed?
400. Prove that air has weight. 401. What is the weight of 100 cubic inches of air?

## Atmospheric Pressure.

402. The particles of air, like those of the other elastic fluids, mutually repel each other. The atmosphere would therefore spread out into space, and become exceedingly rare, if it were not for the attraction of the earth. This prevents it from extending more than fifty miles from the surface, and gives it weight.

403. Since air has weight, it exerts a pressure on all terrestrial bodies. This is known as Atmospheric Pressure. The pressure on any given body is equal to the weight of the column of air resting upon it, and therefore varies according to its size.

Fig. 177.

404. EXPERIMENTS.—The pressure of the atmosphere is proved by experiments.

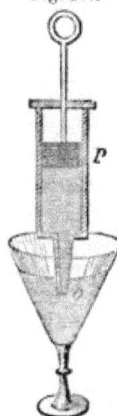

*Experiment* 1.—Take a common syringe, represented in Fig. 177, and let the piston, P, rest on the bottom of the barrel. Insert the nozzle, O, in a vessel of water, and raise the piston. The water enters through O, and follows the piston, as shown in the Figure.

What causes the water to rise? The piston, being air-tight, as it is drawn up, leaves a vacuum behind it; and the pressure of the atmosphere on the water in the vessel drives it into the barrel through O. If the piston does not fit the barrel tightly enough to exclude the air above, no water enters, because the pressure of the air from without is then counterbalanced by that from within the barrel.

*Exp.* 2.—Take a small tube, close one end with the finger, fill it with water, and carefully invert it, as shown in Fig. 178. The water is kept in the tube by atmospheric pressure. Remove the finger, and the downward pressure of the atmosphere, which was before cut off, will counterbalance the upward pressure, and the water will fall by its own weight.

Fig. 178.

*Exp.* 3.—Fill a wine-glass with water, and cover the mouth with a piece of stiff paper. Place the hand over the paper, and invert the glass. On carefully removing

What is the weight of the whole atmosphere? 402. What prevents the atmosphere from spreading out into space? 403. What is Atmospheric Pressure? What causes atmospheric pressure? To what is the atmospheric pressure on any given body equal? 404. Describe the experiment with the syringe that proves the pressure of the atmosphere. What will prevent the water from rising in the syringe? Describe

8

the hand, the water will be found to remain in the glass, supported there by atmospheric pressure.

Fig. 179.

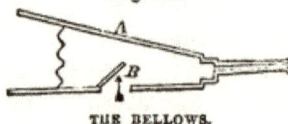

THE BELLOWS.

*Exp.* 4.—When we raise the top board, A, of a common bellows (see Fig. 179), the valve B in the lower board opens. This is because a vacuum is formed within the bellows, and the atmospheric pressure forces the valve up and drives in a portion of the external air.

The same principle is involved in the act of breathing. The cells in the lungs are expanded by muscular action, a vacuum is thus formed, and the pressure of the atmosphere drives in the outer air through the nose or mouth. In a few seconds the muscles contract, and the same air, laden with impurities received from the blood in the lungs, is expelled.

Fig. 180.

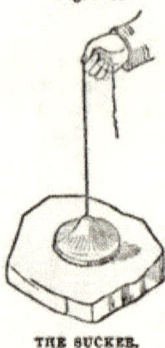

THE SUCKER.

405. The Sucker, a play-thing used by boys, shows the force of atmospheric pressure. It consists of a circular piece of leather with a string attached to the middle. The leather, being first wet so that it may adapt itself to the surface, is pressed firmly upon a flat stone. The string is then gently pulled, so as to form a vacuum between the leather and the stone. On this, the atmospheric pressure from above, not being counterbalanced from beneath, acts on the leather with such force that a stone of great weight may be lifted without the sucker's becoming detached. If a hole is made in the leather, air rushes in, the pressure from above is counterbalanced, and the stone falls by its own weight.

When flies walk on a ceiling, their feet act like suckers. Vacuums are formed beneath them, and they are sustained by atmospheric pressure. It is in the same way that the shell-fish called limpets fasten themselves to rocks.

406. Supported by the pressure of the atmosphere below, while it is cut off from that above, a liquid will not flow from the tap of a barrel unless a small opening is made in the top. As soon as this is done, air is admitted,

the experiment with a small tube that proves the pressure of the atmosphere. How may water be supported in a wine-glass by atmospheric pressure? How is the pressure of the atmosphere exhibited with a common bellows? How do we breathe? 405. Explain the principle involved in the Sucker. How are flies able to walk on a ceiling? 406. Why, when a barrel is tapped, must a hole be made in the top?

# THE BAROMETER. 171

the upward pressure is counterbalanced, and the liquid flows continuously. On the same principle, a small hole is made in the lid of a tea-pot.

407. THE BAROMETER.—The pressure of the atmosphere differs at different times and different places. To measure it, an instrument called the Barometer is used.

The barometer was invented about the middle of the seventeenth century. It was the result of a celebrated experiment performed by Torricelli [to-re-chel'-le], the friend and pupil of Galileo.

408. *Torricellian Experiment.*—The Duke of Tuscany, having dug a well of great depth, and tried to raise water from it with an ordinary pump, found to his surprise that the water would not rise more than 32 feet. Galileo, to whom the fact was referred, was unable to explain it; but shortly before his death he requested Torricelli to investigate the subject. Torricelli, suspecting that the water was raised and supported by atmospheric pressure, proceeded to test the truth of his opinion by experimenting with a column of mercury. Mercury is nearly 14 times as heavy as water; if, therefore, atmospheric pressure supported a column of water 32 feet high, it would support a column of mercury only about one-fourteenth of that height, or 28 inches. Accordingly, he procured a tube 3 feet long, sealed at one end; and having filled it with mercury, and stopped the open end with his finger, he inverted the tube in a vessel of mercury, as shown in Fig. 181. When he removed his finger, the mercury fell, and finally settled, as he had supposed it would, at a height of about 28 inches, leaving a vacuum in the upper part of the tube. This is the famous Torricellian Vacuum.

Fig. 181.

Torricelli did not live to follow up his discovery; but the French philosopher, Pascal, succeeded him with a variety of ingenious experiments. It occurred to Pascal that, if the columns of water and mercury were supported by the pressure of the atmosphere, then at great elevations, where this pressure would necessarily be less, the height of the columns supported would also be less. He tried the experiment on a mountain in Auvergne [o-värn']. At the foot of the mountain, the mercury stood at 28 inches; at the top, it was below 25; and at intervening distances it stood between the two. This proved beyond doubt that the atmosphere exerted a pressure, and that this pressure varied according to the distance above the level of the sea. Perceiving how valuable such an instrument would be for

407. What is the Barometer? When was it invented? Of what was it the result? 408. Relate the circumstances that first directed attention to the subject. Give an account of Torricelli's experiment. What is meant by the Torricellian Vacuum? Who followed up Torricelli's discovery? Give an account of Pascal's experiment.

measuring heights, Pascal soon constructed a Barometer, consisting of a tube and vessel of mercury so attached as to be conveniently carried.

409. *Kinds of Barometers.*—There are several kinds of barometers. The simplest consists of Torricelli's tube and vessel of mercury, with a graduated scale attached to the upper part. The mercury never rises above 31 inches, and seldom falls below 27. The scale is therefore applied only to that part of the tube which lies between these limits.

Fig. 182.

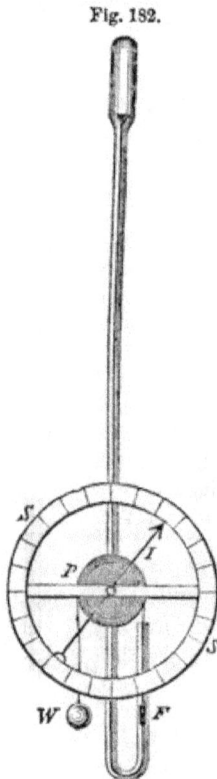

The Wheel Barometer is exhibited in Fig. 182.

· Here the tube, instead of resting in a vessel of mercury, is bent upward at its lower extremity. A float, F, is supported by the mercury in the short arm of the tube. To this float is attached a thread, which passes over the pulley P, and is attached to the ball W. When the mercury falls in the long arm of the tube, it must rise in the short arm, and with it rises the float F. The thread turns the pulley P, and this moves the index I, which is so arranged as to traverse the graduated scale S S.

410. *The Barometer as a Weather-guide.*—The barometer shows that the pressure of the atmosphere at any given place is different at different times. This is because the air is constantly varying in density, on account of a greater or less intermixture of foreign substances. When the air is densest, the mercury stands highest, and we generally have clear weather; but, when the air is rarefied, the mercury falls, and rain not unfrequently follows. Hence, the barometer has been used for predict-

THE WHEEL BAROMETER.

ing changes of weather; and the words FAIR, CHANGE, RAIN, are placed at different points on the scale, to indicate the weather which may be expected when the mercury reaches either of those levels.

411. The only reliable indications, however, afforded by the barometer are *changes* in the level of the mercury. No regard should be paid to the particular point at which it stands at any given time; we should merely ask, is it rising or falling? The following rules generally hold good:—

1. After much dry weather, if the mercury falls steadily, rain will ensue, though it may not begin for several days. The longer it is in coming, the longer it will last.
2. After much wet weather, if the mercury, standing below its medium height, rises steadily, fine weather will ensue, though it may not begin for several days. The longer it is in coming, the longer it will last.
3. A sudden fall of the barometer, in spring or fall, indicates wind; in very hot weather, a thunder-storm; in winter, a change of wind, and rain or snow according to the temperature.
4. Sudden changes of the mercury indicate violent changes of the weather, but not permanent ones.
5. A rise of mercury in autumn, after much wet and windy weather, indicates the approach of cold.

412. At sea, the barometer may be relied on with tolerable certainty, and it is therefore exceedingly useful to navigators. Violent and frequent changes in the mercury almost invariably precede a sudden storm. Warned in time, the prudent mariner furls his sails, and thus escapes the fury of the hurricane which would have proved fatal to his craft had it struck her unprepared.

Dr. Arnott gives the following account of his preservation at sea through the warning of the barometer:—"It was in a southern latitude; the sun had just set with placid appearance, closing a beautiful afternoon; and the usual mirth of the evening watch was proceeding, when the captain's order came to prepare with all haste for a storm: the barometer had begun to fall with appalling rapidity. As yet the oldest sailors had not perceived even a threatening in the sky, and were surprised at the extent and hurry of the preparation; but the required preparations were not completed, when a more awful hurricane burst upon them than the most experienced had ever braved.

to what use has the barometer been applied? 411. What are the only reliable indications afforded by the barometer? What does a steady fall of mercury in the barometer after much dry weather indicate? What does a rise of mercury after much wet weather indicate? What does a sudden fall indicate at the different seasons? What do sudden changes indicate? What does a rise of mercury in autumn indicate? 412. What is said of the barometer at sea? Relate the circumstances of Dr. Arnott's

Nothing could withstand it; the sails, already furled and closely bound to the yards, were riven away in tatters; even the bare yards and masts were in great part disabled, and at one time the whole rigging had nearly fallen by the board. In that awful night, but for the little tube of mercury which had given the warning, neither the strength of the noble ship nor the skill and energies of the commander could have saved one man to tell the tale."

413. DENSITY OF THE AIR AT DIFFERENT LEVELS.—The lowest parts of the atmosphere are the densest, as they have the greatest quantity of air pressing on them from above.

Fig. 183.

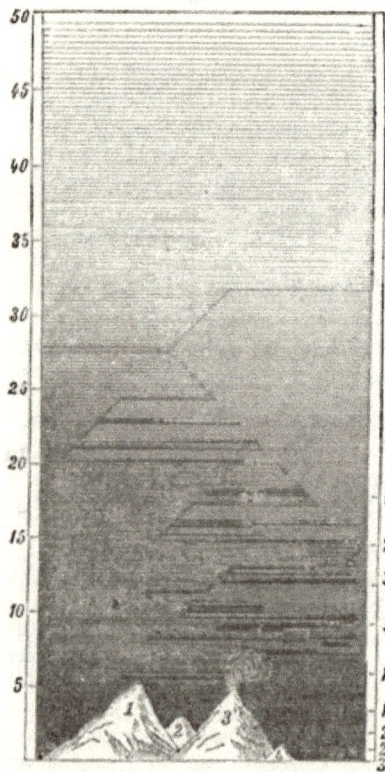

1. *Highest Peak of the Himalayas.*
2. *Highest Peak of the Alps.*
3. *Highest Peak of the Andes.*
4. *Mount Mitchell, N. Carolina.*

414. At the level of the sea, the pressure of the atmosphere on every square inch of surface is 15 pounds. The body of a man of ordinary size has a surface of about 2,000 square inches, and is therefore subjected to the enormous pressure of 30,000 pounds. We do not feel this pressure, because it is counterbalanced by that of the air within our bodies.

415. The higher we go above the level of the sea, the less is the pressure of the atmosphere and the rarer the air. At an elevation of 18 miles, the mercury would fall to 1 inch, — that is, the air above that point is so rare,

preservation at sea by means of the barometer. 413. What parts of the atmosphere are densest, and why? 414. How great is the pressure of the atmosphere at the level of the sea? How great is the pressure on the body of a man of ordinary size? Why

that a column of it 30 miles high weighs no more than an equal column of mercury 1 inch in height.

The shading in Fig. 183 shows the gradual increase in the density of the air as the surface of the earth is approached. The figures in the left margin represent the height of the atmosphere in miles; those on the right, the corresponding height, in inches, at which the mercury stands in the barometer. On the top of Mount Mitchell and Mount Washington, the most elevated peaks in the United States east of the Mississippi, somewhat over a mile high, it stands at 24 inches; on the highest peaks of the Himalayas and Andes, which are about five miles high, at no more than 12.

416. The rarity of the air is painfully felt by those who ascend to great heights on mountains. The pressure of the external air being diminished, that which is in the body expands, the delicate blood-vessels burst, the skin cracks, and blood issues from the nose and ears. Among the Andes, the Indians are subject to a malady called *veta*, which is caused by the rarity of the air. The head aches violently, its veins are swollen, the extremities grow cold, and breathing becomes difficult.

### Effect of Heat on Air.

417. Air is rarefied by heat.

Throw some burning paper into a wine-glass, and before the flame goes out place your hand over the top. The glass will be found to adhere to your hand. This is because the heat rarefies the air within, and thus expels most of it before the top is covered. The pressure of the external air, not being counterbalanced by any pressure from within, fastens the glass and hand together.

418. Cupping-glasses are made to draw on this principle. Incisions having been made in the skin, the sides of the glass are moistened with alcohol, and flame is applied. While the alcohol is burning, the glass is inverted on the skin. The pressure of the air in the body, no longer counterbalanced by the external pressure, causes a flow of blood into the cup.

419. Heated air, being lighter than that which surrounds

do we not feel this pressure? 415. What is said of the air, as we ascend above the sea-level? How would the mercury stand at a height of 18 miles? What does Fig. 183 show? How does the mercury in the barometer stand on the top of Mount Mitchell? On the tops of the Himalayas? 416. What sensations are experienced by persons who ascend to great heights on mountains? Describe the symptoms of the *veta*. 417. What is the effect of heat on air? How may the rarefaction of air by heat be shown? 418. Explain the operation of cupping-glasses. 419. Why does

it, ascends till it reaches a region of the atmosphere as rare as itself.

This is the reason why smoke rises. So, when a fire is kindled in a grate, a draft is produced in the chimney. The air near the fire is rarefied and ascends. A vacuum is thus formed for the instant; cold air rushes in to fill it; this in turn is heated and rises, and thus there is a constant passage of hot air up through the chimney.

Fig. 184.

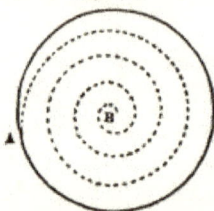

To show the ascent of hot air, take a circular piece of paper, as represented in Fig. 184, and, commencing at any point of the outer edge, as A, cut in the direction of the dotted line. Support it from beneath at B on a piece of wire, and it will hang down, resembling in shape the threads of a cork-screw. If the paper thus suspended be held over a hot stove, it will be carried rapidly round by the ascending currents of heated air.

420. BALLOONS.—By observing the rise of smoke, Stephen and Joseph Montgolfier [*mon-gol-fe-ā'*], paper-manufacturers in France, were led in 1782 to the invention of balloons. The following year, they exhibited their invention to the public.

An immense bag of linen lined with paper was prepared, and brought directly over a fire of chopped straw. In a few minutes, the balloon was filled with rarefied air and released from its fastenings. It rose about a mile, remained suspended ten minutes, and reached the ground a mile and a half from the place of its ascent. The same year, two persons ascended to a height of 3,000 feet in the basket of a smoke balloon, and came down in safety.

On the 1st of January, 1784, a successful ascent was made in a balloon inflated with hydrogen. This gas is now generally used for the purpose, on account of its superior buoyancy. Even when badly prepared, it has but one-sixth of the weight of air, and is three times as light as Montgolfier's mixture of heated air and smoke.

421. Balloons have not as yet been turned to any practical use, from the fact that they are completely at the mercy of the wind, no way of steering them having been devised. A theory has lately been put forth, however, that at a certain height of the atmosphere currents are always setting from

heated air rise? Explain how the kindling of a fire causes a draft in a chimney. How may the ascent of hot air be shown? 420. By whom and when were balloons invented? Describe the Montgolfiers' balloon, and its ascent. When was the first successful ascent made in a balloon inflated with hydrogen? Why is hydrogen now used for

west to east; if this be so, aërial voyages may be made with tolerable certainty, at least in one direction. The theory in question has been in part confirmed by a balloon voyage (the most remarkable on record) made July 1, 1859. Four persons started from St. Louis, and in 19 hours, 40 minutes, landed in Jefferson Co., N. Y., near Lake Ontario,—having travelled about 1,000 miles, at a rate exceeding that of the fastest railroad train.

422. Long before the invention of balloons, attempts were made to navigate the air. At different periods not long after the Christian era, adventurous men launched themselves from the tops of high buildings, and with different sorts of apparatus which they had prepared moved a short distance through the air. Mechanical contrivances resembling wings were more than once resorted to; but several who tried them met with serious accidents, and it was at last proved that wings sufficiently large to support a man in the air would be too heavy for him to move.

## The Air-pump.

423. The Air-pump is an instrument used for removing the air from a vessel called a Receiver. Receivers are made of glass, and are usually of the shape represented in Fig. 185.

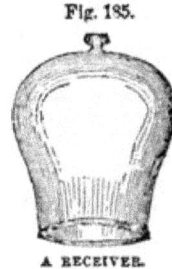

Fig. 185.

A RECEIVER.

424. INVENTION OF THE AIR-PUMP.— The air-pump was invented 1654 A. D., by Otto Guericke [gä'-re-kä], burgomaster of Magdeburg, Germany.

Guericke's first attempt to obtain a vacuum was made with a barrel full of water. Having closed it tightly, he applied a pump to the lower part and commenced drawing off the water. Could he have done this and kept the air out, a vacuum would have been formed; but he had not proceeded far, when the air from without began to force its way with a loud noise through the seams of the barrel. To remedy the difficulty, Guericke substituted a metallic globe for his barrel of water, and the experiment was then successful.

inflating balloons? 421. Why have not balloons been turned to practical use? What remarkable voyage has lately been made? 422. Give an account of the early attempts to navigate the air. 423. What is the Air-pump? Of what are receivers made? 424. By whom and when was the air-pump invented? Give an account of Guericke's first attempt to obtain a vacuum. How did he finally succeed? Describe Gue-

Great improvements have been made on the rude air-pump employed by Guericke; yet, imperfect as his instrument was, it produced results of deep interest to the learned men of that day. His most famous experiment was performed before the Emperor of Germany and his court. Two hollow me-

Fig. 186.

MAGDEBURG
HEMISPHERES.

tallic hemispheres of great size were prepared, fitting each other so closely as to form an air-tight globe. From this globe the air was removed with the pump, and a stop-cock prevented any new air from entering. Fifteen horses were then harnessed to each hemisphere; but their united strength was unable to effect a separation, so tightly were the two parts held together by atmospheric pressure. On turning the stop-cock and readmitting the air, they fell asunder by their own weight.

425. This experiment is often repeated at the present day, on a small scale. The Magdeburg hemispheres, as they are called from Guericke's native city, are represented in Fig. 186. They are fixed to the plate of an air-pump, instead of a receiver; and on exhausting the air they are pressed together so tightly that two men can not pull them apart.

426. SINGLE-BARRELLED AIR-PUMP.—A single-barrelled air-pump is represented in Fig. 187. A is a receiver with its edge carefully ground, resting on a plate near the centre of the stand. In this plate there is a hole leading into a pipe beneath, which connects the receiver with the barrel B.

Fig. 187.

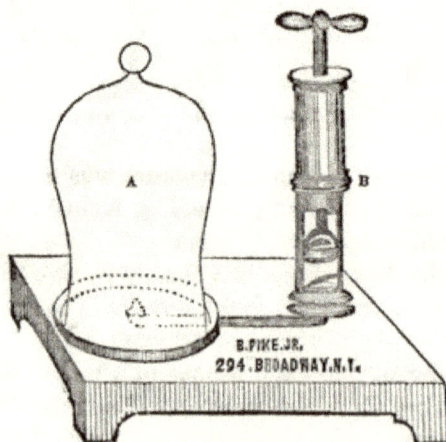

THE SINGLE-BARRELLED AIR-PUMP.

The lower part of the barrel is represented as cut away in the figure, in order to show the interior. A piston is tightly fitted to it, containing a valve opening upward, and connected with a handle, by which it may be worked up and down. At the base of the barrel there is another valve, also opening upward.

427. *Operation.*—The plate having been carefully dusted and rubbed with a little oil, the receiver is placed on it, and the piston is drawn up. A vacuum is thus formed in the lower part of the cylinder, and the air in the receiver, by reason of its elasticity, pushes up the lower valve and enters the barrel. The piston is now in turn driven down; the pressure at once closes the lower valve, while the resistance of the air in the barrel opens the valve in the piston. Through the latter the air passes out, and by the time the piston has reached the bottom, it has all escaped. The piston is then again raised, and the whole operation is repeated,—a barrel-full of air being drawn out from the receiver as often as the piston ascends, and expelled from the barrel as it descends. At last the air in the receiver becomes so rare that it has not sufficient elasticity to open the valve at the base of the barrel. After this the exhaustion can not be carried any further. A perfect vacuum, therefore, is not produced; but the air is rarefied to such a degree that we speak of it as such.

Fig. 188.

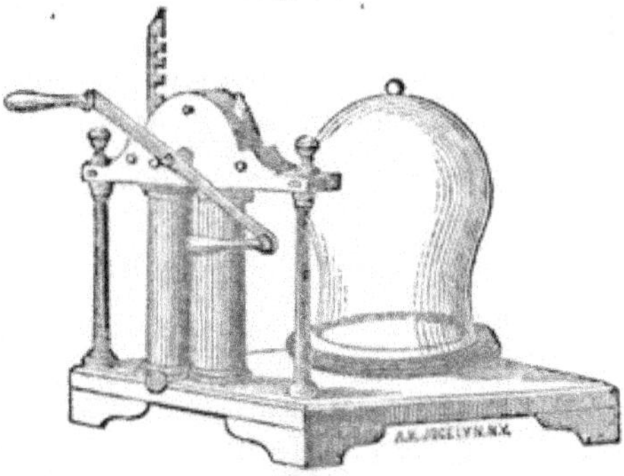

THE DOUBLE-BARRELLED AIR-PUMP.

428. DOUBLE-BARRELLED AIR-PUMP.—The double-barrelled air-pump (see Fig. 188) acts on the same principle as the above, but exhausts the air more quickly in consequence of having two barrels and pistons. A section of the instrument is repsented in Fig. 189, from which its working will be readily understood.

A and B are the barrels, in which the pistons, C, D, work up and down. Each piston is connected with a rack, E, F, the teeth of which work in the cog-wheel G, turned by the handle M. When C is raised, D is lowered; and when C is lowered, D is raised. H I is the passage which

pump, as represented in Fig. 187. Describe the interior of the barrel. 427. How does the single-barrelled air-pump operate? 428. How does the double-barrelled air-pump differ from the single-barrelled? Describe the operation of the double-bar-

Fig. 189.

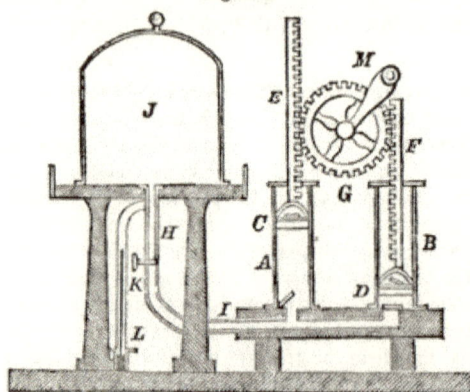

connects the barrels with the receiver J. K is a stop-cock by which the connection may be cut off. L is a tube resting at one end in a small vessel of mercury, and at the other connected with the receiver. This tube is called *a barometer gauge.* As the air in the receiver is rarefied, the external atmospheric pressure on the mercury in the vessel causes it to rise in the tube; the degree of rarefaction is therefore shown by the position of the mercury.

429. EXPERIMENTS WITH THE AIR-PUMP.—With the air-pump and different pieces of apparatus which accompany it, may be performed a variety of experiments, illustrating the properties of air.

Fig. 190.

THE HAND-GLASS.

430. *The Hand-glass.*—The Hand-glass (Fig. 190) is a receiver open at both ends. Set the large end on the plate of the air-pump, and place the hand flat upon the top. As soon as the pump is worked, the pressure of the atmosphere is felt. When the air is exhausted, the hand can hardly be removed from the glass; on readmitting the air through a stop-cock, it is raised without difficulty. The expansion of the air in the palm of the hand is shown by the redness of the flesh, and its puffing out while over the exhausted glass.

431. *The Apple-cutter.*—The Apple-cutter (Fig. 191) is a metallic cylinder with a sharp upper edge. An apple that fits it closely having been placed on its top, the air is exhausted. The pressure of the atmosphere forces the apple down on the sharp edge; the middle part is cut out and falls inside of the vessel.

---

relled air-pump, with the aid of Fig. 189. What is the use of the barometer gauge ? 430. What is the Hand-glass? Describe the experiment with the hand-glass. What causes the redness of the hand? 431. What is the Apple-cutter? Describe the ex

Fig. 191.

THE APPLE-
CUTTER.

432. *The Bladder-glass.*—Over the large end of the hand-glass tie a wet bladder, as shown in Fig. 192. When the bladder has become dry, place the open end on the plate, and exhaust the air from the glass. The pressure of the atmosphere, unsupported from within, soon bursts the bladder with a loud noise. If a piece of thin india rubber be substituted for the bladder, it will be drawn in and distended, till it covers nearly the whole inside of the glass.

Fig. 192.

THE BLADDER-
GLASS.

433. *The Lungs-glass.*—The Lungs-glass (Fig. 193) illustrates the elasticity of air. It is a small glass globe with a metallic stopper. Through this stopper passes a tube, to the lower part of which a bladder is tied. The whole is placed under a receiver, and the air exhausted. The air in the bladder, communicating through the tube with the receiver, is gradually rarefied. The air around it in the glass, having no communication with the receiver, remains of the same density. Owing to its pressure, the bladder becomes shrivelled when the receiver is exhausted; but, on the readmission of the air, it resumes its former dimensions. This movement, regularly repeated, resembles the action of the lungs in breathing, and hence the name given to the apparatus.

Fig. 193.

THE LUNGS-GLASS.

Fig. 194.

VACUUM FOUNTAIN.

434. *Vacuum Fountain.*—Fig. 194 represents a tall glass receiver, terminating at the bottom in a metallic cap, through which a tube passes. This tube is furnished with a stop-cock, and a screw, by means of which it may be fastened to the plate of an air-pump. A jet communicating with the tube rises into the receiver. Screw this apparatus to the plate of the pump, exhaust the air, and close the stop-cock. Then unscrew the whole, place the lower end of the tube in a vessel of water, and open the stop-cock. The pressure of the atmosphere will force the water up through the tube and jet into the vacuum, forming a beautiful miniature fountain.

Another mode of producing a vacuum fountain is with the apparatus shown in Fig. 195. It consists of

periment with the apple-cutter. 432. How is the experiment with the bladder-glass performed? 433. What does the Lungs-glass illustrate? What does it consist of? Describe the experiment. Why is the lungs-glass so called? 434. What does Fig. 194 represent? How is the vacuum fountain produced? Describe another mode of producing

Fig. 195.

a glass vessel with an air-tight stopper, through which a tube extends almost to the bottom. The vessel, nearly filled with water, is placed under a tall receiver, and the air exhausted. The elasticity of the air within the vessel, not being counterbalanced by any pressure from without, forces the water through the tube in the form of a fountain.

435. *Bottle Imps.*—The bottle imps, described in § 397, may be made to dance up and down in a jar of water in an exhausted receiver. These figures are hollow and contain air. When the receiver is exhausted, the pressure on the surface of the water being removed, the air in the figures expands and drives out some of the water. This diminishes their specific gravity, and causes them to

Fig. 196.

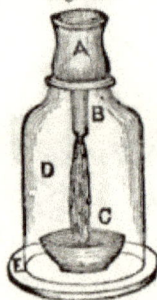

rise. When the air is readmitted, the pressure is restored, the air in the figures is compressed, water enters, their specific gravity is increased, and they sink.

436. *The Mercury Shower.*—On an open-mouthed receiver, D, place the cup A, in the bottom of which is a plug of oak wood, B, projecting downward about two inches. Put some mercury in A, and set the saucer C beneath the oaken plug. Exhaust the air from D, and the mercury will soon be forced by atmospheric pressure through the pores of the oak, and fall into the saucer in a silvery shower.

437. *The Weight-lifter.*—This is an apparatus with which the pressure of the atmosphere is made to lift a heavy weight (see Fig. 197). A is a cylinder attached to a frame, firmly supported by three legs. On the bottom of the cylinder rests a closely fitting piston, to which the platform F is attached. A tube, B C, connects the interior of the cylinder with the plate E of the pump D. When the air is exhausted from A, the pressure of the atmosphere raises the piston, together with the platform and its contents, the whole length of the cylinder. Atmospheric pressure being 15 pounds to the square inch, the number of pounds that can be lifted by a given cylinder may be found by multiplying its area expressed in inches by 15.

438. It has been proposed to transmit mails between distant points, by atmospheric pressure, on the principle of the weight-lifter. A strong metallic tube, perfectly smooth on the inside, would have to be laid between the places, and a piston tightly fitted to it. Large air-pumps, worked by steam, would be placed at both ends of the tube. The mail being attached to the piston at one end of the line, the pumps at the other would be set in motion. A partial vacuum would be produced, and atmospheric pressure would drive the piston through the tube at a rate estimated at 500 miles an

Fig. 197.

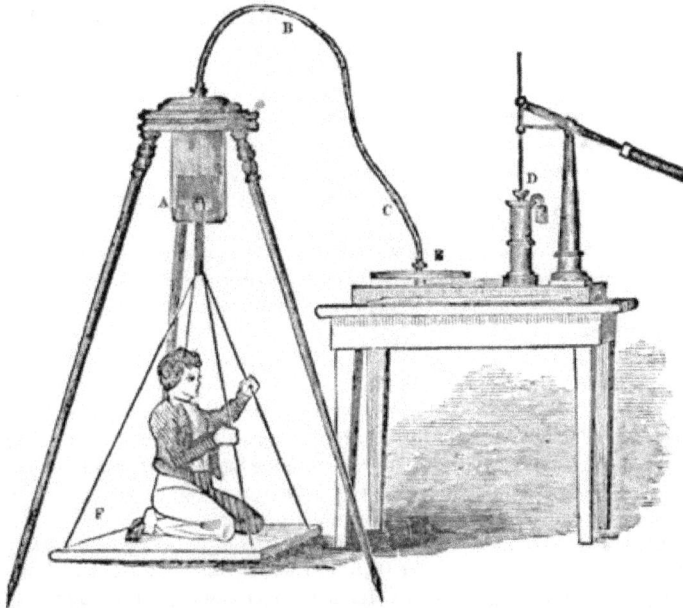

hour. Such is the theory; whether it can be practically applied, remains to be proved.

439. *Vacuum Bell.*—This apparatus is intended to show that air is essential to the production of sound. A bell is so fixed under a receiver that it can be rung by pushing down a sliding-rod which passes through the top. When rung before the receiver is exhausted, the bell is distinctly heard; but, when the air is withdrawn, it is almost inaudible. If a perfect vacuum could be produced, it would not be heard at all.

Fig. 198.

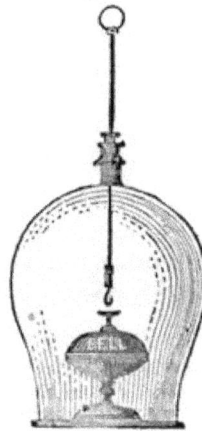

440. *Freezing Apparatus.*—Water may be frozen in a vacuum, with the apparatus shown in Fig. 199.

Fig. 199.

Having placed the liquid in a shallow vessel over a basin containing strong sulphuric acid, set the whole under a receiver and exhaust the air. Under the diminished pressure, the water is rapidly converted into vapor, which is as rapidly absorbed by the acid. The continued evaporation cools the water to such a degree that it is finally covered with ice.

441. *Miscellaneous Experiments.*—In a vacuum, boiling commences at a much lower temperature than in the air. This is shown by placing some hot water under a receiver and exhausting the air. The pressure of the atmosphere being removed from its surface, the water soon boils; but it comes to rest the moment that air is readmitted. For the same reason, water boils at a lower temperature on the top of a mountain than at its base, as has often been observed by travellers.

442. If beer is placed under a receiver and the air exhausted, it begins to foam. This is owing to the elasticity of the carbonic acid in the liquid, rushing out to fill the vacuum. If the air is readmitted, the beer resumes its usual appearance.

443. A shrivelled apple in an exhausted receiver is puffed out to its full size by the expansion of the air within.

444. If a vessel of water containing a piece of wood, a vegetable, or almost any solid substance, is placed under a receiver, and the air is exhausted, minute globules of air can be seen forming on the surface of the solid, and sometimes even bubbling up through the water. This proves the porosity of solids and the presence of air in their pores.

445. A lighted candle in an exhausted receiver is extinguished, and the smoke falls because it is heavier than the rarefied air. If a mouse, rabbit, or other living creature, is placed under a receiver and the air is drawn off, it immediately shows signs of distress, and soon dies.

446. These experiments show that air is everywhere present, and is essential to life and combustion. In a vacuum, animals die, vegetation ceases, and sound can not be produced.

### The Condenser.

447. The Condenser (Fig. 200) is an instrument used for forcing a large quantity of air into a given vessel.

Like the single-barrelled air-pump, the condenser consists of a cylinder, A, with a valve at its base, V, and a piston, P, which also contains a valve, tightly fitted to it.

---

ing apparatus, and the experiment with it. 441. At what temperature does boiling commence in a vacuum, compared with that at which it commences in the air? How is this shown? What is said of the boiling of water on the top of a mountain? 442. What phenomenon is presented when beer is placed under a receiver and the air exhausted? 443. When a shrivelled apple is so placed? 444. How is the presence of air in the pores of solids proved with the air-pump? 445. How is it shown with the air-pump that air is necessary to combustion and animal life? 447. What is the

Instead of opening upward, however, as in the air-pump, these valves open downward.

Fig. 200.

448. *Operation.*—The condenser having been screwed to any strong vessel in which it is desired to condense air, the handle is worked up and down. A vacuum being produced below the piston, as it ascends, its valve is opened and air rushes in; while the valve in the cylinder is closed by the pressure of the air in the vessel. When the piston descends, its valve is closed by the pressure of the air in the cylinder, while the other valve opens and allows this air to be driven into the vessel. With every ascent of the piston, therefore, the cylinder is filled with air, and with every descent this cylinder-full of air is forced into the vessel.

Air is condensed in the chamber of the air-gun (described in § 399) by the use of this instrument.

Fig. 201.

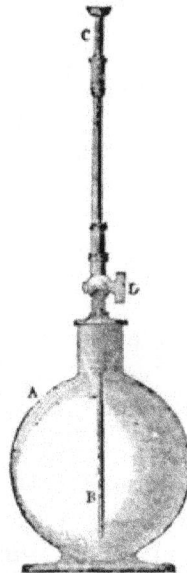

449. *Experiment.*—An interesting experiment may be performed with the condenser and the apparatus represented in Fig. 201. A is a globe half full of water, with a tube, B, reaching nearly to the bottom, and extending upward through an air-tight cap till it terminates in a screw just above the stop-cock D. The condenser, having been screwed on, is worked till a large quantity of air is forced into A. The stop-cock is then closed, the condenser is unscrewed, and a jet-pipe, C, is put on in its place. The stop-cock is now opened, when the pressure of the condensed air, being greater than that of the atmosphere, forces the water in A up through the jet, making a beautiful fountain.—This experiment shows that the elasticity of air is increased by condensing it.

## Pneumatic and Hydraulic Machines.

450. THE SIPHON.—The Siphon, represented in Fig.

202, is a simple instrument for drawing off liquids from a higher to a lower level. It is nothing more than a bent tube, with one leg longer than the other.

Fig. 202.

THE SIPHON.

To use the siphon, fill it with some liquid and then invert it, stopping the long end with the finger, and setting the short one in the liquid to be drawn off. Remove the finger, and the liquid will commence flowing from the long end. The upward pressure of the atmosphere is counterbalanced by its downward pressure on the surface of the liquid to be drawn off, and the liquid in the tube will therefore flow in the direction of its greatest weight. As it flows, a vacuum is formed in the tube, and fresh liquid is constantly forced up into the short leg. The flow continues till the liquid falls below the extremity of the short leg.

451. Some siphons, like that in the figure, have an additional tube, open at the upper end and at the lower communicating with the long leg. This saves the trouble of turning the siphon, every time it is used, to fill it with liquid; for, the long leg being stopped with the finger and the mouth applied to this additional tube, the liquid may by suction readily be made to fill both legs.

452. TANTALUS'S CUP.—Fig. 203 represents Tantalus's Cup, which is simply a goblet containing a siphon, the short

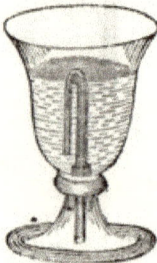

Fig. 203.

TANTALUS'S CUP.

leg of which reaches nearly to the bottom, while its long leg passes through the bottom and extends below. The siphon is concealed by a figure, which seems to be trying to drink. Water is poured in; but, the moment it reaches the lips of the figure, it recedes, because just then it passes the turn of the siphon and begins to be discharged below.

453. THE LIFTING-PUMP.—The Lifting-pump was invented by Ctesibius [te-sib'-e-us], who flourished at Alexandria, in Egypt, 250 B. C. Though the son of a barber and brought up to his father's calling, he attained distinction by his mechanical abilities. Several ingenious

contrivances for raising water are attributed to this philosopher, besides the clepsydra already described.

454. The common Lifting-pump is represented in Fig. 204. It consists of a cylinder, B C, to which is fitted the air-tight piston G, containing a valve opening upward. A is called the suction-pipe; it must be long enough to reach the water that is to be raised. In the top of the suction-pipe is the valve H, opening upward into the cylinder. E is a handle, by which the piston may be worked. F is a spout, from which the water is discharged.

Fig. 204.

455. *Operation.*—To work the pump, raise the piston. As it ascends, it leaves a vacuum behind it, and the water under the pressure of the atmosphere rushes up through A, opens H, and fills the cylinder B C. The piston, having reached the top, is now forced back. Its downward pressure at once closes the valve H, so that the water can not return into the suction-pipe; but the valve in the piston opens, and through it the water rushes above the piston. When the piston has reached the bottom of the cylinder, it is again raised; its valve being now closed by the downward pressure, the water is lifted by the piston into the reservoir D, whence it is discharged by the spout. Meanwhile, the second time the piston rises, a vacuum is formed below it as before, and the whole operation is repeated.

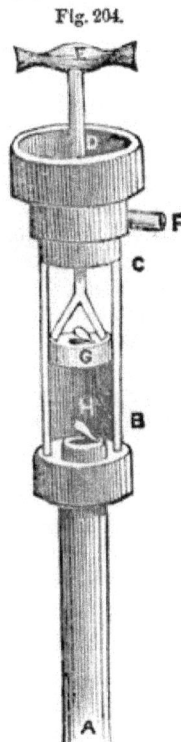

THE LIFTING-PUMP.

456. Thus we see that water is raised in pumps by atmospheric pressure. The air will support a column of water from 32 to 34 feet high. To this elevation, therefore, water can be raised with the lifting-pump; for greater distances, the forcing-pump must be used.

457. THE FORCING-PUMP.—The Forcing-pump, after raising a liquid through its suction-pipe, does not discharge it from a spout above, but by the pressure of the returning piston drives it through an opening in the side below. The

liquid is thus forced, either directly or by means of the
pressure of condensed air, to a greater height than it could
otherwise attain.

458. Fig. 205 represents one form of
the forcing-pump. It has a cylinder, piston, and suction-pipe, like the lifting-pump
just described; but there is no valve in
the piston. Near the bottom of the cylinder enters the pipe M, which communicates with the air-chamber K, by the valve
P, opening upward. The tube I, open at
the bottom and terminating at the upper
end in a jet, passes through the air-tight
top of the chamber K, and extends nearly
to its bottom.

459. *Operation.*—To work the forcing-
pump, raise the piston. A vacuum is
formed; and water, from the reservoir
below, rushes through the suction-pipe,
opens H, and fills the cylinder. The piston is now pushed back, when H at once
closes. The water in the cylinder is forced
into M, raises P, and enters the chamber
K. The water in K soon rises above the
mouth of the tube I, and begins to con-
dense the air in the upper part of the chamber. The higher
the water rises in K, the more the air is condensed, and its
elasticity increases in proportion. Its pressure, therefore,
soon becomes greater than that of the atmosphere, and drives
out the liquid through the jet.

Some forcing-pumps have no air-chamber, but drive out
the liquid by the direct pressure of the descending piston. In
that case, the discharge is by successive impulses; but, when
made from an air-chamber, it is continuous.

Fig. 205.

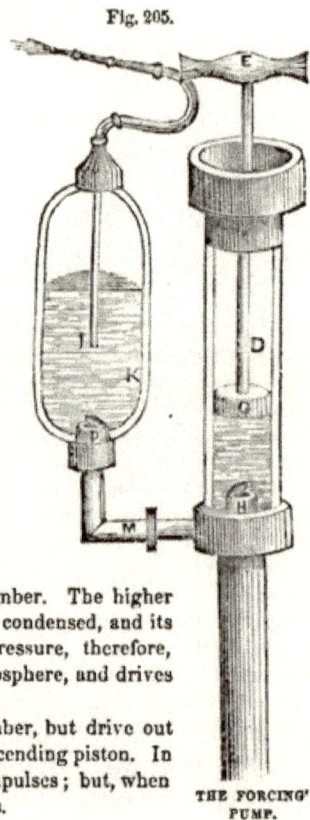

THE FORCING-
PUMP.

460. THE FIRE-ENGINE.—The Fire-engine is a combination of two forcing pumps, with a common air-chamber
and suction-pipe. Its operation will be understood from
Fig. 206.

The pistons, C, D, are attached to a working-beam, A B, turning on the

be used? 457. What is the principle on which the Forcing-pump acts? 458. Describe the form of forcing-pump represented in Fig. 205. 459. Explain its operation.
When there is no air-chamber, how does the forcing-pump drive out the liquid?
460. Of what does the Fire-engine consist? Describe its operation with Fig. 206.

pivot K, so that one rises as the other descends. They are driven up and down by *brakes* attached to the beam and worked by a number of men on each side. F is the suction-pipe. H is the air-chamber, and E a pipe rising from it, to which a flexible leather hose is attached, so that the stream can be turned in any direction. The piston D in Fig. 206 is ascending, followed by a stream of water from the reservoir below, the valve I leading into the air-chamber being closed. The piston C, on the other hand, is descending; its lower valve is closed, and the water drawn into the cylinder during its previous ascent, is now being forced into H, through the open valve J.

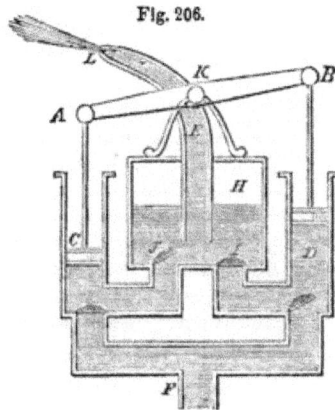

Fig. 206.

461. The fire-engine is one of the most powerful forms of the forcing-pump, since water is being constantly forced into the air-chamber by one of the pistons, and the air is violently compressed. With a good engine, a stream can be thrown more than 100 feet high.

462. THE CENTRIFUGAL PUMP.—The Centrifugal Pump (Fig. 207) is an instrument for raising water by the combined effect of the centrifugal force and atmospheric pressure.

It consists of a vertical axle, A B, and one or more tubes, C, C, fastened to it, extending into a reservoir of water below, and branching off towards the top so as to bring their mouths over the circular trough D. E is a spout for discharging the wa-

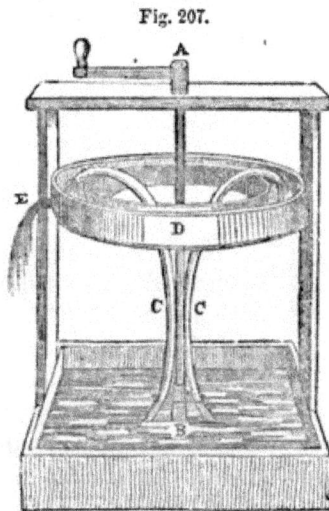

Fig. 207.

THE CENTRIFUGAL PUMP.

ter from the trough. Near the top and bottom of each tube is a valve opening upward.

463. *Operation.*—When the pump is to be worked, the tubes are filled with water, which is prevented from escaping by the lower valves. A rotary motion is then communicated to the tubes by means of a handle attached to the axle. The centrifugal force at once acts on the water within, causing it to open the valves and rush forth from the mouths of the tubes. As it ascends, a vacuum is left behind it, into which water is driven by atmospheric pressure from the reservoir below. Streams are thus kept pouring into the trough as long as the rotary motion is continued.

A large centrifugal pump, worked by steam, has raised no less than 1,800 gallons a minute to a considerable height.

464. THE STOMACH PUMP.—The Stomach Pump is an instrument for injecting a liquid into the stomach of a poisoned person and withdrawing it, without removing the apparatus. The stomach is thus rinsed out, and life is often saved.

Fig. 208.

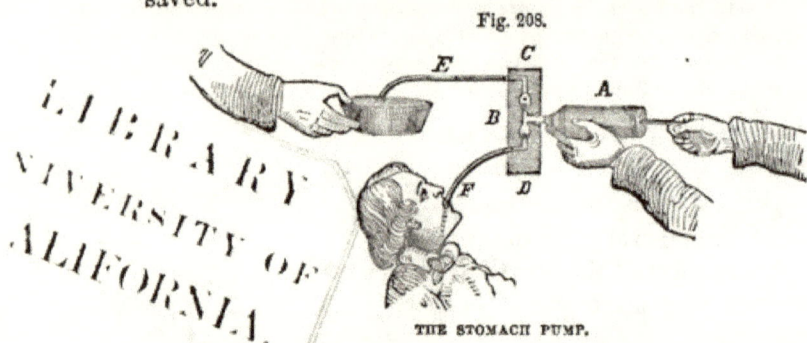

THE STOMACH PUMP.

Fig. 208 represents the stomach pump. A syringe, A, is screwed into a cylindrical box, B, where it communicates with a short metallic tube. This tube leads on either side into a bulb, which is connected with a tube of india rubber. Each bulb contains a movable circular valve of metal, which fits either extremity, and may be made to close either by raising the opposite side of the instrument.

*Operation.*—To work the pump, turn the syringe so as to depress C and elevate D; and then introduce the tube F into the patient's stomach, and E into a basin of warm water. The metallic valves fall to the lowest part of their respective bulbs, which brings them directly opposite where they are in the Figure. Now draw out the handle of the syringe. A vacuum is produced; and the warm water, under atmospheric pressure, rushes up to fill it, all communication with F being cut off by the valve. The syringe being thus charged, the handle is pressed back, and the water, prevented from returning into E by the valve, is forced through F into the stomach. Without removing the india rubber tube from the stomach, now turn the instrument, so as to raise the side C and depress D, as shown in the Figure. The metallic valves are thus thrown to the opposite extremities of their bulbs, and by working the syringe with them in this position, the contents of the stomach 'are drawn off and discharged into the basin. The syringe is thus always charged through the depressed tube and emptied through the elevated one.

465. The consideration of the steam-engine, the greatest of pneumatic machines, is deferred till we shall have treated of the mode of generating steam by heat, a subject which belongs to Pyronomics.

### EXAMPLES FOR PRACTICE.

1. (*See* § 398.) Under a pressure of one atmosphere, a body of oxygen fills 24 cubic inches, and its specific gravity is 1.111. What space will it occupy, and what will be its specific gravity, under a pressure of three atmospheres?

2. Some hydrogen, by a pressure of 20 pounds to the square inch, is forced into a space of one cubic foot. How great a pressure will compress it into half a cubic foot, and how will its density then compare with what it was before?

3. Into what space must we compress 10 cubic inches of air, to double its elastic force?

4. (*See* § 401.) What is the weight of 600 cubic inches of air? What is the weight of the same bulk of water?

5. A vessel, full of air, weighs 1,061 grains; exhausted, it weighs but 1,000 grains. How many cubic inches does it contain?

6. (*See* § 414.) What is the downward atmospheric pressure on the roof of a house containing 115,200 square inches? What is the upward atmospheric pressure on the same roof?

7. What amount of atmospheric pressure is supported by a boy whose body contains 1,000 square inches of surface?

8. (*See* § 409.) When the mercury in the barometer stands at 29 inches, at what height will a column of water be supported by the atmosphere?
   [*Solution.—The specific gravity of water is* 1; *that of mercury,* 13.568. *A column of water will be supported at the height of* 29 × 13.568 *inches.*]

9. When the atmosphere supports a column of water 32 feet high, how high a column of mercury will it support?

10. (*See Fig.* 183.) How far above the earth's surface would the mercury stand only two inches high in the barometer?

# CHAPTER XIII.

## PYRONOMICS.

466. PYRONOMICS is the science that treats of heat.

### Nature of Heat.

467. Heat is the sensation experienced on approaching a warm body.

The invisible agent that produces this sensation is also called Heat. Another name for it is Ca-lor'-ic.

468. Cold is the opposite of heat. It is not a positive agent, but merely implies a greater or less deficiency of heat. There is heat in all substances; but in those which we call *cold*, it exists in an inferior degree.

469. There are two kinds of heat; Free, or Sensible, and Latent.

Free or Sensible Heat is heat that can be felt. Latent Heat is heat that can not be felt. The heat of a fire is Free, or Sensible; the heat in ice is Latent.

470. The Temperature of a body is the amount of sensible heat that it contains.

We can not always judge correctly of a body's temperature by the sensation it produces when we touch it. In the same room, for instance, are a bar of iron and a piece of cloth; they must be of the same temperature, but the iron is cold to the touch while the cloth is not. This is because the iron carries off the heat more rapidly from the part that touches it. So, if one hand be cold and the other warm, a substance which to the former seems hot, to the latter may appear just the reverse. Our sensations, therefore, are not proper criterions by which to judge of a body's temperature.

---

466. What is Pyronomics? 467. What is Heat? What other signification has the term *heat?* What other name is there for it in this sense? 468. What is Cold? 469. How many kinds of heat are there? What is Free or Sensible Heat? What is Latent Heat? Give examples. 470. What is the Temperature of a body? Can we judge of a body's temperature by the sensation it produces when we touch it? State

### 471. What heat is, we do not know.

Some think that it is not a material substance, but results from the vibrations of the particles of bodies. Others believe it to be an exceedingly subtile substance, whose particles repel each other, and thus give it a tendency to diffuse itself, while they have a strong affinity for other matter. This substance, they think, enters into every body, and keeps its particles from coming into absolute contact. As long as it remains at rest, it may be latent; but, when a colder body approaches, there is a tendency to equalize the temperature; a series of vibrations are produced in the subtile atmosphere around each particle, and the heat which was before latent becomes sensible.

Heat seems to be closely connected with light. The one is generally accompanied by the other; and to some extent, as will appear hereafter, they are governed by the same laws.

### 472. Heat has no weight.

Balance a piece of red-hot iron with weights in a sensitive pair of scales; the same weights will exactly balance it when it has become cold. Heat, therefore, must be imponderable; or the loss of so much of it would occasion a perceptible difference in the weight of the iron. So, if a piece of ice is balanced and then allowed to melt, the water formed will weigh precisely the same as the ice.

## Sources of Heat.

473. The principal Sources of Heat are four in number: —the Sun, Chemical Action, Mechanical Action, and Electricity.

474. THE SUN, A SOURCE OF HEAT.—The Sun is the great source of heat, as well as light, to the earth.

What the sun is composed of, that it has thus for thousands of years poured forth undiminished supplies of heat, astronomers can not determine. Some think that the whole of its immense mass is heated to such a degree as to make it luminous. According to others, the great body of the sun is not luminous, but its surface is covered with flames from which rays of heat and light are constantly emitted. In either case, it is hard to explain how combustion can be continued so long without exhausting the material by which it is supported.

475. The heat at the sun's surface is supposed to be more intense than any with which we are acquainted. By

some facts to prove this. 471. What is heat? What do some think it results from? What do others believe it to be ? How do the latter account for its being sometimes latent and sometimes sensible? With what is heat connected? 472. What is the weight of heat? Prove this. 473. What are the principal sources of heat? 474. What is the great source of heat to the earth? What two theories have been advanced to account for the sun's heat? 475. How great is the heat at the sun's surface supposed

9

the time it reaches us, modified by the immense distance it has traversed, it is just sufficient to warm the earth into fertility.

The sun does not heat all parts of the earth alike. This is because its rays strike some portions perpendicularly and others obliquely. The perpendicular rays are absorbed more than the oblique ones, and therefore produce a greater degree of heat in the parts on which they strike. For the same reason, it is hotter about noon than any other time of day, the sun being then more directly over head.

The variety of productions in different parts of the earth is owing to the difference in the amount of heat received from the sun. The trees and plants of the tropics are quite different from those of the temperate regions, and these again are unlike those of cold climates. In the far north and south, so little heat is received that vegetation entirely ceases.

Fig. 209.

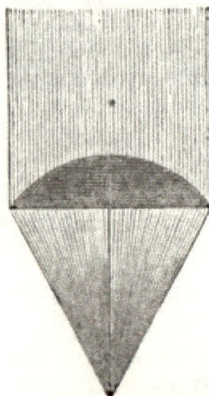

476. The sun's heat may be increased by collecting a number of its rays into one point called a Focus. This may be done with a convex lens, or glass of the shape represented in Fig. 209. With such a lens, three feet in diameter, the metals have been melted.

A similar effect may be produced with concave mirrors, so arranged as to reflect the rays that strike them to one and the same focus. When the Romans were besieging Syracuse, 213 B. C., Archimedes is said to have used a number of metallic mirrors with such effect as to set fire to their fleet. The experiment has been repeated in modern times. Buffon, with a combination of 168 mirrors, showed that tarred planks could be set on fire at a distance of 150 feet, and that at 60 feet silver could be fused.

477. *Heat below the Earth's Surface.*—The sun's heat, even when it falls perpendicularly on the surface, does not penetrate into the earth farther than 100 feet. Beyond this depth, all the heat that is felt, comes, not from the sun, but from the interior of the earth.

to be? Why is it less intense when it reaches us? Why does not the sun heat all parts of the earth alike? To what is the variety of productions in different parts of the earth owing? 476. How may the sun's heat be increased? In what other way may a similar effect be produced? What did Archimedes accomplish with a number of metallic mirrors? Give an account of Buffon's experiment. 477. What is the greatest

As we descend below the earth's surface, the temperature increases about one degree for every 45 feet. At this rate, water would boil at a depth of less than two miles, and at 125 miles all known substances would be melted. It is thought, therefore, that the great mass of the interior of the earth is in a state of fusion. The discharge of melted earthy matter, called *lava*, during the eruption of volcanoes, goes to prove this; while the hot springs in different parts of the world (particularly numerous in Iceland) show that a high temperature prevails at no very great depth. At the surface this internal heat is not perceptible, because the outer crust of the earth is a bad conductor.

478. CHEMICAL ACTION, A SOURCE OF HEAT.—When, by combining two or more substances, we produce a new substance totally different in its properties from either, we say that Chemical Action has taken place. Such action is always accompanied with an increase of temperature. If, for instance, we mix equal quantities of sulphuric acid and water, chemical action takes place, a new substance is formed, and heat is given out. The heat produced by chemical action is sometimes sufficient to ignite inflammable substances. Thus a drop of sulphuric acid will set fire to a mixture of sugar and chlorate of potassa.

479. *Combustion.*—One of the commonest processes in which chemical action is exhibited, is Combustion, or Burning. This is the great source of artificial heat, as the sun is of natural heat.

Combustion is nothing more than a chemical union of the oxygen of the air with the combustible body or some of its elements. Latent heat is given out, by which the gases or vapors produced are rendered luminous; and hence what we call Flame. The rise of temperature is proportioned to the rapidity with which the chemical union takes place; and this depends in a great measure on the amount of oxygen supplied.

If we wish to make a fire hotter, we have only to bring more air in contact with the fuel. This may be done with a bellows, or in the case of grates with a blower. To fill the vacuum produced by the ascent of the heated air through the chimney, cold air must enter; by putting on the blower, we pre-

distance to which the sun's heat penetrates? Beyond this depth, whence is the heat derived? Descending below the earth's surface, at what rate does the temperature increase? At what depth would water boil? How great would the temperature be at a depth of 125 miles? In what state is the interior of the earth supposed to be? What phenomena support this opinion? 478. When does Chemical Action take place? With what is chemical action always accompanied? Give an example. 479. In what common process is chemical action exhibited? What is Combustion? What is the cause of flame? To what is the rise of temperature proportioned? What must be

vent it from entering anywhere except at the bottom of the grate, and cause what does enter to pass through the ignited coals, thus increasing their supply of oxygen.

480. *Animal Heat.*—To Chemical Action is attributable Animal or Vital Heat,—that is, the heat generated in all organic beings that possess life.

Different living creatures have different degrees of animal heat. Birds have the most; beasts come next; then fish and insects. In the same class of animals, however, the amount of vital heat is nearly uniform; and under ordinary circumstances it remains the same, whether the surrounding medium be warm or cold. Other things being equal, the heat of the human body is as great in winter as in summer, in the frigid as in the torrid zone. We do not feel equally hot, to be sure; but, as already explained, we must not judge of temperature by our feelings.

481. Animal heat is produced by a process similar to combustion. When we breathe, air is taken into the lungs, where it comes in contact with particles of carbon contained in the blood. This carbon unites chemically with the oxygen of the air inhaled, and, as in the case of combustion, latent heat is evolved. The heat is less than that produced by combustion, because the particles of carbon are extremely small.

As in combustion, whatever increases the supply of oxygen increases the animal heat. Running or bodily exertion of any kind makes us hotter, because it quickens the circulation of the blood, obliges us to breathe faster, and thus brings more air (and consequently more oxygen) into the lungs.

482. The carbon consumed comes from the food we eat. Greasy food generates it most plentifully. In winter, therefore, when we need an abundance of carbon, we eat meat more freely than in summer, when we seek to reduce our vital heat as much as possible. So, the inhabitants of cold regions consume more greasy food than those of warmer climates. The Esquimaux thrive on fish-oil and seals' fat, which to the people of the tropics would be neither palatable nor wholesome.

483. MECHANICAL ACTION, A SOURCE OF HEAT.—Mechan-

---

done, if we wish to make a fire hotter? 480. What is Animal or Vital heat? To what is it attributable? What is said of animal heat in different living creatures? In the same class of animals? Does it differ in different seasons? 481. How is animal heat produced? Why is it less than the heat produced by combustion? How is animal heat increased? Give examples. 482. How is the carbon consumed, produced? What sort of food generates carbon most plentifully? What follows, with respect to our diet at different seasons? How does the diet of the inhabitants of cold regions compare with that of tropical nations? 483. What is the third source of heat?

ical Action is a familiar source of heat. Under this head are embraced Friction or Rubbing, and Percussion or Striking. By compressing the particles of a body, mechanical action forces out its latent heat and makes it sensible.

**484.** *Heat from Friction.*—Touch a row-lock, in which an oar has been rapidly plying, or a gimlet that has just been vigorously worked, and you will feel the heat produced by friction. Rub a metallic button to and fro on a dry board, and you will soon make it so hot that you can not bear your finger on it. By drawing a match across a rough surface, you develop heat enough to ignite it. By rubbing two pieces of ice together, in a freezing temperature, latent heat is liberated in sufficient quantities to melt them.

Machinery has been ignited by the rubbing of its parts on each other. Savages kindle a fire by rubbing two dry sticks violently together. In boring a brass cannon, immersed in water by way of experiment, sufficient heat has been generated to boil the water in two hours and a half. The friction of two large iron plates has even been employed as a practical source of heat.

It is to be observed that in all the above cases heat is produced by the friction of solids. The friction of fluids is insufficient to generate heat.

**485.** *Heat from Percussion.*—By striking flint and steel together, we develop sufficient heat to ignite the minute fragments broken off, and produce sparks. In like manner, the hammer of a gun, descending on a percussion-cap, sets fire to the fulminating mixture of which the cap is made.

A nail may be made red-hot by hammering it rapidly on an anvil. Before lucifer matches were invented, blacksmiths used to ignite sulphur matches and kindle their forge-fires with a nail hammered to a red heat.

By violent and quick compression, enough heat can be set free from air to ignite tinder. This is done with the Fire Syringe (see Fig. 210). In the extremity of the piston is a small cavity, in which some tinder is placed. When the piston is driven rapidly down, the air in the barrel is compressed,

What are included under this head? How is it that mechanical action produces heat? 484. State some familiar cases in which heat is produced by friction. What is sometimes the effect of friction on machinery? How do savages kindle their fires? How great a heat has been produced by boring a brass cannon? How has friction been turned to practical use? What is said of the friction of fluids? 485. Give some familiar examples of the production of heat by percussion. How did blacksmiths formerly kindle their forge-fires? Describe the Fire-syringe, and the experiment

Fig. 210.

latent heat is evolved, and on withdrawing the piston the tinder will be found ignited.

If a body is compressed by violent percussion more than once, the heat produced is less each time, until at last all the latent heat is forced out, and it may be struck or hammered without any material increase of temperature. Iron, when thus deprived of its latent heat, becomes stiff and brittle. The metals generally lose their ductility, and can not be drawn out into wire till their latent heat is restored by subjecting them to the action of fire.

486. ELECTRICITY, A SOURCE OF HEAT.—The passage of electricity is sometimes attended with intense heat. Lightning, for instance, sets fire to trees and houses, and melts metallic bodies that it strikes. The heat produced by the galvanic battery ignites or fuses every known substance.

THE FIRE
SYRINGE.

## Diffusion of Heat.

487. Heat tends to diffuse itself equally among bodies of different temperature. So strong is this tendency, that, unless fresh supplies are received, the hottest body soon becomes cool, in consequence of parting with its heat to surrounding objects cooler than itself.

488. Heat is diffused in three ways:—

1. By CONDUCTION, when it passes from one particle of a body to another in contact with it. If one end of a poker is placed in a fire, the other becomes heated by Conduction.

2. By CONVECTION, when it is conveyed by the actual motion of some of the particles of a body. When a pot of water is placed over a fire, the particles at the bottom are first heated, and ascend, carrying heat with them and diffusing it by Convection.

performed with it. What is found, when a body is violently struck more than once? What change is produced in iron thus treated? In the metals generally? 486. What is the fourth source of heat? Give examples. 487. What is the tendency of heat? 488. In how many ways is heat diffused? Name, describe, and give an example of

3. By RADIATION, when it passes from one body to another not in contact with it, leaping over the intervening space. A joint of meat, placed before the fire, is roasted by Radiated Heat.

489. CONDUCTION.—Some substances allow heat to pass freely through their particles; others do not. The former are called Conductors of heat; the latter, Bad Conductors, or Non-conductors.

As a general rule, dense solids are conductors of heat; porous and fibrous solids, as well as liquids, gases, and vapors, are bad conductors.

490. *The Conductometer.*—The metals are all good conductors of heat, but some are better than others. This is shown by the Conductometer, represented in Fig. 211.

Fig. 211.

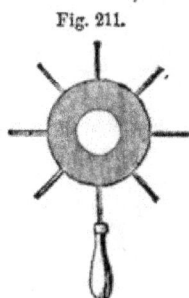

THE CONDUCTOMETER.

The conductometer consists of a circular plate of brass, in the outer edge of which are inserted rods of different metals, of the same size and length, each having a small cavity in its extremity for holding a piece of phosphorus. When the plate is brought over the flame of a lamp, the heat passes along the different rods and ignites the pieces of phosphorus, but not all at the same time. It first reaches the end of the rod that is the best conductor; and thus the order in which the pieces of phosphorus take fire indicates the order in which the metals that the rods are made of rank as conductors of heat.

491. *Conducting Power of different Substances.*—Gold is the best conductor among the metals. The conducting power of gold being set down at 1,000, that of some other common substances compares with it as follows:—

| | | |
|---|---|---|
| Platinum .... 981 | Iron............ 374 | Lead ........ 180 |
| Silver ....... 973 | Zinc............ 363 | Marble ...... 24 |
| Copper ...... 898 | Tin ............ 304 | Clay......... 11 |

Platinum and silver, it will be seen, are nearly as good conductors as gold.

each. 489. What are Conductors of heat? What are Bad Conductors, or Non-conductors? As a general rule, what substances are good conductors of heat, and what not? 490. How do the metals rank in conducting power? Describe the Conductometer, and its mode of operation. 491. Among the metals, what is the best conductor? The next? The next? Which is the better, iron or lead? How may

A silver spoon containing water, with a piece of muslin wrapped smoothly around it, may be held in the flame of a lamp till the water boils without the muslin's burning, so rapidly does the metal carry off the heat.

492. Wood is a bad conductor of heat. A log blazing at one end may be handled at the other without inconvenience. Hence metallic tea-pots, sauce-pans, &c., are often provided with wooden handles. Dense wood and coal are better conductors than porous wood. This is one reason why they are harder to kindle; they conduct the heat away before a sufficient amount is collected in them to produce combustion. Earthen-ware of all kinds ranks far below the metals in conducting power.

493. Fibrous substances, like wool, hair, and fur, are bad conductors. The finer and closer their fibres, the less their conducting power. Thus we see why Providence has clothed the animals of cold climates with a shaggy covering, from which those of the tropics are free; and why the coats of many animals in temperate regions change with the seasons, being closer and longer in winter, thinner and shorter in summer.

494. The best non-conductors among solids are straw, saw-dust, pow-dered charcoal, and plaster of paris. Recourse is had to these articles when it is desired to protect an object from extremes of temperature. Straw is bound round tender plants in winter, to prevent their warmth from being drawn off. It is also used for thatching the roofs of houses, preventing the external heat from entering in summer, and the heat within from being with-drawn in winter. Ice shipped to warm climates is packed in saw-dust, to keep out the heat of the atmosphere. For the same reason, the hollow apart-ments that constitute the sides of refrigerators are filled with powdered char-coal. Plaster of paris is used for filling in the sides of fire-proof safes. So impervious to heat does it render them that they may be exposed to flames for hours without injury to the papers within.

495. If we bare our feet, and place one of them on a carpet and the other on oil-cloth, the latter feels much colder than the former. This is not because the oil-cloth is colder than the carpet, for being in the same room their temperature must be the same; but oil-cloth is a good con-ductor, whereas carpet is not. A good conductor, brought in contact with the body, carries off our animal heat and makes us feel cold. A bad conductor, on the other hand, prevents our animal heat from escaping. Hence the differ-

the conducting power of silver be proved? 492. Why are metallic tea-pots often pro-vided with wooden handles? Why is dense wood hard to kindle? How does earthen-ware rank in conducting power? 493. How do fibrous substances rank? As re-gards the coats of animals, how is the goodness of Providence shown? 494. What are the best solid non-conductors? For what are these substances severally used, and what is the effect in each case? 495. If we bare our feet, and place one on a carpet and the other on oil-cloth, what do we feel? Explain the reason of this. Of the

ence of warmth in different kinds of clothing. That fabric feels the warmest, which is the worst conductor.

Of the materials used for clothing, wool is the worst conductor and linen the best; cotton and silk rank between the two. Linen is therefore the most comfortable fabric for summer clothing, and woollen for winter. A linen under-garment is cooler than a silk or muslin one, and these in turn are much cooler than flannel.

496. The heat of our bodies is generally greater than that of the atmosphere surrounding them. If we were placed in an atmosphere warmer than our bodies, woollen would be the coolest dress that could be worn, because, being a bad conductor, it would not transmit the external heat. Hence firemen and others exposed to a high degree of heat, always wear flannel. Hence, also, a blanket is wrapped round ice, to keep it from melting.

497. *Conducting Power of Liquids.*—Liquids (except mercury, which is a metal) are very bad conductors of heat. This may be shown by several experiments.

Freeze some water in the bottom of a tube, and on the ice pour some more water. Inclining the tube, apply the flame of a lamp to the liquid till it boils. The ice remains for a long time unmelted. If mercury is used instead of water, the ice begins to melt almost immediately on the application of heat.

Again, in a funnel-shaped glass vessel (represented in Fig. 212) fix a thermometer, or instrument for measuring heat, with its bulb uppermost. Cover the bulb with water to the depth of half an inch; then pour on some ether, and set fire to it. The burning of the ether generates a great heat; yet the thermometer, only half an inch below it, indicates little or no increase of temperature.

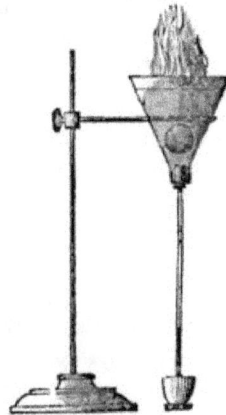
Fig. 212.

498. *Conducting Power of Gases and Vapors.*—Gases and vapors are still worse conductors of heat than liquids. The less their specific gravity, the less appears to be their conducting power.

499. Air is one of the worst conductors known. If we

materials used for clothing, which is the worst conductor? Which, the best? How do cotton and silk rank? What fabric, then, is the most appropriate for summer wear, and what for winter? 496. Why do firemen wear flannel? Why is a blanket wrapped round ice? 497. How do liquids rank in conducting power? Prove that water is a bad conductor. Prove it by an experiment with the apparatus represented in Fig. 212. 498. How do gases and vapors rank in conducting power? 499. What

9*

could keep a body of air perfectly still, it would take a long
time for heat applied to one portion of it to be transmitted
throughout the whole.

In summer, when there is no breeze, we feel oppressively warm, because
the air does not carry off the heat generated within us.  Fanning cools us,
because it drives off the air heated by contact with our bodies and brings up
a fresh supply, which, after withdrawing more or less heat, is in turn driven
away.  In this case it will be observed that the heat is carried off by *convection*, and not by *conduction*.  If air were a good conductor, it would soon
take so much heat from animals and plants that their vital action could not
make up the deficiency, and they would be chilled to death.

Closed cellars are cooler than the surrounding air in summer, and warmer in winter.  If air were a good conductor, this would not be the case.  As
it is, the doors being kept closed, currents of air are excluded; and, since
heat passes very slowly from particle to particle, extremes of temperature
without are not felt within.

It is the air in fibrous and porous solids that makes them bad conductors.
Drive out this air by compression, and you increase their conducting power.
Let wool, or cotton, for instance, be twisted into rolls, and it will carry off
heat faster than it did when loose.  Accordingly, clothing that allows some
air to remain in contact with the body is warmer than that which fits very
tight.  So, double sashes and double doors, confining a body of non-conducting air, protect apartments from extremes of heat and cold.

500. The uses of air as a non-conductor are seen in the operations of nature.  Filling the pores and interstices in the bark of plants, it protects the
tender parts within from sudden falls of temperature.  In cold climates, vegetation is further protected by snow, which, owing to the air imprisoned
among its particles, is a very bad conductor.  A mantle of snow on a field
has very much the same effect that a covering of wool would have.  Hence
we are told in Scripture that God " giveth snow like wool ".—The Esquimaux
shield themselves from the excessive cold of their climate in huts of snow.

501. CONVECTION.—Fluids, as we have just seen, are
bad conductors, but they are readily heated by convection.
Heat being applied beneath, the lower particles become
expanded and rarefied.  They therefore ascend, carrying
up their heat, while cooler and heavier particles from above

---

is said of the conducting power of air?  Why do we feel oppressively warm in summer, when there is no breeze?  What is the effect of fanning?  If air were a good
conductor, what would be the consequence to animals and plants?  Why are closed
cellars exempt from extremes of temperature?  What makes fibrous and porous solids bad conductors?  Prove this.  Compare the warmth of loose clothing with that
which fits very tight.  On what principle do double sashes operate?  500. Show the
uses of air as a non-conductor in the economy of nature.  What is the effect of snow?
What use is made of it by the Esquimaux?  501. How are fluids readily heated?

take their place. This process is repeated till heat is diffused throughout the whole,—not conducted from one stationary particle to another, but actually conveyed by the particles receiving it.

The process of convection is exhibited when water is set over a fire to boil. The particles soon begin to move, as may be shown by throwing in some powdered amber, which is seen to rise and descend, more and more rapidly as the temperature increases. Heat is thus diffused throughout the whole body of liquid, till ebullition, or boiling, commences.

502. In cooling, this process is reversed. The particles at the top yield their heat to the air in contact with them. Being thus made heavier, they descend, while warmer and lighter particles take their place. The greater the surface exposed to the air, the sooner the liquid loses its heat; hence we pour our tea into a saucer, to cool it.

503. To heat a body of liquid by convection, the fire must be applied beneath. A pot of water can not be made to boil by a fire kindled on its lid. The particles at the top may be heated, but they will remain there on account of their superior lightness, and there will be no diffusion of heat.

504. Thin liquids, like water, are heated and cooled more quickly than thick ones, like tar, because their particles move more freely among themselves, and thus diffuse heat more readily.

505. Heat is diffused through gases and vapors, as through liquids, by convection. Heated air, like heated water, ascends, carrying its heat with it. Consequently, to make the temperature of a room uniform, a fire-place should be set as near the floor as possible.—With the same temperature, we feel colder on a windy day than on a still one; because the heat is more rapidly withdrawn from our bodies by the fresh currents of air constantly brought in contact with them.

506. Solids can not be heated by convection, because their particles cohere.

507. RADIATION.—A body not in contact with the source of heat can not be heated by conduction or convection. If it receives heat, it is by a third process, called Radiation.

Describe the operation. In what familiar process is convection exhibited? Describe the process of boiling. 502. Describe the process of cooling. 503. To heat a liquid, where must the fire be applied? Why can not a pot of water be made to boil by a fire kindled on its lid? 504. What kind of liquids are heated and cooled most quickly? Why? 505. What, besides liquids, are heated by convection? Where should a fire-place be set, and why? Why do we feel colder on a windy day than on a still one? 506. Can solids be heated by convection? Why not? 507. What bodies are heated

If we place our hands under a fire in a grate, we at once feel a sensation of heat. This heat can not reach our hands by conduction, for air is a bad conductor,—nor by convection, for heated currents ascend. It is transmitted in rays sent forth from the fire through the intervening space. Heat thus diffused is called Radiant Heat. All the heat that we receive from the sun, and much of that from fire, is radiant heat.

508. All substances radiate heat, but not equally well. Much depends on the character of the surface. Rough and dull surfaces radiate better than smooth and bright ones.

Lamp-black is the best radiator known. Rating its radiating power at 100, that of crown-glass is 90; black lead, 75; tarnished lead, 45; clean lead, 19; bright metals generally, 12. The radiating power of metals is increased by scratching their surface, or letting them become tarnished.

509. A heated body confined in a covered vessel parts with its heat more or less rapidly according to the radiating power of the vessel containing it. For tea-pots, therefore, bright silver is preferable to earthen-ware, because it is a worse radiator and keeps the tea warm for a longer time. Stoves, on the contrary, should be made of a good radiator, so that the heat of the fire may be freely diffused. Cast-iron is better for this purpose than sheet-iron, because its surface is rough; the radiating power of both is increased by rubbing in black lead. When heat is to be conveyed from one room to another, a pipe should be used of bright tin, which is a bad radiator and prevents the escape of heat by the way.

The atmosphere receives its heat, not directly from the sun, but by radiation from the earth; hence, as we ascend from the earth's surface, the heat diminishes.

510. *Law of Radiant Heat.—Radiant heat diminishes in intensity as the square of the distance from the radiating body increases.*

A body 10 feet from a fire will receive from it only $1/100$ of the heat that a body 1 foot from it receives.

511. Radiant heat, striking different bodies, is reflected

by radiation? What is heat diffused by radiation called? Give a familiar example of radiant heat. 508. By what is a body's radiating power affected? What surfaces radiate heat the best? What is the best known radiator? Rating the radiating power of lamp-black at 100, what is that of crown-glass? Black lead? Tarnished lead? Clean lead? Bright metals generally? How may the radiating power of the metals be increased? 509. Why is bright silver preferable to earthen-ware for tea-pots? Of what should stoves be made? When heat is to be conveyed from one room to another, what should be employed? Why? How does the atmosphere receive its heat? What follows? 510. State the law of radiant heat. Give an example.

by some, absorbed by others, and transmitted by a third class.

512. *Reflection of Radiant Heat.*—Radiant heat is reflected by polished and light-colored surfaces. Polished gold reflects about three-fourths of the radiant heat it receives, and looking-glass about one-fifth ; whereas metallic surfaces blackened reflect only one-twentieth.

513. White and light-colored clothes are worn in summer, because they reflect heat. For the same reason, it is harder to heat water in a new tin vessel than in one that has been blackened over the fire.

514. The reflection of radiant heat may be illustrated with the apparatus represented in Fig. 213. A and B are concave metallic mirrors, highly pol-

Fig. 213.

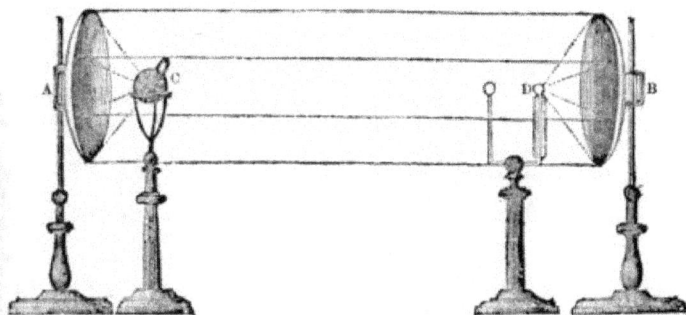

ished. In the focus of A is placed a red-hot ball C. This ball radiates hea* in all directions, and some of its rays strike the mirror A, from which they are reflected in parallel lines to B. By B they are again reflected and brought to a focus at D, where a thermometer indicates a rise of temperature. Sufficient heat may thus be concentrated at D to set fire to phosphorus or gun· powder.

515. When radiant heat is reflected by a plane surface, the angle of reflection (see § 96) is always equal to the angle of incidence. If it strikes the surface perpendicularly, it is reflected perpendicularly, back to the radiating body. If the line in which it approaches the surface forms an angle

with the perpendicular, it glances off at an equal angle on the other side.

516. *Absorption of Radiant Heat.*—Radiant heat is absorbed by dull and dark-colored surfaces. Good reflectors are bad absorbents and radiators; bad reflectors are good absorbents and radiators.

Of the colors, black is the best absorbent of heat, and violet the next best; white is the worst, and yellow next to the worst.

Lay two pieces of cloth, one white and the other black, on a snow-bank, in the sunshine. Under the black piece, which absorbs the heat that strikes it, the snow melts rapidly; not so under the white cloth, for by it the heat is reflected. Dark-colored clothing is therefore best adapted to winter.

Dark mould absorbs the sun's heat; hence one cause of its fertility. White sand reflects the hot rays; hence it burns our faces when we walk over it in summer. Hoar-frost remains longer in the morning on light than dark substances: this is because light colors reflect the sun's heat, while dark colors absorb it, and thus melt the hoar-frost, which is nothing more than frozen dew.

517. *Transmission of Radiant Heat.*—Transparent substances, or such as allow light to pass through them, for the most part transmit heat also. The sun's rays, for instance, falling on the atmosphere of the earth, which is a transparent medium, are transmitted through it to objects on the surface. More or less heat is absorbed in the act of transmission.

518. Substances that transmit heat freely are called Di-a-ther'-ma-nous. Those that absorb the greater part and transmit little or none are called A-ther'-ma-nous.

519. All transparent substances are not diathermanous. Water, for example, which offers but little obstruction to rays of light, intercepts nearly all the heat that strikes it. Alum is another instance in point.

---

equal? 516. By what surfaces is radiant heat absorbed? What is said of good reflectors? What, of bad reflectors? What color is the best absorbent of heat? What, the next best? What color is the worst absorbent? What, the next worst? Prove by an experiment the difference in absorbing power between white and black. Why is dark-colored clothing best adapted to winter? What is the difference between dark mould and white sand in absorbing power? Why does hoar-frost remain longer in the morning on light than dark substances? 517. What substances, for the most part, transmit heat? Give an example. 518. What are Diathermanous substances? What are Athermanous substances? 519. Name a transparent substance that is not dia-

All diathermanous substances are not transparent. Quartz, though it may intercept light almost entirely, transmits heat quite freely.

As a general rule, the rarer transparent substances, such as gases and vapors, transmit heat the best; the denser ones, such as rock-crystal, transmit it the least freely. The farther the rays have to pass through a given substance, the more heat is intercepted.

## Effects of Heat.

520. The effects of heat are five in number: Expansion, which changes the size of bodies; Liquefaction and Vaporization, which change their form; Incandescence, which changes their color; and Combustion, which changes their nature.

521. EXPANSION.—Heat expands bodies.

Insinuating itself between the particles of bodies, it forces them asunder, and thus makes them occupy a greater space. Heat, therefore, opposes cohesion. Solids, in which cohesion is strongest, expand the least under the influence of heat; liquids, having less cohesion, expand more; gases and vapors, in which cohesion is entirely wanting, expand the most. Heat converts solids into liquids, and liquids into gases and vapors, by weakening their cohesion. It turns ice, for example, into water, and water into steam.

522. *Expansion of Solids.*—All solids except clay are expanded by heat; but not equally. Of the metals, tin is among those that expand most. Clay is contracted by baking, and ever afterwards remains so; this is supposed to be owing to a chemical change produced in it by heat.

The expansion of solids is illustrated with the apparatus represented in Fig. 214. A brass ball is suspended from a pillar, to which is also attached a ring just large enough to let the ball pass through it at ordinary temperatures. Heat the ball with a lamp placed beneath, and it will expand to such a degree that it can not pass through the ring. Let it cool, and it will go through as before.

523. A sheet-iron stove in which a hot fire is quickly kindled or put out, sometimes makes a cracking noise, in consequence of the rapid ex-

thermanous. Name a diathermanous substance that is not transparent. As a general rule, what transparent substances transmit heat the best, and what the worst? 520. State the effects of heat. 521. What is the first of these? How is it that heat expands bodies? What force does it oppose? Which expand the most under the influence of heat, solids, liquids, or gases,—and why? Into what does heat convert solids? Into what, liquids? 522. What solids are expanded by heat? What metal is expanded more than most of the others? What is the effect of heat on clay? Illustrate the expansion of solids with the apparatus represented in Fig. 214. 523. Why

Fig. 214.

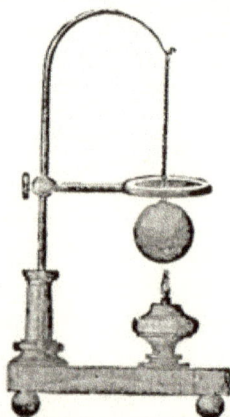

pansion or contraction of the metal. A blower
placed on or taken from a hot fire produces a similar noise for the same reason. New furniture
standing in the sun or near a fire is apt to warp
and crack in consequence of the expansive effects
of heat.

When boiling water is poured into china cups
and glass vessels, they often crack. This is because the inner surface is expanded by heat,
while the outer is not, china-ware and glass being bad conductors. The unequal expansion
cracks the vessel. Cold water poured on a hot
glass or stove produces the same effect. On the
same principle, glass chimneys are apt to crack,
when brought too suddenly over the flame of a
lamp or gas-burner. A cut made in the bottom
with a diamond allows an opportunity for expansion, and prevents the chimney from breaking.

When a glass stopper becomes fastened in a bottle, it may often be withdrawn by placing the neck of the bottle in warm water. The neck is expanded before the heat reaches the stopper.

524. The force with which a body expands when heated and contracts when cooling, is very great. In iron
bridges, therefore, and other structures in which long bars
of metal are employed, there is danger of the parts' separating, unless provision is made for the expansion caused
by a rise of temperature. The middle arch of an iron
bridge has been known to rise an inch in the summer of a
temperate climate. So, when great lengths of iron pipe
are laid for conveying steam or hot water, sliding joints
must be used, or the apparatus will burst in consequence
of the expansion of the metal.

525. The fact that heat expands bodies and cold contracts them, is often
turned to practical account. Coopers, for instance, heat their iron hoops,
and while they are thus expanded put them on casks which they just
fit. As they cool, they contract and bind the staves tightly together. The

do a sheet-iron stove and a blower sometimes make a cracking noise? What causes
new furniture to warp? What makes glass vessels crack when boiling water is poured
into them? When are glass chimneys apt to crack? How may their cracking be
prevented? When a glass stopper becomes fastened in a bottle, how may it be withdrawn? 524. What is said of the force with which bodies expand and contract?
What precautions must be taken in consequence? 525. What practical use is made
of the fact that heat expands bodies and cold contracts them? What ingenious appli-

wheel-wright fastens the tire, or outer rim of iron, on his wheel in the same way.

The contraction of iron, when cooling, has been ingeniously used for drawing together the walls of buildings that have bulged out and threaten to fall. Several holes are made opposite to each other in the walls, into which are introduced stout bars of iron, projecting on both sides and terminating at each end in a screw. To each screw a nut is fitted. The bars are then heated by lamps placed beneath, and when they have expanded the nuts are screwed up close to the walls. As the bars cool, they gradually contract, and with such force as to bring the walls back to a perpendicular position.

526. *Expansion of Liquids.*—Liquids, when heated, expand much more than solids, but not all alike. Thus water, raised from its freezing-point to the temperature at which it boils, has its bulk increased one-twenty-second; alcohol, between the same limits, increases one-ninth.

The higher the temperature, the greater the rate at which liquids expand.

527. In proportion as heat expands liquids, it rarefies them, the same quantity of matter being made to occupy a larger space. This fact is shown in the process of boiling, described in § 501.

528. Water at certain temperatures forms a remarkable exception to the general law that liquids are expanded by heat and contracted by cold. As it cools down from the boiling-point, it contracts, and consequently increases in density, till it reaches 39 degrees, or 7 degrees above its freezing-point. Below this temperature, it expands.

The expansion of water in freezing is proved every winter by the bursting of pipes, pitchers, &c., containing it. The force with which it expands is tremendous. An iron plug weighing three pounds and closing a bomb-shell filled with water, has been thrown 15 feet by the freezing and expansion of the liquid within. Immense masses of rock are sometimes split off by the freezing of water which has insinuated itself into minute fissures.

The expansion and consequent rarefaction of water in freezing, afford a

cation has been made of the contraction of iron when cooling? Give an account of the process. 526. How does the expansion of heated liquids compare with that of solids? Compare the expansion of water with that of alcohol. On what does the rate at which liquids expand depend? 527. Besides expanding liquids, what does heat do to them? 528. What exception is there to the law that liquids are contracted by cold? How is the expansion of water in freezing proved? What cases are cited, to show the great force with which water expands in freezing? How does the expansion

striking proof of the goodness of Providence. The great body of a large mass of water never becomes cold enough to freeze; it freezes only on the top, where it comes in contact with very cold air. As it is, the ice formed on the surface remains there on account of its superior rarity, and protects the water below and the fish that inhabit it from further cold. If water continued to contract and increase in density as it approached the freezing-point, the ice first formed would sink; the fresh surface exposed to the air would in its turn freeze, and another layer of ice would sink; and this would go on till even in a mild winter every body of water would be converted into a solid mass, and all living things therein destroyed.

529. Iron, zinc, and several other metals, when cooling down from a melted to a solid state, expand like freezing water. This is because the particles assume a crystalline arrangement, by which greater interstices are left between them.

530. *Expansion of Gases and Vapors.*—Aëriform bodies expand equally under a given increase of temperature. At the boiling-point of water, their bulk is one-third greater than at the freezing-point.

531. Fill a bladder with air, tie its neck, and place it before a fire; the heat will soon expand the confined air to such a degree as to burst the bladder.

The popping of grains of corn, the bursting open of chestnuts when roasting, and the crackling of burning wood, are caused by the expansion of the air within them. Porter-bottles have to be kept in a cool place in summer, lest the heat expand the carbonic acid in the porter and break the bottles.

532. LIQUEFACTION.—Heat melts solids. This process is called Liquefaction.

Some solids, such as wax and butter, require but little heat to melt them. Others, like metals and stones, melt only at the highest temperatures that can be produced. Such substances are called *refractory*.

Even substances that are liquid at ordinary temperatures may be looked upon as melted solids, for they can be reduced by cold to the solid state.

533. When a solid is converted into a liquid, sensible heat is absorbed. When a liquid is converted into a solid,

---

of water in freezing exhibit the goodness of Providence? 529. How do we account for the expansion of several of the metals, when cooling down from a melted state? 530. What is said of the expansion of aëriform bodies? How great is their expansion, when they are raised from the freezing-point to the boiling-point of water? 531. How may we illustrate the expansion of air by heat with a bladder? What familiar examples are given of the expansion of air by heat? 532. What is Liquefaction? What difference is there in solids, as regards their capability of being melted? How may substances that are liquid at ordinary temperatures be looked upon? 533. By what

latent heat is given out. This is another merciful provision, for thus extremes of temperature and their effects are modified.

When a solid is rapidly melted, so much heat is absorbed by the liquid that intense cold is produced. This is the principle on which freezing mixtures operate. Ice cream, for instance, is frozen with a mixture of salt and snow or pounded ice; the latter is rapidly melted, and so much heat is absorbed in the process that the cream is brought to a solid form.

534. VAPORIZATION.—Heat converts liquids into vapors. This process is called Vaporization.

Heat, applied to a solid, first expands it, then melts it, and finally turns it into vapor. Some solids pass at once into vapor, without becoming liquids.

535. A great degree of heat is not essential to vaporization. At ordinary temperatures, wherever a surface of water is in contact with the air, vapor is formed. This process is known as Spontaneous Evaporation. By its means the atmosphere becomes charged with moisture, and clouds and dew are formed. The drier the air, and the more it is agitated, so as to bring fresh currents in contact with the liquid, the more rapidly does evaporation take place.

536. A drop of water let fall on a cold iron moistens its surface; let fall on a very hot iron, it hisses and runs off without leaving any trace of moisture. In the latter case, the water does not touch the iron at all, but is separated from it by a thin layer of vapor into which part of the drop is converted by the heat radiated from the iron. Laundresses try their irons in this way, to see if they are hot enough for use. On the same principle, jugglers plunge their hands into melted metal with impunity, by first wetting them. The moisture on their hands is converted into vapor, which keeps the seething metal from their skin.

537. When vapor is formed, sensible heat is absorbed, and cold is produced.

Hence when the skin is moistened with a volatile liquid (that is, one that readily passes into vapor) like alcohol, a sensation of cold is soon experienced. So, a shower or water sprinkled on the floor cools the air in sum-

merciful provisions are extremes of temperature modified? On what principle do freezing mixtures operate? 534. What is Vaporization? What are the successive effects of heat on solids? 535. What is Spontaneous Evaporation? What are the effects of evaporation on the earth's surface? To what is the rapidity of evaporation proportioned? 536. Explain the principle on which laundresses try their irons. What use do jugglers make of this principle? 537. With what phenomena is the formation of vapor accompanied? Give some examples of cold produced by the for-

mer.—Green wood does not make so hot a fire as dry, because, when the moisture it contains is converted into vapor, a large amount of sensible heat is absorbed and carried off.

538. CONDENSATION.—The turning of vapor back into a liquid state is called Condensation.

539. *Distillation.*—Some substances are converted into vapor at lower temperatures than others. This fact is taken advantage of in Distillation.

Distillation is the process of separating one substance from another by evaporating and then condensing it. It was known to the Arabians at an early date. Fig. 215 represents a Still, or apparatus for distilling.

Fig. 215.

A STILL.

540. A is a *boiler*, resting on a furnace. In its *head*, B, is inserted a pipe, *b c*, which enters the *worm-tub*, R, and there terminates in a *worm*, represented by the dotted lines. The substance to be distilled having been placed in the boiler and a fire kindled beneath, vapor soon rises. Passing through the pipe *b c*, it enters the worm, in which it is to be condensed. The worm is surrounded with cold water, with which the vat is filled, and the vapor is soon cooled down into a liquid form, and issues from the lower extremity of

mation of vapor. Which makes the hotter fire, green wood or dry,—and why? 538. What is meant by the Condensation of vapor? 539. What is Distillation? On what fact is the process based? To whom was distillation early known? What is an apparatus for distilling called? 540. With the aid of Fig. 215, describe the still,

the worm, falling into a vessel prepared to receive it. To condense the vapor, the water in the vat must be kept cold. For this purpose, a stream is
kept flowing into it through the pipe *p p*, while a similar stream of water
partially warmed by the hot vapor as constantly escapes at *q*. By this process water may be obtained perfectly pure, as the earthy matter dissolved in
it is not converted into vapor, but remains behind in the boiler. With a
similar apparatus, spirituous liquors are distilled from grain.

541. INCANDESCENCE.—When a body is raised to a certain very high temperature, it begins to emit light as well
as heat. This state is called Incandescence, or Glowing
Heat.

An incandescent body becomes successively dull red,
bright red, yellow, and white. All solids and liquids, not
previously converted into vapor by heat, become incandescent. The temperature at which incandescence commences is the same for all bodies, and may be set down at
977 degrees of Fahrenheit's Thermometer (see § 544).

## Instruments for measuring Heat.

542. The expansion of bodies by heat furnishes us the
means of measuring changes of temperature. Liquids,
which are easily affected, are used for measuring variations
in moderate temperatures. Solids, which require a higher
degree of heat to expand them perceptibly, are used for
measuring variations in elevated temperatures. Hence we
have two instruments, the Thermometer and the Pyrometer.

543. THE THERMOMETER.—The Thermometer is an instrument in which a liquid, usually mercury, is employed
for measuring variations that occur in moderate temperatures.

The thermometer (see Fig. 216) consists of a tube closed at one end and
terminating in a bulb at the other. The bulb and part of the tube contain
mercury, above which is a vacuum, all air having been excluded before the
top of the tube was closed. Expanded by heat, the mercury rises in the

and its mode of operation. 541. What is Incandescence? What colors mark the
successive stages of incandescence? What substances become incandescent? At
what temperature does incandescence commence? 542. What means have we of
measuring changes of temperature? In what cases are liquids used? In what, solids? Name the instruments used for measuring changes of temperature. 543. What

Fig. 216.

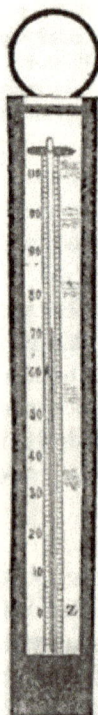

tube; when the temperature falls, the mercury, contracting, falls also. The tube is fixed in a stand or case, and has a graduated scale beside it for measuring the rise and fall of the mercury. This scale is formed in the following way :—The thermometer is brought into contact with melting ice, and the point at which the mercury stands is marked. It is next plunged in boiling water, and the point to which the mercury rises is also marked. The interval is then divided into a number of equal spaces, called *degrees*.

544. As the thermometer does not indicate the amount of heat in a body, but merely its changes of temperature, the number of degrees into which the interval between the freezing and the boiling mark is divided is arbitrary. Three different divisions are in use: Fahrenheit's, in the United States, Great Britain, and Holland; Reaumur's [ro'-murz], in Spain and parts of Germany; and the Centigrade, the most convenient of the three, in France, Sweden, &c.

In Fahrenheit's scale the freezing-point is called 32, the boiling-point, 212; when, therefore, the mercury stands at 0, or zero, it is 32 degrees below the freezing-point. In Reaumur's scale the freezing-point is called 0, the boiling-point 80. In the Centigrade the freezing-point is 0, the boiling-point 100. When degrees of the thermometer are mentioned, it is usual to indicate the scale referred to by the letters F., R., or C., as the case may be. Thus 40° F. means 40 degrees on Fahrenheit's scale; 15° R., 15 degrees on Reaumur's scale, &c. In this country, when no scale is mentioned, Fahrenheit's is meant.

THE THERMOMETER.

545. Imperfect thermometers were in use at the beginning of the seventeenth century. It is uncertain whether the honor of their invention belongs to Sanctorio, an Italian physician,—Drebbel, a Dutch peasant,—or Galileo. Various liquids have been tried; the astronomer Roemer was the first to use mercury, the advantages of which are such that it has superseded all others.

546. *The Differential Thermometer.*—This instrument,

is the Thermometer? Of what does it consist? How is the scale of the thermometer formed? 544. What is said of the number of degrees into which the scale is divided? Name the three principal scales, and tell where each is used. What are the freezing-point and the boiling-point respectively called in Fahrenheit's scale? What, in Reaumur's scale? In the Centigrade scale? How are the different scales indicated? 545. When were thermometers first used? To whom does the honor of their invention belong? What liquid has superseded all others in the thermometer? Who

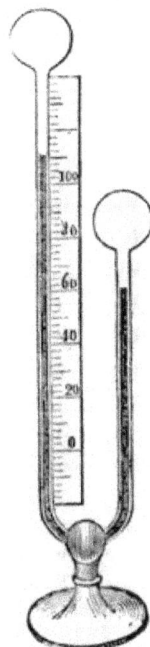

represented in Fig. 217, measures minute differences of temperature.

Fig. 217.

It consists of a long glass tube, bent twice at right angles, somewhat in the form of the letter U.  One arm is furnished with a scale of 100 degrees, and each terminates in a bulb.  The tube contains a small quantity of sulphuric acid, colored red, and so disposed that when both bulbs are of the same temperature it stands at 0 on the scale.  Let either bulb be heated ever so little more than the other, and the expansion of the air within will drive the liquid down and cause it to ascend the opposite arm to a distance measured by the scale.  Ordinary changes of temperature do not affect the instrument, because both bulbs are acted on alike.

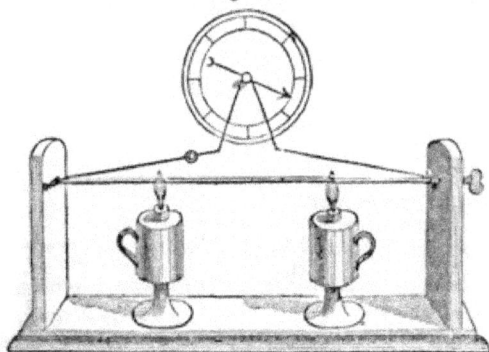

547. THE PYROMETER.—The Pyrometer (see Fig. 218) is used for measuring variations in elevated temperatures, and comparing the expansive power of different metals for a given degree of heat.

Fig. 218.

THE PYROMETER.

THE DIFFERENTIAL THERMOMETER.

A metal bar is fixed in an upright at one end by means of a screw, and left free to expand at the other. It there touches a pin projecting from a rod which rests against an opposite upright, in a circular support at each side.  This rod terminates at one end in an arm bent at right angles, which is connected by a cord and pulley with an index traversing a scale marked with degrees. Near its extremity is a ball, the weight of which, under ordinary circumstances, keeps the index at the highest point of the scale.  When lamps are placed beneath and the bar expands, it pushes against the pin, turns the rod

more or less around, and thus raises the arm containing the ball and moves
the index along the scale. The relative degree of heat applied to the bar is
thus indicated. By keeping the heat the same, and using rods of different
metals, we can ascertain their relative expansive power.

## Specific Heat.

548. Put a pound of water and a pound of olive oil in
two similar vessels, and apply heat. It will take twice as
long to raise the water to a given temperature as it will the
oil. Let them cool, and the water will be twice as long in
parting with its heat as the oil. Water, therefore, must
receive twice as much heat as olive oil in reaching a given
temperature.

The relative amount of heat which a body receives in
reaching a given temperature is called its Specific Heat, or
its Capacity for Heat.

549. In estimating the specific heat of bodies, that of water is taken as a
standard. Reckoning the specific heat of water as 1, that of iron is about
$1/9$, and mercury only $1/33$. As a general thing, the densest bodies have the
least specific heat; solids have less than liquids, and liquids less than gases
and vapors.

550. As the elastic fluids expand, they are rarefied, and their specific heat
becomes greater,—that is, it requires more heat to raise them to a given temperature. This is one reason why the upper regions of the atmosphere are
colder than the lower, as is found by those who ascend mountains.

## Steam.

551. GENERATION OF STEAM.—Water is rapidly turned
into steam at its boiling-point, which in an open vessel at
the level of the sea is 212° F. After it commences boiling,
water can not be raised to any higher temperature, because
all the heat subsequently applied is absorbed by the steam
and passes off with it.

used? Describe the Pyrometer. 548. How is it proved that water must receive twice
as much heat as olive oil in reaching a given temperature? What is meant by Specific Heat? 549. In estimating the specific heat of bodies, what is taken as a standard? What is the specific heat of iron? Of mercury? As a general thing, what
bodies have the least specific heat? 550. Under what circumstances is the specific
heat of elastic fluids increased? What fact is thus explained? 551. How is steam
generated? Why can not water, after it commences boiling, be raised to any higher

If the water is in a close vessel, the steam first formed, being confined, presses on the water and prevents it from boiling as soon as before. It may now be raised to a more elevated temperature, for heat is not withdrawn by the formation of steam till it reaches a higher point.

552. Steam has the same temperature as the water from which it is formed, the heat absorbed in the process of formation becoming latent. When it is generated from water in an open vessel, its temperature is 212°; in a confined vessel it will be higher, according to the pressure on the surface of the water.

553. Steam is colorless and invisible. When cooled by contact with the atmosphere, it begins to turn back into a liquid state, and assumes a grey mist-like appearance. Look at the spout of a tea-kettle full of boiling water. For half an inch from the extremity nothing can be seen; beyond that, the steam, cooling and beginning to condense, becomes visible.

Fig. 219.

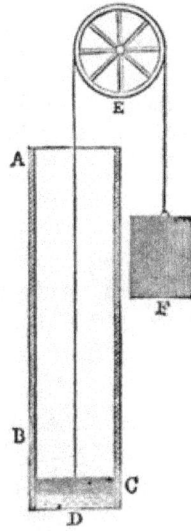

554. The generation and properties of steam may be understood from Fig. 219. A B represents the inside of a tall glass tube, the section of which has an area of one square inch. The tube is closed at its lower end, and contains a cubic inch of water, D, and resting on it a tightly-fitting piston, C. A cord, fastened to the piston, is carried round the wheel E, and attached to the weight F. F is made just heavy enough to counterbalance the piston and its friction against the tube. Suppose a thermometer to be placed in the water, and apply heat at the bottom of the tube. As soon as the thermometer indicates a temperature of 212°, the piston begins to rise, leaving a space apparently empty between it and the water. The fire continues to impart heat to the water, but the mercury in the thermometer remains stationary at 212°; the piston keeps rising, and the water begins to diminish. If the process were continued and the tube were long enough, the piston would at last reach a

temperature? Under what circumstances may water be raised to a higher temperature than 212°? 552. What is the temperature of steam? 553. What is the color of steam? Explain the mist-like appearance a short distance from the spout of a boiling tea-kettle. 554. With the aid of Fig. 219, show the process of generating steam, and

10

height of nearly 1,700 inches, by which time the water would entirely disappear. If the tube were then weighed, though nothing could be seen in it but the piston, it would be found to have exactly the same weight as at first. The water would simply be converted into steam, and thus increased in volume 1,700 times. The piston, with the pressure of the atmosphere on it (which is 15 pounds, the area of the piston being one square inch), would be raised 1,700 inches.

All the time steam is forming, a uniform amount of heat is applied to the tube. As the mercury in the thermometer rises no higher than 212°, it is evident that the heat imparted after it reaches that point is absorbed by the steam and becomes latent. To determine the amount of this latent heat, we must compare the time required to raise the water from the freezing to the boiling point with the time that elapses from the commencement of boiling till the water disappears. We shall find that the latter interval is $5\frac{1}{2}$ times as great as the former; and, since from the freezing-point (32°) to the boiling-point (212°) is 180°, we conclude that the amount of heat absorbed is $5\frac{1}{2}$ times 180°, or nearly 1,000 degrees. That is, the heat applied would have raised the water to a temperature of nearly 1,000°, if it could have remained in the liquid state.

555. If, besides the pressure of the atmosphere on P, a weight of 15 pounds were placed on it, it would be said to have a pressure of *two atmospheres*. Steam, in this case, would not commence forming till the water reached a temperature of $251\frac{1}{2}$ degrees; and, when the whole was evaporated, the piston would stand only about half as high as before. Under a pressure of three atmospheres, the piston would be raised about one-third as high, &c.; the mechanical force developed in the evaporation of a given quantity of water remaining nearly the same. This force, for a cubic inch of water, is sufficient to raise a ton a foot high.

556. Steam has a high degree of elasticity and expansibility. Under a pressure of two atmospheres, or 30 pounds to the square inch, it would raise the piston in the above experiment about 850 inches; if 15 pounds were removed from the piston, the expansive force of the steam would drive it up 850 inches farther.

557. CONDENSATION OF STEAM.—Steam retains its form only as long as it retains the latent heat absorbed. The

describe some of its properties. When water is converted into steam, how many times is its volume increased? How is this proved with the apparatus just described? Prove that heat becomes latent in the steam. How can the amount of latent heat be determined? 555. When is steam said to have a pressure of two atmospheres? How high would the piston then be raised? How high would the piston be raised under a pressure of three atmospheres? How great is the mechanical force developed in evaporating a cubic inch of water? 556. Prove the expansibility of steam. 557. How long does steam retain its form? When is it condensed? Show

moment it is forced to part with this heat, it is turned back into the liquid form, or *condensed*.

In the above experiment, after the piston has been raised 1,700 inches, let the fire be removed, and cold water be applied to the surface of the tube. The latent heat will be abstracted, and the steam will be condensed and form once more a cubic inch of water at the bottom of the tube. As the steam condenses, successive vacuums are produced; and the piston, forced down by the pressure of the atmosphere, descends, and finally rests on the water as at first.

By applying heat again, the process may be repeated. An up-and-down motion may in this way be communicated to the piston; and the piston may be connected with machinery, which will thus be set in motion by the alternate evaporation of water and condensation of steam. This was the principle of the Atmospheric Engine, which was once extensively used, but has now been superseded.

## The Steam-Engine.

558. HERO'S ENGINE.—Steam and some of its properties appear to have been known to the ancients centuries before the Christian era. Hero, of Alexandria, who flourished about 200 years B. C., has left us a description of a steam-engine by which machinery could be set in motion.

Fig. 220 represents Hero's engine. A hollow metallic globe is supported by pivots, and provided with a number of jets equally distant from the pivots, and bent at right angles near their outer end. As soon as steam is introduced into the globe, it issues violently from the mouth of each jet, while on the opposite side of each it presses without being able to escape. This unbalanced pressure makes the globe revolve. Machinery may be set in motion by means of a band connected with this apparatus.

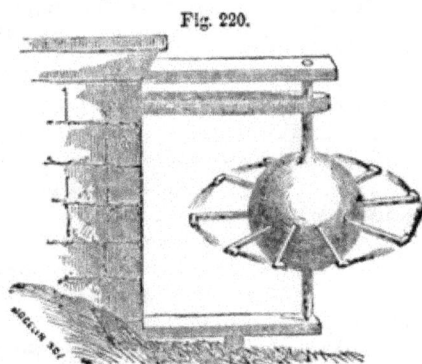

Fig. 220.

HERO'S STEAM-ENGINE.

559. Hero's was a simple rotatory engine. No use was made of it for

2,000 years; but the principle involved has been revived, and is applied in rotatory engines at the present day.

560. DE GARAY'S ENGINE.—In 1543, a Spaniard, by the name of De Garay, undertook to propel a vessel of 200 tons in the harbor of Barcelona by the force of steam. He kept his machinery a secret, but it was observed that a boiler and two wheels constituted the principal part of his apparatus. The experiment succeeded. The vessel moved three miles an hour, and was turned or stopped at pleasure; but the Emperor Charles V., by whose order the trial was made, never followed the matter up, and De Garay and his invention were forgotten.

561. ENGINES OF DE CAUS AND BRANCA.—In 1615, De Caus, a French mathematician, devised an apparatus by which water could be raised in a tube through the agency of steam. A few years afterwards, an Italian physician, named Branca, ground his drugs by means of a wheel set in motion by steam. The steam was led from a close vessel, in which it was prepared, and discharged against flanges on the rim of the wheel.

562. THE MARQUIS OF WORCESTER'S ENGINE.—The Marquis of Worcester, by many regarded as the inventor of the steam-engine, greatly improved on the imperfect attempts of those who had preceded him.

Some say that Worcester derived his ideas from De Caus. Others claim that his invention was purely original, and the result of reflections to which he was led during his imprisonment in the Tower of London, in 1656, for plotting against the government of Cromwell. Observing how the steam kept moving the lid of the pot in which he was cooking his dinner, he could not help thinking that this power could be turned to a variety of useful purposes, and set about devising an engine in which it might be applied to the raising of water.

The Marquis of Worcester generated his steam in a boiler, and led it by pipes to two vessels communicating on one side with the reservoir from which it was to be drawn, and on the other with the cistern into which it was to be discharged.

---

559. What sort of an engine was Hero's, and what is said of it? 560. Give an account of De Garay's engine, and the experiment made with it. 561. Give an account of De Caus's engine. Of Branca's. 562. Whom do many regard as the inventor of the steam-engine? What claim has he to the honor? How was he led to reflect on the subject?

563. PAPIN'S ENGINE.—The next step was taken by Papin, who devised the mode of giving a piston an up-and-down motion in a cylinder by alternately generating and condensing steam below a piston.

564. SAVERY'S ENGINE.—Captain Thomas Savery, in 1698, constructed an engine superior to any before invented. He was led to investigate the subject by the following occurrence. Having finished a flask of wine at a tavern, he flung it on the fire, and called for a basin of water to wash his hands. Some of the wine remained in the flask, and steam soon began to issue from it. Observing this, Savery thought that he would try the effect of inverting the flask and plunging its mouth into the basin of cold water. No sooner had he done this than the steam condensed, and the water rushing into the flask nearly filled it. Confident that he could advantageously apply this principle in machinery, Savery rested not till he invented an engine which was employed with success in drawing off the water from mines.

565. The principle on which Savery's engine worked, may be understood from Fig. 221. S is a pipe connecting a boiler in which steam is generated (and which does not appear in the Figure) with a cylindrical vessel, C, called *the receiver*. I is known as *the injection-pipe*, and is used for throwing cold water into the receiver to condense the steam. The steam-pipe, S, and the injection-pipe, I, contain the stop-cocks, G, B, which are moved by the common handle, A, so arranged that when one is opened the other is closed. F is a pipe which descends to the reservoir whence the water is to be drawn, and is commanded by the valve V, opening upward. E D is a pipe leading from the bottom of the receiver up to the cistern, into which the water is to be discharged. This pipe contains the valve Q, opening upward.

Fig. 221.

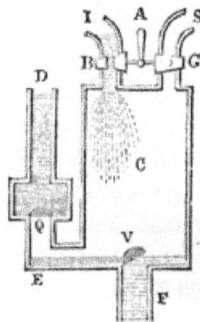

*Operation.*—To work the engine, open the stop-cock G, which of course involves the shutting of B. The steam rushes in through S, and fills the receiver C, driving out the air through the valve Q. When C is full, shut G and open B. Cold water at once enters through the injection-pipe and condenses the steam in C. A vacuum is thus formed, and the water in the reservoir or mine, under the pressure of the atmosphere, forces open the valve V, and rushes up through F into G, till the receiver is nearly filled. G is then opened and B closed; when the steam again enters through S, and by its expansive force opens the valve Q, and drives the water up through E D into the cistern.

566. NEWCOMEN'S ENGINE.—Savery's engine was employed only for raising water; but Newcomen, an intelligent blacksmith, extended its sphere of usefulness, by connecting a piston, worked up and down on Papin's principle, with a beam turning on a pivot, by means of which machinery of different kinds could be set in motion.

567. About this time, also, the engine was made self-acting through the ingenuity of Humphrey Potter, a lad employed to turn the stop-cocks Preferring play to this monotonous labor, he contrived to fasten cords m the beam to the handle of the stop-cocks, in such a way that the latter were opened and closed at the proper times, while he was away, enjoying himself with his companions. His device was after a time found out, and saved so much labor that it was at once adopted as an essential part of the machine.

568. WATT'S ENGINE.—The genius of James Watt brought the steam-engine to such perfection that but little improvement has since been made in it. Gifted with remarkable mathematical powers and a reflective mind, he commenced his experiments in 1763. Having been employed to repair one of Newcomen's engines, he soon perceived that there was a great loss in consequence of having every time to cool down the receiver from a high degree of heat before the steam could be condensed. This difficulty he remedied by providing a separate chamber called a *condenser*, to which the steam was conveyed and in which it was condensed. He also made the movement of the piston more prompt and effective by introducing steam into the cylinder alternately above and below it. The Double-acting Condensing Steam-engine, as improved by Watt, and

566. What was the only purpose for which Savery's engine was employed? Who extended its usefulness, and how? 567. Give an account of Humphrey Potter's improvement, and the circumstances under which it was devised. 568. Who brought the steam-engine to comparative perfection? When did Watt commence his experiments? What disadvantage did he perceive that Newcomen's engines labored under? How did he remedy the difficulty? What other improvement did he make?

now generally constructed for manufacturing establishments, is represented in Fig. 222.

569. *Description of the Parts.*—A is the *cylinder*, in which the *piston* T works. This piston is connected by the *piston-rod* R with the *working-beam*

Fig. 222.

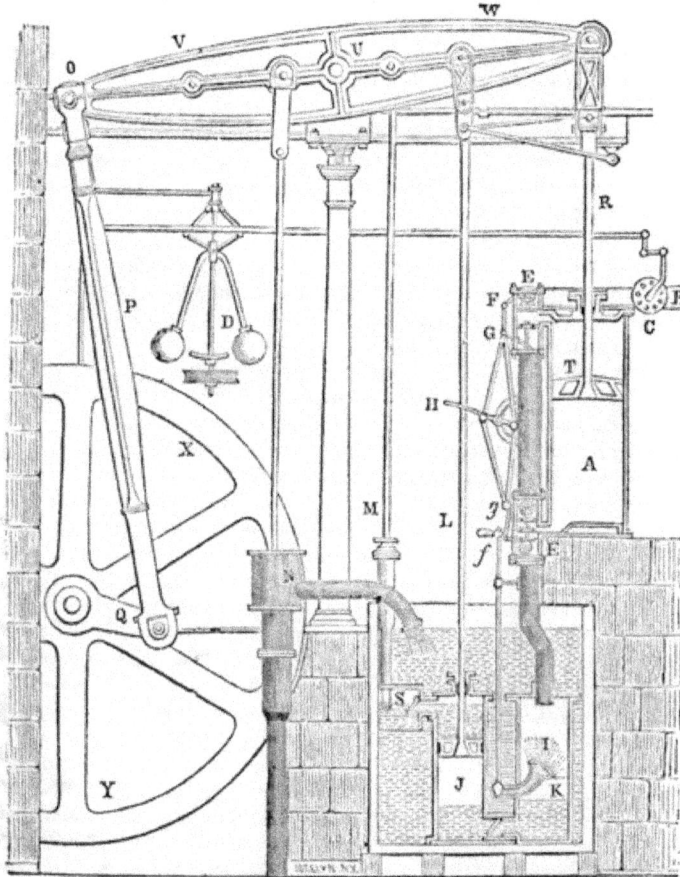

THE DOUBLE-ACTING CONDENSING STEAM-ENGINE.

V W, which turns on a pivot, U. The other end of the working-beam, O, imparts a rotary motion to the heavy *fly-wheel* X Y, by means of the *connecting-rod* P and the *crank* Q. The fly, as explained on page 125, regulates the motion, and is directly connected with the machinery to be moved. Steam

569. Describe the parts of Watt's Double-acting Condensing Engine. Show how the

is conveyed to the cylinder A from the *boiler* (which is not seen in the figure), through the *steam-pipe* B, which is commanded by the *throttle-valve* C. This valve is connected with the *governor* D, in such a way as to be opened when the supply of steam is too small and closed when it is too great.

Communicating with the cylinder at its top and bottom on the left, are two hollow *steam-boxes*, E, E, each of which is divided into three compartments by two valves. F is called the *upper induction-valve*, and opens or closes communication between the steam-pipe and the upper part of the cylinder, so as to admit or intercept a supply of steam. G, called the *upper exhaustion-valve*, opens or closes communication between the upper part of the piston and the *condenser* K, so that the steam may either be allowed to escape into the latter or confined in the cylinder. The *lower induction-valve g*, and the *lower exhaustion-valve f*, stand in the same relation to the lower part of the cylinder, the former connecting it with the steam-pipe, and the latter with the condenser K. These valves are connected by a system of levers with a common handle, H, called a *spanner*, which is made to work at the proper intervals by a pin projecting from the rod L, which is moved by the working-beam. The spanner works so as to open and close the valves by pairs. When it is pressed up, it opens F and *f*, and closes G and *g*; when pressed down, it closes F and *f* and opens G and *g*.

Below is the condensing apparatus, consisting of two cylinders, I and J, immersed in a cistern of cold water. A pipe, K, having an end like the rose of a watering-pot, conveys water from the cistern to the cylinder I (the supply being regulated by a stop-cock), and thus condenses the steam which is from time to time admitted into I. The other cylinder, J, called the *air-pump*, contains a piston with a valve in it opening upward, which works like the bucket of a common pump, and draws off the surplus water that collects at the bottom of the cylinder I into the *upper reservoir* S. The *hot-water pump* M then conveys this water to the cistern that supplies the boiler. To keep the water around the condensing apparatus at the right temperature, a fresh supply is constantly introduced through the *cold-water pump* N; which, like the hot-water pump and the air-pump, is kept in operation by rods connected with the working-beam.

570. *Operation.*—The working of the engine is as follows:—Let the piston be at the top of the cylinder, and all the space below be filled with steam. The upper induction-valve and the lower exhaustion-valve are then opened by the spanner, while the upper exhaustion-valve and the lower induction-valve are closed. By this means steam is introduced above the piston, while the steam beneath is drawn off into the condenser, where it is converted into water. The pressure of the steam above at once forces the piston to the bottom of the cylinder. Just at this moment the spanner is moved in the opposite direction, and the valves that were before opened are closed, while those that were previously closed are opened. The steam is now admitted beneath the piston, and the steam above is drawn off into the condenser and converted into water as before. While this action is going on, the cold-water pump

valves work. Describe the condensing apparatus.   570. How is the engine worked?

is constantly supplying the cistern in which the condenser is immersed; while the air-pump is drawing off the hot water from the condenser to the upper reservoir, whence it is conveyed by the hot-water pump to the cistern that supplies the boiler. An up-and-down motion is thus communicated to the piston, and by it to the working-beam, which causes the fly to revolve, and moves the machinery with which it is connected.

571. *The Governor.*—The Governor, an ingenious piece of mechanism, by which the throttle-valve in the steam-pipe is opened and closed, and the supply of steam regulated as the machinery requires, is worthy of further description.

The governor and its connection with the throttle-valve are represented in Fig. 223. It consists of two heavy balls of iron, E, E, suspended by metallic arms from the point *e*. At *e* they cross, forming a joint, and are continued to *f, f,* where they are attached by pivots to other bars, *f h, f h.* These bars are joined to one end of a lever, the other end of which, H, is connected at W with the handle of the

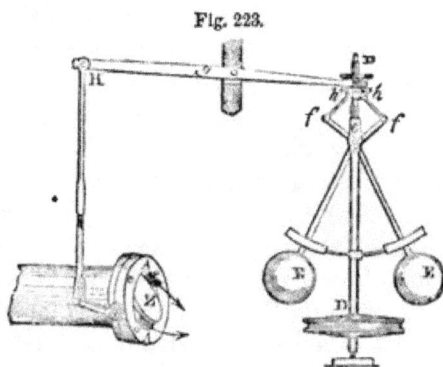

Fig. 223.

THE GOVERNOR.

valve Z. The spindle D D, to which the balls are attached, turns with the fly-wheel. When the fly-wheel revolves very rapidly, the balls E E, under the influence of the centrifugal force, fly out from the spindle, and with the aid of the bars *f h, f h,* pull down the end of the lever *g*. The other end, H, is of course raised, and with it the handle of the valve Z, which is thus made to close the mouth of the steam-pipe A and cut off the supply of steam. On the other hand, when the motion of the fly diminishes, the centrifugal force of the balls E E also diminishes, and they fall towards the spindle. The nearer end of the lever *g* is thus raised, while the end H is depressed. The valve Z is by this means opened, and admits a full supply of steam. The governor thus acts almost with human intelligence, now admitting, and now cutting off the steam, just as is required.

572. *The Boiler.*—The boiler is made of thick wrought-iron or copper plates, riveted as strongly as possible, so as to resist the expansive force of the steam generated within.

How are the cisterns supplied? 571. What is the Governor? Describe the governor, and its connection with the throttle-valve. Show the workings of the gov-

10*

The fire is applied in an apartment beneath or within the boiler called the Furnace.

Boilers are made of different shapes, but are generally cylindrical, because this form is one of the strongest. Watt made his concave on the bottom, in order to bring a greater extent of surface in contact with the flame.

573. *The Safety Valve.*—The pressure on the boiler, in consequence of the expansive force of steam, is immense. If it is allowed to become too great, the boiler bursts, often with fatal effects. To prevent such catastrophes, a Safety Valve is fixed in the upper part of the boiler, which is forced open and allows some of the steam to escape whenever the pressure exceeds a certain amount. A lever, with a weight which slides to and fro on its arm, is attached to the valve; and the engineer, by placing the weight at different distances, can determine the amount of pressure which the boiler shall sustain before the valve will open.

574. KINDS OF ENGINES.—Engines are divided into two kinds, Low Pressure and High Pressure.

In the Low Pressure Engine, one form of which has been described above, the steam is carried off and condensed; while in the High Pressure Engine it is allowed to escape into a chimney, and thence into the open air. The latter, having no condensing apparatus, is much the simpler in its construction. It is noisy when in operation, in consequence of the puffing sound made by the steam as it escapes.

575. As regards their use, engines may be divided into three classes; Stationary Engines, employed in manufacturing, Marine Engines, for propelling boats, and Locomotive Engines, for drawing wheeled carriages.

576. THE LOCOMOTIVE ENGINE.—The Locomotive is a high pressure engine. The principle on which it works may be understood from Fig. 224.

---

ernor. 572. Of what is the boiler made? Where is the fire applied? What is the usual shape of boilers? What shape did Watt make his, and why? 573. What is the use of the Safety Valve? How is it worked? 574. How are engines divided? What constitutes the difference between Low Pressure and High Pressure Engines? Which are the simpler? Which are the more noisy, and why? 575. As regards their use,

Fig. 224.

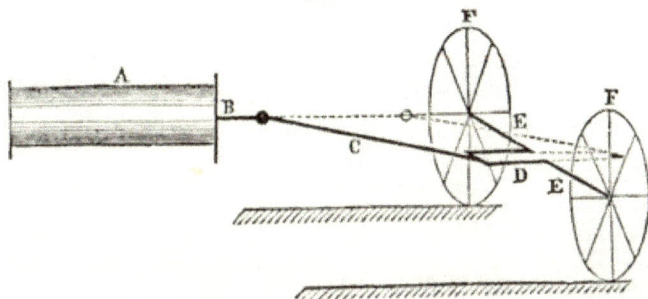

The cylinder A in this engine is horizontal instead of vertical, and the piston works horizontally. B, the piston-rod, is connected by a crank, D, with the axle E E of the wheels, F, F. The piston, moving alternately in and out of the cylinder, with the aid of the crank causes the axle and wheels to revolve; and the wheels, by their friction on the rails, move forward the engine and whatever may be attached to it. The heavy line represents the position of the parts when the piston is at the remote extremity of the cylinder; the dotted line shows their position, when the piston has reached the other end. Steam is first introduced on one side of the piston, and then on the other, being allowed to escape as soon as it has done its work,—that is, driven the piston to the opposite extremity. The rest of the machinery consists of arrangements for boiling the water, for regulating the admission of steam into the cylinder and its discharge, for providing draught for the fire, and for giving the driver the means of starting and stopping the engine, and reversing the direction of its motion.

577. *History.*—Watt seems to have been the first to conceive the idea of propelling wheeled carriages by steam; but he was so engaged in perfecting the stationary engine that he did not attempt to carry out his idea. William Murdoch, in 1784, first constructed a locomotive. Though little more than a toy, it worked successfully, and travelled so fast that on one occasion its inventor in vain tried to keep pace with it.

Eighteen years passed before any use was made of Murdoch's invention; at the end of that time, in 1802, Richard Trevithick publicly exhibited a locomotive engine, so con-

Into what three classes may engines be divided? 576. With Fig. 224, show the principle on which the locomotive engine works. What does the rest of the machinery consist of? 577. Who first conceived the idea of the locomotive engine? Who first carried out the idea? What is said of Murdoch's engine? Who exhibited an im-

structed that it could be used for transporting cars. Important modifications and improvements have since been made, for many of which the world is indebted to George Stephenson, who shares with Trevithick the honor of this great invention.

### EXAMPLES FOR PRACTICE.

1. (*See* § 510.) A joint of meat stands 2 feet from a fire, a fowl 4 feet; how does the heat which strikes the former compare with that received by the latter?

2. How does the heat which my finger receives from the blaze of a candle, when held an inch from it, compare with what it receives when held a foot from it?

3. If we were but one-fifth of our present distance from the sun, how many times as much heat would we receive from it?

4. The planet Neptune is about 30 times as far from the sun as the earth is; how does its solar heat compare with ours?

5. To receive a certain amount of heat from a fire, an object is placed 3 feet from it; to receive only one-fourth as much heat, how far from the fire must it be placed?

6. (*See* § 526.) A quantity of water at the freezing-point measures 22 gallons; how much will it measure when its temperature has increased to the boiling-point?

7. I have a vessel which holds 46 gallons; how much water at a temperature of 32° must I put in it, to exactly fill the vessel when it boils?

8. What will be the increase in measure of 18 gallons of alcohol, when raised from 32° to 212°? What will be the increase in weight?

9. (*See* § 554.) Under a pressure of one atmosphere, how many cubic inches of steam will be generated from 2 cubic inches of water? From 10 cubic inches of water?

10. If 3,400 cubic feet of steam (under a pressure of one atmosphere) be condensed, how much water will it make?

11. (*See* § 555.) Under a pressure of two atmospheres, about how many cubic inches of steam will two inches of water generate? How many, under a pressure of three atmospheres?

12. About how many cubic inches of steam will be required, to raise 10 tons 10 feet high? If the steam were condensed, how many cubic inches of water would it make?

---

proved locomotive in 1802? Who subsequently made important improvements in the locomotive?

# CHAPTER XIV.

## OPTICS.

578. OPTICS is the science that treats of light and vision.

### Nature of Light.

579. Light is an agent, by the action of which upon the eye we are enabled to see.

Light is imponderable; for it moves with great velocity, and if it had any weight, though it were ever so little, its striking force would be felt by every object with which it comes in contact. Yet it does not affect even the most sensitive balance.

580. With respect to the nature of light, two theories have been advanced, the Corpuscular and the Undulatory.

581. *Corpuscular Theory.*—The Corpuscular Theory teaches that light consists of extremely minute particles of matter, thrown off from luminous bodies, which strike the eye and produce the sensation of light, just as particles thrown off by an odoriferous substance affect the organ of smell. This theory, held as long ago as the days of Pythagoras, was received by Newton; but, failing to account for many of the facts more recently discovered in connection with light, it has now but few supporters.

582. *Undulatory Theory.*—According to the Undulatory Theory, light is produced by the undulations of an exceedingly subtile imponderable medium, known as Ether, with which space is filled; just as sound is produced by the vibrations of air. A luminous object millions of miles away causes the ether in contact with it to move in minute waves, like the surface of a pond rippled by throwing in a stone. These undulations are transmitted with inconceivable rapidity, till they reach the eye, strike the sensitive membrane that lines it, and produce the phenomena of vision. This theory, advanced by Descartes [dā-kart'], but first definitely laid down by Huygens, explains most of the phenomena of optics, and is now generally received.

---

578. What is Optics? 579. What is Light? How is it proved that light is imponderable? 580. What two theories have been advanced with respect to the nature of light? 581. State the chief points of the Corpuscular Theory. By whom was it held? 582. According to the Undulatory Theory, how is light produced? By whom was the Undulatory Theory advanced? Which of these theories is now generally received?

583. *Rays.*—Rays are single lines of light, the smallest distinct parts into which light can be resolved.

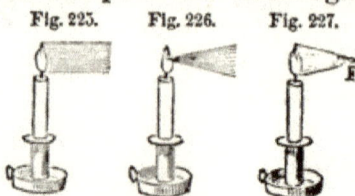

Fig. 225.  Fig. 226.  Fig. 227.

Rays of light from the same body either move in parallel lines, as in Fig. 225; or *diverge*, that is, separate from each other, as in Fig. 226; or *converge*, that is, come together at a point called the Focus, as in Fig. 227.

A Beam of light is a collection of parallel rays.

A Pencil of light is a collection of rays not parallel.

A Diverging Pencil is a collection of diverging rays.

A Converging Pencil is a collection of converging rays.

### Division of Bodies.

584. SELF-LUMINOUS AND NON-LUMINOUS BODIES.—As regards the production of light, bodies are divided into two classes, Self-luminous and Non-luminous.

Self-luminous bodies are those which are seen by the light that they themselves produce; as, the sun, the stars, a lighted candle.

Non-luminous bodies are those that produce no light of their own, but are seen only by that of other bodies. The moon is non-luminous, its light being borrowed from the sun. The furniture in a dark room is non-luminous, being invisible until the light of the sun, a lamp, or some other luminous body, is admitted.

Many non-luminous bodies, when exposed to a heat of 977° F., become incandescent, and grow brighter and brighter with every increase of temperature beyond that point, till they reach a white heat. This is a striking proof of the connection between light and heat.

585. TRANSPARENT, TRANSLUCENT, AND OPAQUE BODIES.

583. What are Rays? How may rays move? What is a Beam of light? What is a Pencil of light? What is a Diverging Pencil? What is a Converging Pencil? 584. As regards the production of light, how are bodies divided? What are Self-luminous bodies? What are Non-luminous bodies? Give examples. What striking proof have we of the connection between light and heat? 585. As regards the transmission

—As regards the transmission of light, bodies are divided into three classes; Transparent, Translucent, and Opaque.

Transparent bodies are such as allow light to pass freely through them; air, water, glass, are transparent.

Translucent bodies are such as allow light to pass through them, but not freely; ground glass, thin horn, paper, are translucent.

Opaque bodies are such as do not allow light to pass through them; wood, stone, the metals, are opaque.

*Transparent* and *opaque* are relative terms. No substance transmits light without intercepting some by the way. It is computed that the sun's rays lose nearly one-fourth of their brilliancy by passing through the earth's atmosphere; and that, if this atmosphere extended fifteen times as far from the surface as it now does, we should receive no light at all from the sun, but should be plunged in perpetual night. On the other hand, an opaque substance, if made very thin, may become transparent. Gold leaf, for instance, held in the sun's rays, transmits a dull greenish light.

586. MEDIA.—By a Medium (plural, *media*) is meant any substance through which a body or agent moves in passing from one point to another. Air is the medium in which birds fly; water, the medium in which fish swim; ether, the medium in which the planets move. In connection with light, any substance through which it passes is a medium; as air, water, glass, &c.

587. A Uniform Medium is one that is of the same composition and density throughout.

### Sources of Light.

588. The principal sources of light are nearly the same as those of heat; viz., the Sun and Stars, Chemical Action, Mechanical Action, Electricity, and Phosphorescence.

Most of our artificial light is produced by chemical action, as exhibited in the process of combustion (see § 479). To this is due the light of lamps, can-

of light, how are bodies divided? What are Transparent bodies? What are Translucent bodies? What are Opaque bodies? What is said of the terms *transparent* and *opaque*? How much of their brilliancy do the sun's rays lose in passing through the atmosphere? What would be the consequence if the atmosphere extended fifteen times as far as at present? How may an opaque substance be made transparent? 586. What is a Medium? Give examples. 587. What is a Uniform Medium? 588. Name the principal sources of light. How is most of our artificial light pro-

dles, gas, fires, &c.—The mechanical action involved in percussion is also a source of light. Sparks are produced when flint and steel are struck violently together.—Lightning and the sparks given off from the electrical machine are examples of light produced by electricity.—Phosphorescent light is unaccompanied with heat. It is seen in decayed wood, fire-flies, glow-worms, and certain marine animals. Vast tracts of ocean are sometimes rendered luminous by myriads of phosphorescent creatures.

589. THE SUN AND STARS, SOURCES OF LIGHT.—The sun has already been mentioned (§ 474) as the great natural source of heat and light to the earth. Notwithstanding the loss of some of its brightness in consequence of passing through our atmosphere, its light is more intense than any other with which we are acquainted. The most dazzling artificial lights look like black specks, when held up between the eye and the sun, so much more brilliant is the latter. It would require the concentrated brightness of 5,563 wax candles at the distance of a foot, to equal the light which we receive from the sun at a distance of 95,000,000 miles.

The fixed stars are the suns of other systems. Like our sun, they are self-luminous, and therefore sources of light, though unimportant to us as such by reason of their great distance. The light we get from Sirius, one of the brightest of the fixed stars, is only one twenty-thousand-millionth of what we receive from the sun. When the sun shines, the stars are invisible, their light being lost in his superior brightness.

The light of some of the stars is so faint, that it is entirely absorbed by the atmosphere before it reaches the eye of an observer at the level of the sea. This is the reason why more stars are visible from the top of a mountain than from its base.

590. The moon and planets are non-luminous, receiving from the sun the

duced? Give an example of light produced by mechanical action. Of light produced by electricity. What is the peculiarity of phosphorescent light? In what is it seen? 589. What is the great natural source of light to the earth? How does the sun's light compare with other lights with which we are acquainted? Prove this. To how many wax candles is the light received from the sun equal? What are the fixed stars? What renders them unimportant to us, as sources of light? How does the light of Sirius compare with that of the sun? Why are the stars invisible in the day-time? Why can more stars be seen from the top of a mountain than from its base? 590. What heavenly bodies are non-luminous? What follows with respect to

light with which they shine. This light, reflected to the earth, is much inferior in brightness to that received directly from the sun. The latter body, for example, gives us 800,000 times as much light as the moon.

## Propagation of Light.

591. DIRECTION.—*Light radiates from every point of a luminous surface in every direction.*

The flame of a candle can be seen by thousands of persons at once, because a ray from the flame meets the eye of each. Within the immense space belonging to the solar system, there is no point at which an observer can be placed without seeing the sun, provided no opaque body intervenes. From the sun, therefore, and from every luminous body, an infinite number of rays proceed.

592. *In a uniform medium, light is propagated in straight lines.*

Look through a straight tube at the sun, and you see it; not so, if you look through a bent or curved tube. Place a book between your eye and a gas-burner; the latter is not visible, because, to reach your eye, the light from it would have to deviate from a straight line. Darken a room, and admit a sunbeam through a small hole in a shutter. Its path, marked out by the floating dust that it illuminates, is seen to be a straight line.

593. The rays proceeding in straight lines from different particles of a luminous body cross at every point within the sphere of its illumination, but without at all interfering with each other; just as different forces may act on an object, and each produce the same effect as if it acted alone. A dozen candles will shine through a hole in the wall of a dark room, and each with the same intensity and direction as if no other rays than its own traversed the narrow passage.

594. VELOCITY.—Light travels with the enormous velocity of 192,000 miles in a second. While you count one, it goes eight times round the earth; it would take the swiftest bird three weeks to fly once around it. Light traverses the space between the sun and the earth in about 8 minutes; a cannon-ball would be seventeen years in going the same distance.

---

their light? How does the moon's light compare with the sun's? 591. What is the law for the direction of radiated light? Show the truth of this law in the case of a candle and the sun. 592. In a uniform medium, how is light propagated? Prove this by some familiar experiments. 593. What is said of the rays proceeding in straight lines from different particles of a luminous body? Illustrate this with candles shining through a hole. 594. What is the velocity of light? How does it compare with that of the swiftest bird? With that of a cannon-ball? By whom was the

The velocity of light was discovered accidentally, by Roemer, an eminent Danish astronomer, when engaged in a series of observations on one of the moons of the planet Jupiter. This moon, in a certain part of its path, becomes invisible to an observer on the earth, in consequence of getting behind its planet. Knowing that the revolutions of the moon must be performed in the same time, Roemer supposed that the intervals between these invisible periods would of course be uniform. To his surprise, he found that they differed a little every time; increasing for six months (at the expiration of which, the eclipse was sixteen minutes later than at first), and then decreasing at the same rate for a similar period, till at the end of a year he found the interval precisely the same as at first. The conclusion was inevitable. The discrepancy was caused by the difference in the earth's distance. If the first observation was made when the earth was at that point of her orbit which was nearest to Jupiter, six months afterwards she would be at the most distant point; and the light from Jupiter's moon, to reach the observer's eye, would have to travel the whole distance across the orbit (about 190,000,000 miles) farther than before. Here was the key to a grand discovery. If light was sixteen minutes, or 960 seconds, in travelling 190,000,000 miles, it was easy to find how far it travelled in one second.

595. INTENSITY AT DIFFERENT DISTANCES.—*The intensity of light diminishes according to the square of the distance from the luminous body that produces it.*

Let several objects be placed respectively 1 foot, 2 feet, 3 feet, &c., from a luminous body; they will then receive different degrees of light proportioned to each other as 1, $\frac{1}{4}$, $\frac{1}{9}$, &c.—A planet twice as far from the sun as the earth is, would receive from it only $\frac{1}{4}$ as much light; one three times as far, $\frac{1}{9}$ as much; one ten times as far, $\frac{1}{100}$ as much.

Fig. 228.

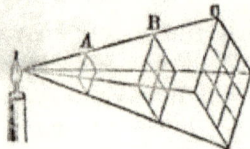

596. This is illustrated with Fig. 228. A square card placed at A, a distance of 1 foot from the candle, receives from a given point in the flame a certain amount of light. This same light, if not intercepted at A, goes on to B at a distance of 2 feet; it there illuminates four squares of the same size as the card, and has, therefore, but one-fourth of its former intensity. If allowed to proceed to C, 3 feet, it illuminates nine such squares, and has but one-ninth of its original intensity, &c.

## Shadows.

597. Light falling on an opaque body is intercepted.

The darkness thus produced behind the opaque body is called its Shadow.

598. Shadows are not all equally dark. They may be more or less illumined by reflected light or by rays from some luminous body that are not intercepted. Thus, if there are two lighted candles in different parts of a room, the shadow cast by either is less dark than if it were burning alone. Again, the brighter the light that produces a shadow, the darker it appears by contrast. Hence, to compare the intensity of different lights, observe the shadows respectively cast at equal distances; the one that throws the darkest shadow is the brightest light.

599. When the luminous body is larger than the opaque body it shines on, the latter throws a shadow smaller than itself; and this shadow diminishes according to the distance of the surface on which it is thrown.

In Fig. 229, let A be a luminous, and B an opaque, body. B's shadow, no matter how near the surface on which it is thrown, must be smaller than B itself; and, as the surface is removed from B, the shadow diminishes, till it is reduced to a point at C.

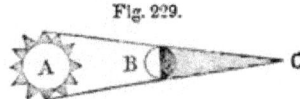

Fig. 229.

If, on the contrary, the opaque body is the larger of the two, it throws a shadow greater than itself; and this shadow increases according to the distance of the surface on which it is thrown.

600. THE PENUMBRA.—Every luminous body has an infinite number of points, from each of which proceeds a pencil of rays. When an opaque body is interposed, some of the space behind it is cut off from all the rays of the luminous body, and this constitutes the shadow proper. Part of the space, however, while it is cut off from some of the rays, is illumined by others; this is called the Penumbra.

Fig. 230.

In Fig. 230, let O P be the flame of a candle, and AB an opaque object placed before it. The space A B C D is not reached

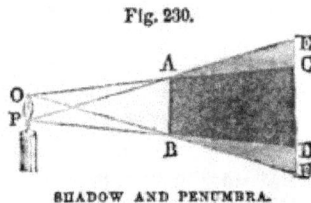

SHADOW AND PENUMBRA.

meant by a body's Shadow? 598. Why are not all shadows equally dark? How may we compare the intensity of different lights? 599. When does a body throw a shadow smaller than itself? Illustrate this law with Fig. 229. When does a body throw a shadow larger than itself? 600. What is meant by the Penumbra? How is

by any ray from O P, and is therefore the Shadow of A B. The space A E C, while it is cut off from the rays produced by the lower extremity of the flame, is illumined by its upper extremity; hence it is nowhere so dark as the shadow, and becomes lighter and lighter as the line AE is approached. So the space B D F is cut off from the rays produced by the upper part of the flame, but receives those from the lower part, and is therefore partially illuminated. The spaces A C E, B D F, constitute the Penumbra, or imperfect shadow, of A B.

## Reflection of Light.

601. When light strikes an opaque body, some of it is absorbed, and some reflected, or thrown back into the medium from which it came. According to the Undulatory Theory, we should say that some of the undulations that strike the opaque body are brought to rest, while others are reproduced in the same medium with a different direction from what they had before.

The reflection of light is analogous to the reflected motion of an india rubber ball thrown against a solid surface. It is by the light irregularly reflected from their surfaces that all non-luminous bodies are seen.

Transparent surfaces, as well as opaque, reflect some of the light that strikes them; otherwise, they would not be visible. We see overhanging objects mirrored in a stream with great distinctness, because a portion of the rays received from them are reflected by the water to our eyes.

602. That branch of Optics which treats of the laws and principles of reflected light, is called CATOPTRICS.

603. Rays that strike a body are called Incident Rays.

604. REFLECTIVE POWER OF DIFFERENT SURFACES.— Different surfaces reflect the light that strikes them in different degrees. By none is the whole reflected.

If any surface were a perfect reflector,—that is, threw back all the light that struck it,—the eye would fail to distinguish it. Looking at such a surface, we should see nothing but images of the bodies that produced the incident rays. If, for example, the moon reflected all the light it received, it would have the appearance of another sun. It is because there is not a

it produced? Illustrate the mode in which the shadow and penumbra are produced, with Fig. 230. 601. When light strikes an opaque body, what becomes of it? Express this according to the Undulatory Theory. To what is the reflection of light analogous? How are non-luminous bodies seen? Is the reflection of light confined to opaque surfaces? Prove that it is not. 602. What is Catoptrics? 603. What is meant by Incident Rays? 604. What is said of the reflection of light from different surfaces? If any surface were a perfect reflector, what would be the

perfect and regular reflection that the non-luminous bodies which meet the eye every moment are visible.

Though incident light is never wholly reflected, yet from some surfaces it is thrown off with a high degree of regularity, and with its intensity diminished comparatively little. If, for instance, we look at a good plate-glass mirror hung opposite to us at the end of a room, we can hardly persuade ourselves that there is not another apartment beyond, the counterpart of the one which we are in. The surface of the mirror is not seen at all, in consequence of its great reflective power.

605. The proportion of incident light reflected depends on two things :—1. The angle at which it strikes the surface. 2. The character of the surface.

The more obliquely light strikes a surface, the greater is the quantity reflected.

In Fig. 231, let C D be a surface of polished black marble. A and B are incident beams, with an intensity rated at 1,000. Let B strike the marble at an angle of 3 degrees, and a beam having an intensity of 600 will be reflected. Let A strike it at an angle of 90 degrees, and the reflected beam will have an intensity of only about 20.

Fig. 231.

Light-colored and polished surfaces reflect a much greater proportion of incident light than dark and dull ones. Here again the laws of light and heat agree.

A room with white walls is much lighter than one with black or dark-colored walls. A house painted some light color, or a dome covered with polished tin, is more readily seen from a distance than a dark wall or an ordinary roof.

606. MIRRORS.—The laws of reflected light are best investigated and explained with the aid of mirrors.

607. Mirrors are solids with regular and polished surfaces, having a high degree of reflective power. They are made either of some metal susceptible of a high polish, such as silver and steel, or of clear glass covered on the back with silver or a mixture of tin and mercury. A metallic mirror is sometimes called a Speculum (plural, *specula*).

consequence ? What is said of the reflective power of some surfaces, such as a good plate-glass mirror ? 605. On what does the proportion of incident light reflected depend ? At what angle is the most incident light reflected ? Illustrate this with Fig. 231. What sort of surfaces reflect the most incident light ? 606. With what are the laws of reflected light best investigated ? 607. What are Mirrors ? Of what are they

From glass mirrors there are two reflections; one from the surface first struck, the other from the back coated with mercury. Hence two images of an object before the mirror are presented, the distance between them being equal to the thickness of the glass. But the image produced by the front surface is always faint; and, when the back is well coated, the other image is so much superior that the faint one is entirely lost.

608. *Kinds of Mirrors.*—As regards shape, mirrors are divided into three classes; Plane, Concave, and Convex.

A Plane Mirror (A B, in Fig. 232) is one that reflects from a flat surface, like a common looking-glass.

A Concave Mirror (E F, in Fig. 233) is one that reflects from a curved surface hollowing in like the inside of the peel of an orange.

A Convex Mirror (C D, in Fig. 234) is one that reflects from a curved surface rounding out like the outside of an orange.

A concave mirror polished on both sides becomes a convex mirror when its opposite side is presented to the incident rays.

609. GREAT LAW OF REFLECTED LIGHT.—The law of reflected light is like that of reflected motion :—*The angle of reflection is always equal to the angle of incidence.* This law holds good whether the reflecting surface is plane, concave, or convex.

Fig. 232.     Fig. 233.     Fig. 234.

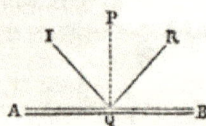 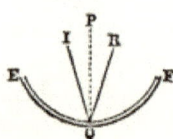 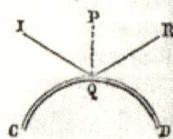

Figs. 232, 233, 234, illustrate this law. In each Figure, I represents the incident ray, R the reflected ray, and P a perpendicular. I Q P, the angle which the incident ray makes with the perpendicular, is called the angle of incidence. R Q P, the angle which the reflected ray makes with the same perpendicular, is the angle of reflection. From every surface, whatever its form, the incident ray is thrown off in such a way as to make the angle of reflection equal to the angle of incidence.

made? What is a Speculum? How many reflections are there from glass mirrors? How are they produced? What is said of the images formed? 608. As regards shape, how are mirrors divided? What is a Plane Mirror? What is a Concave Mirror? What is a Convex Mirror? How may a concave mirror polished on both sides be made a convex mirror? 609. State the law of reflected light. Illustrate this with

610. From these Figures it is obvious that an object which would not otherwise be visible can be seen by reflection from a mirror. Thus, let the upper part of P Q represent an opaque screen, I an object on one side of it, and R the eye of an observer on the other. I is not visible to a person at R looking directly at it, on account of the interposition of the screen; but, as the angle of reflection is always equal to the angle of incidence, it can be seen from R by looking at the mirror.

611. IMAGES.—By the Image of an object is meant a luminous picture of it formed by rays proceeding from its different points. An image is said to be inverted when it represents its object as upside down,—that is, with its lowest part uppermost.

Fig. 235.

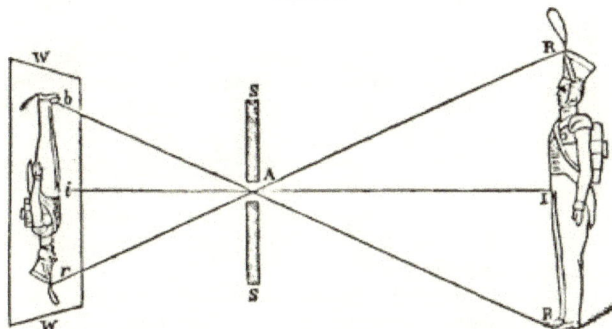

Fig. 235 illustrates the formation of an image. R B represents a soldier with a red coat and blue trowsers standing in strong sunlight opposite the white wall W. Let the shutters S S be thrown open, and not only the light reflected from the person of the soldier, but also other rays, enter the apartment, making its light a mixture of all colors, or white, in which the red and the blue tinge of the dress are lost, and no image is formed. Now let the shutters S S be closed, leaving at A an exceedingly small aperture, through which the rays reflected from the figure are allowed to reach the wall. As light is propagated in straight lines, the ray R will strike the wall at r, B at b, and I at i. The image will therefore be inverted; and, as each ray retains its color, the coat will remain red and the trowsers blue. This experiment confirms two principles already stated:—1. That every ray moves in a straight line; 2. That an infinite number of rays may cross each other without interfering with the effect which each would separately have.

612. Images formed by apertures are always inverted.

**613. Reflection from Plane Mirrors.**—Plane mirrors do not alter the relative direction of incident rays. If the incident rays are parallel, they will remain parallel after reflection; if divergent, they will continue to diverge; if convergent, they will continue to converge.

614. Objects seen in a plane mirror seem to lie in the direction of the reflected rays that meet the eye, and to be as far behind the mirror as they really are in front of it. These principles are illustrated with Fig. 236.

Fig. 236.

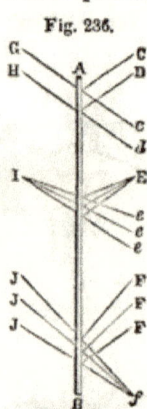

A B is a plane mirror. C, D, are parallel rays striking its surface. They are reflected in parallel lines to c, d; and to an observer at those points will appear to come from G, H, as far behind the mirror as C, D, are in front of it.

E is a diverging pencil. After reflection, its rays continue to diverge to e, e, e; and to an observer there they appear to diverge in unbroken straight lines from the point I, as far behind the mirror as E is before it.

F, F, F, represent converging rays. After reflection, they continue to converge, and meet at the point f. An observer at f would suppose them to come in unbroken lines from J, J, J, as far behind the mirror as F, F, F, are in front of it.

615. When we walk towards a looking-glass, our image seems to advance towards us; and when we recede from it, the image also recedes. The image always appears to be the same distance from the mirror that the object is.

616. The angle of reflection being equal to the angle of incidence, it follows that a person may see his whole figure

Fig. 237.

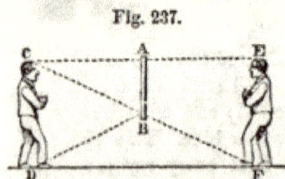

reflected from a mirror whose length is but half his own height. In Fig. 237, C D represents a man standing before the mirror A B. The incident ray from the head C strikes the mirror perpendicularly, is reflected in the same line, and appears to come

612. What kind of images are formed by apertures? 613. What effect have plane mirrors on the relative direction of incident rays? 614. How do objects seen in a plane mirror seem to lie? With Fig. 236, illustrate the reflection of parallel, diverging, and converging rays from a plane mirror. 615. When we approach and recede from a looking-glass, what phenomena are presented? 616. How is it that a person can see his whole figure reflected from a mirror whose length is but half his height?

from E. The ray from his foot D strikes the mirror at B, is reflected at an equal angle to his eye, and appears to come in an unbroken line from F. The extremities of his person being seen, the intermediate parts are also visible, forming a complete image.

617. *Images formed by Plane Mirrors.*—The size of images formed by plane mirrors is not changed, except so far as they seem smaller in consequence of their apparent distance behind the mirror.

618. As the image faces the opposite way from the object, if the mirror is vertical (that is, perpendicular to the floor), the right side of the object will be the left of the image, and the left side of the object the right of the image. If a person stands before a mirror with a book in his right hand, the book seems to be in the left hand of his image; and, if he brings the printed page near the mirror, he can not read it, for the reflection turns about both letters and words, side for side.

Place the same plane mirror in a horizontal position (that is, lay it on the floor with its face up), and the image, which before simply had its sides transposed, now becomes inverted, or seems to stand on its head. On the same principle, a tree or other object reflected from the surface of a pond, is inverted.

619. *The Kaleidoscope.*—When an object is placed between two parallel plane mirrors, each produces an image of its own, and reproduces the image reflected to it from the other. This image of an image is again reflected by each to the other, and thus a series of images is produced, till the rays become so faint by successive reflections as to be no longer discernible.

When the mirrors are placed at right angles to each other, an object between them forms three images,—one produced by each separately, and one by a twofold reflection from both. Placed so as to form with each other an angle of 60 degrees, the two mirrors will produce five images; at 45 degrees, seven.

This principle is applied in the Kaleidoscope [*ka-li'-do-scope*], a beautiful toy invented by Sir David Brewster.

617. What is said of the size of images formed by plane mirrors? 618. If the mirror is vertical, how does the image differ from the object? How, if the mirror is horizontal? 619. What takes place when an object is placed between two parallel plane mirrors? How many images are formed when the mirrors are placed at right angles

620. The kaleidoscope consists of two narrow strips of glass running lengthwise through a tube, and forming with each other an angle of 60 or 45 degrees. One end of the tube, to which the eye is to be applied, is covered with clear glass. The other end terminates in a cell formed by two parallel pieces of glass an eighth of an inch apart, the outer one of which is ground to prevent external objects from marring the effect. This cell contains beads or small pieces of glass of different colors, free to move among themselves. On applying an eye to the tube, we see the objects in the cell multiplied by repeated reflections from the mirrors, and symmetrically arranged, with their images, around a common centre. By shaking the tube, we bring the objects into new relative positions, and have new combinations presented.

621. *The Magic Perspective.*—By arranging four plane mirrors as represented in Fig. 238, a person is enabled to see an object by looking directly towards it, though an opaque screen is interposed.

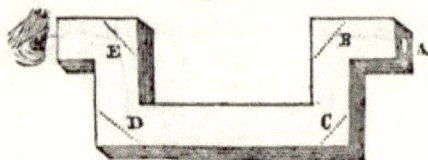

Fig. 238.

THE MAGIC PERSPECTIVE.

A rectangular box is bent four times at right angles; and in each of these angles is placed a piece of looking-glass, B, C, D, E, at such an inclination that the incident ray may strike it at an angle of 45 degrees. Any object opposite the aperture A is visible to an eye applied at the other extremity, though an opaque screen be placed between the arms of the instrument. The rays from the object first strike B at an angle of 45 degrees, and are reflected at the same angle to C, thence to D, thence to E, and finally to the observer's eye. The inventor of this instrument recommended its use in time of war, for discovering an enemy's movements without any exposure of the observer's person. It is more commonly used, however, by itinerant showmen, who for a penny allow the curious to read through a brick.

622. REFLECTION FROM CONCAVE MIRRORS.—In general, the effect of concave mirrors is to make incident rays more convergent or less divergent. In most cases, the images they produce appear in front of them.

623. Parallel rays striking a concave mirror are made to converge to a point called the Principal Focus. This

to each other? How many, when they form an angle of 60 degrees? Of 45 degrees? In what is this principle applied? 620. Describe the Kaleidoscope. 621. How is a person enabled to see an object by looking towards it, though an opaque screen is interposed? Describe the Magic Perspective. By whom is it commonly used? 622. What is the general effect of concave mirrors? What is said of the images they

point is half way between the surface of the mirror and the
centre of the sphere which the mirror would form if it were
extended with uniform curvature.

In Fig. 239, let A E B be a concave mirror, forming part of the surface of a sphere,
of which C is the centre. The parallel rays
$d, e, f, g, h$, are reflected to the principal focus F, midway between the surface and the
centre C.

Fig. 239.

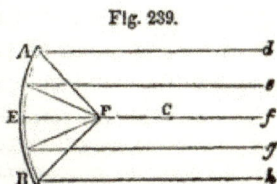

Not only is light concentrated at the focus, but also heat, as we had occasion to
note in § 476. Tinder, wood, or any other combustible material, is readily
ignited, and with a combination of such mirrors the most intense heat can be
produced. Hence concave mirrors are sometimes called Burning Glasses.

624. Converging rays reflected from a concave mirror
are made to converge more.

625. Diverging rays reflected from concave mirrors are
differently affected according to the position of the point
from which they diverge.

626. Diverging rays starting from the principal focus
are made parallel. This is obvious from Fig. 239. The
rays diverging from F, after striking the mirror, are reflected in parallel lines to $d, e, f, g, h$.

This principle is turned to account in light-houses. The light is placed in
the focus of a concave mirror, and its rays are reflected in parallel lines from
every point of the mirror's surface. No image of the light is produced, but
the whole surface of the mirror appears illuminated.

627. Diverging rays coming from a point between the
principal focus and the mirror, become less divergent after
reflection. An object in such a position forms an image
larger than itself, which seems to be situated behind the
mirror.

628. Diverging rays coming from a point between the

produce? 623. What effect has a concave mirror on parallel rays that strike it?
How is the principal focus situated? Illustrate this effect with Fig. 239. What are
concave mirrors sometimes called, and why? 624. What is the effect of concave mirrors on converging rays? 626. What is the effect of concave mirrors on diverging
rays starting from the principal focus? How is this principle turned to account?
627. What effect have concave mirrors on diverging rays coming from a point between the principal focus and the mirror? What kind of an image is formed?
628. What effect have concave mirrors on rays diverging from a point between the

principal focus and the centre, converge, after reflection, to a focus on the other side of the centre. An inverted image will there be visible, suspended in the air. This image is made more distinct, and its effect greatly increased, by causing a cloud of thin bluish smoke to rise about the spot from a chafing-dish placed beneath.

By concealing with screens the mirror, the object, and the light that illumines it, and allowing the reflected rays to pass through an aperture, we may give the image all the appearance of reality. The observer beholds delicious fruit hanging in the air without any visible support, and can hardly convince himself that it is a delusion, even when he tries to grasp it without success. He sees a pail full of water standing bottom upward without spilling its contents, and men with every semblance of life walking on their heads. It was with apparatus of this kind that the pretended magicians of the Middle Ages wrought many of their miracles, terrifying the uninitiated with sudden apparitions of skulls, drawn swords, skeletons, ghosts, &c.

629. Diverging rays coming from the centre are reflected by a concave mirror back to the same point. Here, as in all other cases, the angle of reflection is equal to the angle of incidence. Striking the surface at right angles, they are reflected at right angles back to the centre.

630. Diverging rays coming from a point beyond the centre, after reflection by a concave mirror, converge to a point on the other side of the centre. In this case, the image is inverted and smaller than the object.

631. REFLECTION BY CONVEX MIRRORS.—In general, the effect of convex mirrors is to make incident rays more divergent or less convergent. The images they produce, like those of plane mirrors, seem to stand behind them, and are generally smaller than the objects they represent.

632. Parallel rays striking a convex mirror are made to diverge, as if they proceeded from a point on the opposite side of the mirror, called the Virtual Focus. This point is

principal focus and the centre? What sort of an image is formed? How is the image made more distinct? How may wonderful effects be produced with this mirror? By whom was apparatus of this kind employed? 629. What is the effect of concave mirrors on diverging rays coming from the centre? 630. What is their effect on diverging rays coming from a point beyond the centre? In this case, what kind of an image is produced? 631. What is the general effect of convex mirrors? What is said of the images they produce? 632. What is the effect of a convex mirror on parallel

half way between the mirror and the centre of the sphere which the mirror would form, if it were extended with uniform curvature.

In Fig. 240, let A B represent a convex mirror forming part of the surface of a sphere, of which C is the centre. The parallel rays $a$, $b$, $c$, $d$, $e$, diverge after reflection to $f$, $g$, $c$, $h$, $i$, as if they had come from the virtual focus F on the other side of the mirror. F is half way between the mirror and its centre C.

Fig. 240.

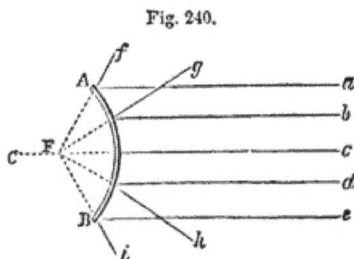

633. Diverging rays falling on a convex mirror are made more divergent by reflection. Converging rays are made less convergent, in some cases even becoming parallel.

## Refraction of Light.

634. When light strikes a transparent body, some of it is reflected and makes the body visible. The rest enters the body, and is partly absorbed and partly transmitted through it. According to the undulatory theory, we should say that some of the undulations that strike the transparent body are reproduced in the same medium with a change of direction, while others are brought to rest within the body, and others again are transmitted through it with certain modifications.

We have treated of that portion of the light which is reflected; we must now look at that which enters the transparent body.

635. When a boy rowing a boat brings his oar into the water, it no longer looks straight, but broken at the point where it enters. The same appearance is presented when he plunges a spoon or cane obliquely in a pail of water. On taking out the oar, the spoon, and the cane, they look perfectly straight again. It is evident, therefore, that the rays coming from the parts

rays? Where does the virtual focus lie? Illustrate the effect of convex mirrors on parallel rays, with Fig. 240. 633. What is the effect of convex mirrors on diverging rays? On converging rays? 634. When light strikes a transparent body, what becomes of it? Express this according to the Undulatory Theory. 635. Give some familiar examples which prove that rays are bent on passing from one medium to an-

immersed are turned from their course on entering the air, so that the points from which they come appear to lie where they do not really lie. Rays thus turned from their course are said to be *refracted*.

636. Refraction is that change of direction which a ray of light experiences on passing obliquely from one medium to another.

For an example, see the ray A in Fig. 241. If there were no water in the vessel, it would go on in a straight line to B; when the vessel is filled, it is refracted to C.

637. That branch of Optics which treats of the laws and principles of refracted light, is called Dioptrics.

638. REFRACTIVE POWER OF DIFFERENT MEDIA.—All media do not have the same refractive power. Rays of light falling from the air on water, alcohol, glass, and ice, are turned from their course in different degrees by each.

A medium that has great refractive power is said to be *dense;* one that has but little, is called *rare.*. The terms *dense* and *rare*, therefore, applied to media in Optics, have a different meaning from that which they convey in other departments of Natural Philosophy.

As a general rule, those media are the densest that have the greatest specific gravity; and, of media having about the same specific gravity, the most inflammable is the densest. The following substances are arranged according to their refractive power, chromate of lead, a transparent solid, being the densest :—Chromate of lead, diamond, phosphorus, sulphur, mother-of-pearl, quartz, amber, plate-glass, olive oil, alcohol, water, ice, air, oxygen, hydrogen.

639. LAWS OF REFRACTED LIGHT.—1. *In a uniform medium, there is no refraction. It is only on passing from one medium (or stratum of a medium) to another, that a ray is turned from its course.*

2. *Only such rays as enter a medium obliquely are refracted,—not such as enter at right angles.* .

3. *When a ray passes obliquely from a rarer to a denser*

other. What term is applied to such rays? 636. What is Refraction? Illustrate this definition with Fig. 241. 637. What is Dioptrics? 638. What is said of the refractive power of different media? What is a Dense Medium? What is a Rare Medium? What is said of the meaning of the terms *dense* and *rare* in Optics? As a general rule, what media are the densest? Mention some substances in the order of their refractive power? 639. What is the first law of refracted light? The second? The

*medium, it is refracted towards a line perpendicular to the surface.* In Fig. 241, let the ray A pass from air, a rarer medium, into water, a denser medium, and instead of going on in a straight line to B, it will be refracted to C, nearer the perpendicular.

Fig. 241.

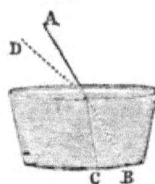

4. *When a ray passes from a denser medium into a rarer, it is refracted from the perpendicular.* In Fig. 241, let the ray B pass obliquely from water into air, and instead of going on in a straight line to A, it will be refracted to D, farther from the perpendicular.

640. An interesting experiment which every pupil may perform for himself, admirably illustrates refraction, and proves the last law to be true. Place a coin on the bottom of an empty vessel (see Fig. 242), and fix the eye in such a position that it just misses seeing it on account of the vessel's side coming between. Keep the eye there, and let water be poured in; the coin will then become visible, the rays from its surface being refracted so as to meet the eye. The coin will appear to lie at N, some distance above the bottom of the vessel; because the rays from it that last meet the eye, if continued in straight lines, would go on to that point.

Fig. 242.

The change caused by refraction in the apparent position of an object often misleads persons standing on the bank of a sheet of water as to its depth. Objects on the bottom seem to be several feet nearer the surface than they are, and bathers, deceived by the appearance, venture beyond their depth and are drowned.

641. ATMOSPHERIC REFRACTION.—Rays from the heavenly bodies, on entering our atmosphere obliquely from a rarer medium, are refracted towards the perpendicular. Hence we never see these bodies in their real position, except when they are directly over head.

The sun is visible to us some time before he really rises above the horizon, and remains visible at night after he has sunk below it. We owe our twilight to successive reflections and refractions of his rays by atmospheric strata of different densities, after he has disappeared.

third? The fourth? Illustrate the third and the fourth law with Fig. 241. 640. What interesting experiment illustrates refraction? How are persons standing on the bank of a sheet of water often deceived? 641. When do we see the heavenly bodies in their real position? Why, at other times, do we not see them in their real position?

642. *Mirage.*—Different strata of the atmosphere differ in their refractive power. Accordingly, rays from an object below the horizon (that is, concealed from us by the roundness of the earth) may, under peculiar circumstances, by successive refractions through different strata, be made to describe a curve to our eyes, and will in that case appear to come from a distant point in the air lying in the direction of the line described by the ray as it entered the eye. Such is the origin of the phenomenon called Mirage [*me-rahzh'*].

Mirage is the appearance in the air of an erect or inverted image of some distant object which is itself invisible. It is most frequently seen on the water, but has also appeared to persons travelling through deserts, with such vividness as to make them believe that they saw trees and springs before them in the distance.

Mirage is sometimes remarkably distinct at sea. Captain Scoresby, on one occasion, in a whaling-ship, recognized his father's vessel, when distant from him more than 30 miles (and consequently below the horizon), by its inverted image in the air, though he did not previously know that it was cruising in that part of the ocean. Another notable case occurred on the coast of Sussex, England. Cliffs were distinctly seen in the air; and the sailors, crowding to the beach, recognized different parts of the French shore, distant from 40 to 50 miles. These phenomena are comparatively frequent in the Strait of Messina, and as there exhibited have been called Fata Morgana [*fah'-tah mor-gah'-nah*].

643. REFRACTION BY PRISMS AND LENSES.—Prisms and lenses are much used in experimenting on light and in the construction of optical instruments.

Fig. 243.

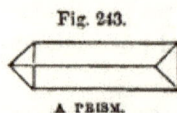

A PRISM.

644. *Prisms.*—A Prism (see Fig. 243) is a solid piece of glass, having for its sides three plane surfaces and for its ends two equal and parallel triangles.

645. A ray of light falling on a prism must pass through two of its surfaces. If it strike both of them obliquely, it

To what do we owe our twilight? 642. Explain how an object below the horizon is rendered visible. What phenomenon is thus produced? What is Mirage? Where is it seen? What case of mirage is recorded by Captain Scoresby? What other notable case is mentioned? Where are these phenomena frequent? 643. What are much used in experimenting on light? 644. What is a Prism? 645. What is the effect of

will be twice refracted; if it strike one surface perpendicularly and the other obliquely, it will be refracted but once. In either case, the object from which it comes will appear to lie in a position more or less removed from its real one.

Fig. 244 shows the refractive effect of a prism. A ray from E, entering the prism A B C, from air, a rarer medium, is refracted to D, and on passing back into the rarer medium, at that point is refracted to the eye. The object from which it comes appears to lie at F, in the direction from which the ray entered the eye. Had there been but one refraction, it would still have appeared elevated above its real position, but not so much.

Fig. 244.

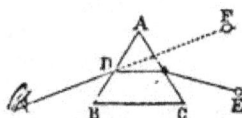

646. *Lenses.*—A lens is a transparent body which has two polished surfaces, either both curved or one curved and the other plane. The general effect of lenses is to refract rays of light, and magnify or diminish objects seen through them. They are generally made of glass; but in spectacles rock crystal is sometimes used instead of glass, because it is harder and less easily scratched.

647. *Classes of Lenses.*—Lenses are divided into six classes according to their shape. Fig. 245 shows these six classes. The name of each is given on one side, and a description of it on the other.

Fig. 245.

| DOUBLE CONVEX LENS. | | Both sides convex. |
|---|---|---|
| PLANO-CONVEX LENS. | | One side convex, the other plane. |
| MENISCUS. | | One side convex, the other concave. Thickest in the middle. |
| DOUBLE CONCAVE LENS. | | Both sides concave. |
| PLANO-CONCAVE LENS. | | One side concave, the other plane. |
| CONCAVO-CONVEX LENS. | | One side concave, the other convex. Of uniform thickness, or thickest at the ends. |

a prism on a ray of light? Show this effect with Fig. 244. 646. What is a lens? What is the general effect of lenses? Of what are they made? 647. Into how many classes are lenses divided? Name them. Describe the Double Convex Lens. The Plano-convex. The Meniscus. The Double Concave Lens. The Plano-concave.

The first three of the above lenses, which are thickest in the middle, are called Convex Lenses, and their effect is to make rays passing through them incline more towards each other. The next two (the double concave and plano-concave) which are thinnest in the middle, are called Concave Lenses, and their effect is to make rays passing through them incline farther from each other.

The concavo-convex lens, when its two surfaces are parallel (as in the above Figure) does not change the direction of rays passing through it, for the convergent effect of the convex surface is nullified by the divergent effect of the concave surface. When the convex surface has a greater curvature than the concave, this lens becomes a meniscus. When the concave surface has the greater curvature, it becomes a concave lens, and participates in the properties of that class.

648. *Refraction by Convex Lenses.*—The general effect of convex lenses is threefold :—1. They make rays passing through them incline more towards each other than before. 2. They enable us to see objects which are invisible to the naked eye on account of their distance. 3. They magnify objects seen through them.

649. A double convex lens of glass, with sides equally convex, brings parallel rays passing through it to a focus at the centre of the sphere, of which the surface of the lens first struck by the rays forms a part. This is shown in Fig. 246. Converging rays would be brought to a focus between the centre and the lens ; diverging rays, on the other side of the centre.

Fig. 246.                              Fig. 247.

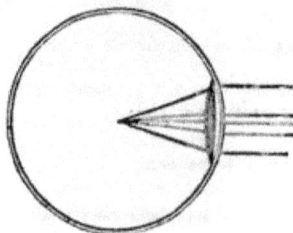 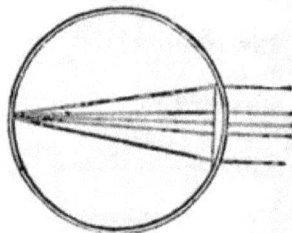

The Concavo-convex. What are the first three of these lenses called ? What is their effect ? What are the double concave and the plano-concave lens called ? What is their effect ? What is the effect of the concavo-convex lens, when its two surfaces are parallel ? When the convex surface has a greater curvature than the concave ? When the concave surface has a greater curvature than the convex ? 648. What is the general effect of convex lenses ? 649. What is the effect of a double convex glass lens on parallel rays passing through it ? On converging rays ? On diverging rays ?

A plano-convex lens brings parallel rays to a focus at a distance from the lens about equal to the diameter of the sphere of which the convex surface of the lens forms a part. This is shown in Fig. 247.

650. Convex lenses collect heat as well as light at their focus. Hence they are sometimes called Burning Glasses. Hold an old person's eye-glass in the sun-shine a short distance from your hand. A bright spot of light marks the focus, and the heat at that point soon becomes too great to be borne. All the rays that fall on the surface of the lens being concentrated in this one point, *the heat at the focus is as many times greater than the heat of ordinary sun-light as the area of the lens is greater than the area of the focus.* If the area of the lens be 100 square inches, and that of the focus ¼ of an inch, the ordinary heat of the sun will be increased 400 times.

651. The second effect of convex lenses follows from the first. Light, it will be remembered, diminishes in intensity according to the square of the distance from the luminous body; hence rays from exceedingly remote stars become so faint by the time they reach the eye as not to produce the sensation of vision. A convex glass concentrates a great number of these faint rays, and thus renders the distant object visible to an eye placed at its focus.

652. The third effect of convex lenses is to magnify objects seen through them. Hence they are sometimes called Magnifying Glasses. The glasses used by old persons, as well as by engravers and others who have to deal with minute objects, are convex lenses.

653. *Refraction by Concave Lenses.*—The effects of concave lenses are opposite to those of convex. 1. They make rays passing through them incline farther from each other. 2. They diminish objects seen through them.

654. All the above laws relating to prisms and lenses apply to rays passing into them from a rarer medium, such as air. If they come from a denser medium, the results will be reversed,—convex lenses will have a diverging and diminishing effect, while concave lenses will have a converging and magnifying effect.

What is the effect of a plano-convex lens on parallel rays? 650. What are convex lenses sometimes called, and why? How may their concentration of heat be shown? How does the heat at the focus compare with that of ordinary sun-light? 651. Show how a convex lens enables us to see distant heavenly bodies that would otherwise be invisible. 652. What is the third effect of convex lenses? What are they sometimes

**655.** *Glasses with Parallel Surfaces.*—When rays pass through a refracting medium having parallel surfaces, they leave it, not exactly in the same line, but in a direction parallel to that in which they entered it. The last refraction nullifies the change of direction produced by the first. Hence we see objects through a pane of window-glass very nearly in their real position. Irregularities in the glass cause objects seen through it to look distorted.

**656.** *The Multiplying Glass.*—If a plano-convex lens have its convex surface ground into several flat surfaces, an object seen through it will be multiplied as many times as there are flat surfaces.

Fig. 248.

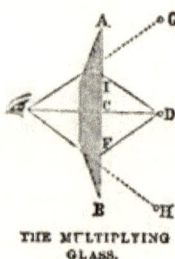

THE MULTIPLYING GLASS.

In Fig. 248, A B represents a multiplying glass, and D an object viewed through it. The ray D C, striking both surfaces perpendicularly, reaches the eye without refraction; but D I and D F, falling obliquely, suffer two refractions, which bring them also to the eye at the focus. As objects are always seen in the direction in which their rays enter the eye, three objects like D will be visible: one at D, in its real position; the others, in the direction of the dotted lines, at G and H.

**657. DOUBLE REFRACTION.** — Certain substances (chiefly minerals) have the property of causing rays which pass through them to take two distinct paths, and thus produce two images. This phenomenon is called Double Refraction.

Fig. 249.

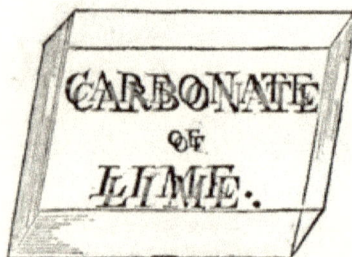

A crystal of carbonate of lime, commonly called Iceland Spar, is one of the best substances for exhibiting double refraction. Let it be placed over a piece of paper containing lines, and each line will be seen double, as shown in Fig. 249.

Keeping the same side on the paper, and turning the crystal round on its axis, we find that the double lines continue parallel, but that the distance between them varies,—diminishing till they coincide, then increasing; then diminishing till they coincide again, and then once more increas-

called in consequence? 653. What are the general effects of concave lenses? 654. In what case do the above laws relating to prisms and lenses apply? Suppose the rays pass into them from a denser medium, what will be the result? 655. What effect has a refracting medium with parallel surfaces on incident rays? How do we see objects through a pane of window-glass? 656. How is the multiplying glass formed? How many times is an object seen through it multiplied? Show this with Fig. 248. 657. What is Double Refraction? How is it exhibited with Iceland spar? What pho-

ing. During each revolution of the crystal, the lines will coincide twice. A single pencil of rays is thus refracted into two distinct pencils, one of which, following the usual law of refraction, is called the Ordinary Pencil, while the other, deviating from that law, is called the Extraordinary Pencil.

## Polarization of Light.

658. Light is said to be *polarized*, when, on being reflected or refracted by a surface which it strikes at a certain angle, it is absorbed by a similar surface perpendicular to the former one, though it is reflected or transmitted by one forming any other angle with it.

Let A and B (Fig. 250) be two tubes open at both ends, and so adjusted to each other that B turns stiffly within A. In each tube fix a piece of polished glass, M, N, roughened and blackened on the back, so as to form an angle of 33 degrees with the axis of the tubes. Bring the instrument into such a position

Fig. 250.

that the light from a luminous body, falling on M, may be reflected along the axis and strike N. Now, keeping the tube A stationary, turn within it the tube B, carrying the reflector N. The reflection from N, if observed, will be seen to keep varying in intensity. In the two positions in which N is parallel to M, the reflection will be brightest; at the points midway between these, —that is, when N is perpendicular to M,—there is no reflection at all. We express this by saying that the light reflected from M is *polarized*.

659. The polarizing angle,—that is, the angle which the incident ray must make with a perpendicular to the first reflecting surface, in order to be polarized,—is different in the case of different substances. For glass, it is about 57 degrees.

660. If a polarized ray be received on a crystal of Iceland spar, there will be but a single refraction.

661. Light is polarized by reflection at a certain angle, as we have just seen; by transmission through substances that have the property of double refraction,—through some imperfectly crystallized substances, such as agate, mother-of-pearl, &c.,—and also through a sufficient number of uncrystallized plates. However produced, polarized light always has the same properties. Its phenomena are striking, and seem to prove the truth of the undulatory

nomena are presented as the crystal is turned around? What are the two pencils presented to the eye called? 658. When is light said to be polarized? Illustrate the polarization of light with Fig. 250. 659. What is meant by the polarizing angle? What is this angle in the case of glass? 660. If a polarized ray is received on a crystal of Iceland spar, what follows? 661. Mention the different ways in which light is polarized. What is said of the properties and phenomena of polarized light, how-

theory. It is thought that the undulations of ether ordinarily take place in planes perpendicular to the direction in which they are propagated; but that, when light is polarized, they take place in planes parallel to this direction. At certain angles, the undulations, thus changed from their usual direction, are reproduced or transmitted by the second reflecting or refracting surface, and reach the eye; but, when the two surfaces form an angle of 90 degrees, they are stopped, and the sensation of vision is not produced.

662. The mineral called Tourmaline [*toor'-ma-leen*] possesses the property of polarizing light in a high degree. It is cut into plates one-twentieth of an inch thick, which are fixed between plates of glass for convenience of use. If we look at the sun through such a plate, we shall find that most of the light is transmitted. Place a second plate behind the first and parallel to it, and the light will still be transmitted; but turn the second plate so as to bring it at right angles to the first, and no light will pass through.

663. Some crystals viewed by polarized light, exhibit systems of beautiful rings, like those shown in Fig. 251. Plates of the mineral called Selenite,

Fig. 251.

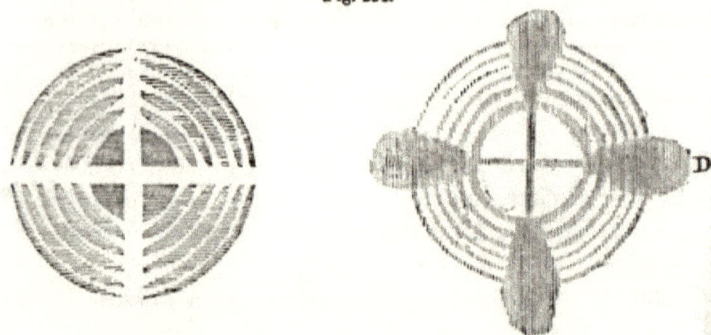

bearing different designs, placed so as to be seen by polarized light, display the most gorgeous coloring, and may be made to undergo remarkable and beautiful changes by causing one of the reflecting surfaces to revolve.

## Chromatics.

664. Chromatics is that branch of Optics which treats of colors.

---

ever it is produced? Explain the polarization of light according to the undulatory theory. 662. What mineral possesses the property of polarizing light in a high degree? How is tourmaline prepared? What experiment may be performed with tourmaline plates? 663. What phenomena are seen when certain crystals are viewed

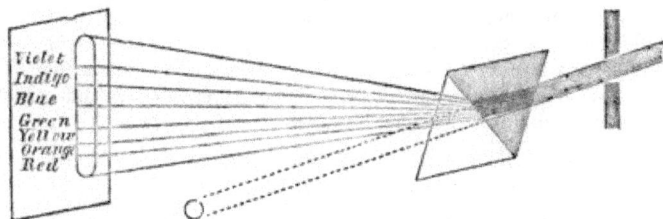

665. THE SOLAR SPECTRUM.—If a ray from the sun be admitted into a dark room through a small aperture, it will form a circular spot of white light on the surface receiving it. But if, after entering the room, it be received on a prism, as shown in Fig. 252, it will be decomposed into

Fig. 252.

Violet
Indigo
Blue
Green
Yellow
Orange
Red

THE SOLAR SPECTRUM.

seven different colors. When made to fall on a white surface, these seven colors are distinctly seen, covering an oblong space, which is called the Solar Spectrum (plural, *spectra*). They are known as the Primary Colors, and in every spectrum they are arranged in the order shown in the Figure. By combining the primary colors in different proportions, other colors are produced.

The seven colors, it will be observed, do not occupy equal spaces of the spectrum. Violet covers the greatest part, more than one-fifth of the whole; and orange the least, less than one-thirteenth of the whole.

666. Ordinary sun-light (and all white light) is therefore composed of seven colors combined in different proportions. In further proof of this, we may re-unite the seven primary colors of the spectrum, and we shall have simply a small circular spot of white light. To re-unite the colors, we may receive the spectrum on a concave mirror or double convex lens, which brings together at its focus the parts of the decomposed ray. Or, we may receive the spectrum on another prism placed in contact with the first, as shown in Fig. 252. In either case, we have the same circular spot of white light that would have been formed if the ray had not been decomposed at all.

by polarized light? When plates of selenite are viewed by polarized light? 664. What is Chromatics? 665. Describe the solar spectrum, and the way in which it is formed. Name the seven primary colors in order. How are the other colors produced? Which color occupies most of the spectrum, and which the least? 666. Of what, then, is all white light composed? What further proof have we

We may produce white light by combining the seven primary colors in another way. Divide the surface of a circular card into seven parts proportioned to each other as the spaces which the different colors occupy in the spectrum, and paint them the corresponding shades. Then cause the card to revolve rapidly. No separate color will be visible, but the whole card will look white.

667. A prism decomposes white light into its seven component parts, because these parts are refracted differently, some more and some less. It will be observed that red, which occupies the lowest part of the spectrum, is turned from its course the least; orange, a little more; yellow, still more; then green; then blue; then indigo; while violet, which is at the top of the spectrum, is refracted the most. The colors, therefore, have different degrees of refrangibility. This fact was discovered by Sir Isaac Newton.

668. DIFFERENCE OF COLOR, EXPLAINED. — According to the Undulatory Theory, the color of light depends on the size of the minute waves that produce it. The undulations that excite in the eye the sensation of red light are each $\frac{1}{39000}$ of an inch in breadth; those that produce violet, $\frac{1}{60000}$; while the intermediate colors are produced by undulations varying between these limits.

669. Color is not a property inherent in bodies, but in the light that they reflect. A non-luminous body seems to be whatever color it reflects to the eye.

An object lying in green light, looks green; in red light, red, &c. This is because green or red is the only light that falls upon it, and therefore it can reflect no other to the eye. A body seen by ordinary light looks green, when it absorbs all or most of the other colors of the spectrum, and reflects or transmits green alone. It looks red when it absorbs the other colors, and reflects or transmits red, &c. It looks white, when it does not decompose the light that falls on it, but reflects all the colors combined. It looks black, when it absorbs nearly all the light that falls on it, and does not reflect any particular color in preference to the rest.

670. What colors a substance absorbs and what it reflects, depends chiefly on its structure. The particles of some bodies are so arranged as to have a peculiar affinity for certain colors; these they absorb, reflecting the rest.

---

of this? How may we re-unite the seven primary colors? What other mode is there of doing this? 667. To what is it owing that a prism decomposes white light into its seven component parts? By whom was this fact discovered? 668. According to the Undulatory Theory, on what does the color of light depend? What is the difference in the undulations that respectively produce red and violet light? 669. In what is the property of color inherent? Why does an object lying in green light look green? When does an object seen by ordinary light look green? When does it look white? When, black? 670. What is it that determines what colors a

Changes of color are caused by changes of structure. We may show this by an experiment with a substance called iodide of mercury. This mineral is a bright scarlet; when heated and allowed to cool undisturbed, it becomes yellow; but, the moment the surface is scratched, the particles rearrange themselves, and the color turns back to scarlet. Here the same particles undergo a marked change of color by simply being made to assume a different arrangement.

671. COMPLEMENTARY COLORS.—Any two colors are said to be Complementary, when, if combined in due proportion, they will produce white. Those colors are complementary to each other which are distant half the length of the spectrum; as,

| | |
|---|---|
| Red and green, | Orange and blue, |
| Yellow and violet, | White and black. |

It is a curious fact that if we look intently at a bright object of any given color and then close our eyes, we shall still see it, but tinged with the complementary color. After gazing a few moments at a bright fire, everything we look at seems to have a greenish hue. If we place a red wafer on a piece of white paper and look at it intently, we shall soon see a circle of light green playing around it. A blue wafer will have a similar circle of orange, and a yellow wafer one of a violet tinge.

672. A color appears to the best advantage, when placed beside its complementary color.

Thus red is set off by green; blue, by orange, &c. A pale face appears paler still when a black dress is worn. On white paper, black ink is plainer and pleasanter to the eye than ink of any other color. In arranging bouquets, and selecting different articles of dress that are to be worn together, the effect of each individual color is heightened by bringing it in immediate contrast with its complementary color.

673. PROPERTIES OF THE SPECTRUM.—Every ray of ordinary sun-light appears to have three distinct properties:—1. Brightness. 2. Heat. 3. Power of producing chemical effects. This last property is called Actinism.

674. The chemical effects of sun-light are shown in various ways. Phosphorus and nitrate of silver undergo a marked change when exposed to the

substance absorbs, and what it reflects? By what are changes of color caused? Prove this with an experiment. 671. When are two colors said to be Complementary? Name four pairs of complementary colors. What curious fact is stated with respect to complementary colors? Give examples. 672. When does a color appear to the best advantage? Give examples. 673. How many distinct properties has every ray of ordinary sun-light? Name them. 674. Instance some of the chemical

solar rays. Daguerreotypes and photographs are taken by means of the action of light on sensitive chemical preparations. Almost all the colored vegetable juices, when exposed to sun-light, undergo a change of hue. Hydrogen and chlorine, which may be mixed without danger in the dark, combine with a loud explosion in the light. Light, also, is essential to the chemical changes which result in the healthy growth of plants. Hence plants kept in a dark room become pale and sickly. A similar effect is produced on persons kept away from the light of the sun.

675. Ordinary sun-light combines these three properties, but the seven colors into which it is decomposed by the prism do not possess them alike. Brightness belongs particularly to yellow; heat, to red; actinism, to violet and indigo.

An object that is bright yellow makes a more vivid impression on the eye than one of any other color. Hence soldiers dressed in yellow are more distinct objects of aim to an enemy and more apt to be shot than those dressed in dark green or gray.

The red portion of the spectrum has the most heat. This is shown by placing the bulb of a thermometer successively in each of the colors of the spectrum. It will be most affected by the red, but will show a still higher temperature, if brought a short distance below the red end of the spectrum, where no light falls at all. This shows that the heat of a solar ray is refracted as well as its light, but in a less degree.

Actinism is strongest in violet and indigo rays. If a seed be placed under a dark blue glass, so that all the light that strikes it will be tinged with that color, it will germinate in one-fourth of the time that it usually takes. Placed under a red glass, it will hardly germinate at all, because red, although it contains more heat than the other colors, has little or no actinism.

676. DARK LINES IN THE SPECTRUM.—If the solar spectrum be viewed through a telescope, a great number of dark lines, parallel to each other but differing in breadth, will be seen crossing its surface. Seven of these are particularly distinct, but with a powerful telescope as many as 2,000 have been counted.

The position of these lines is always the same in the solar spectrum; but, when a ray of star-light is decomposed, their number and arrangement

effects of sun-light. 675. Do the seven primary colors possess these three properties in equal degrees? To which does brightness particularly belong? To which, heat? To which, actinism? What follows from the peculiar brightness of yellow? How is it proved that the red portion of the spectrum has the most heat? How does the refraction of solar heat compare with that of solar light? Prove this. How may it be shown that actinism is strongest in violet and indigo rays? 676. Describe the dark lines in the spectrum. What is said of the lines found in spectra produced from star-

are different, nor do they correspond in spectra formed by rays from different stars. When rays produced by electricity or combustion are decomposed with the prism, bright lines are found crossing the spectrum instead of dark ones.

677. DISPERSION OF LIGHT.—By the Dispersion of light is meant the formation of a spectrum from a single ray. Spectra formed by different refractive media are of different lengths. Thus flint-glass forms a spectrum about twice as long as crown-glass forms, and four times as long as water. Flint-glass is therefore said to have twice the dispersive power of crown-glass, and four times that of water.

678. ACHROMATIC LENSES.—Lenses, like prisms, refract light, and produce spectra. Rays passing through a convex lens, therefore, instead of coming to a focus at a single point, are more or less dispersed, and form colored fringes about the focus. This defect is called Chromatic Aberration. It was long a serious drawback in the use of optical instruments; but the difficulty is now remedied by combining two lenses of such different materials that the dispersive power of the one may nullify that of the other. Lenses combined on this principle are called Achromatic Lenses.

*Achromatic* means colorless, and the lenses are so called because they do not fringe their images with the colors of the spectrum. A double convex lens of crown glass may be united with a plano-concave lens of flint glass. The latter corrects the chromatic aberration of the former, without entirely nullifying its converging effect.

679. THE RAINBOW.—The Rainbow is an arch composed of the seven primary colors, which is visible in the sky when the sun shines during a shower. It appears in the opposite quarter to the sun,—in the west in the morning, and the east in the afternoon.

When the sun is in the horizon, the rainbow is a circle; but the lower part of it is intercepted by the earth's surface, and therefore we do not gen-

light? In spectra produced from the light of electricity or combustion? 677. What is meant by the Dispersion of light? When are different media said to differ in dispersive power? 678. What is Chromatic Aberration? How is it corrected? What does *achromatic* mean? Why are achromatic lenses so called? How may an achromatic lens be formed? 679. What is the Rainbow? Where is it seen? What is the

erally see more than a semi-circle. From the mast-head of a vessel or the top of a mountain, more than a semi-circle is visible.

680. The rainbow is caused by the refraction and reflection of the sun's rays by drops of falling rain. Each drop operates like a prism, decomposing the light that strikes it. The observer's eye is so placed as to receive but one of the colors from one drop, but from other drops it receives the other colors, and thus has an arched spectrum formed complete. As no two persons occupy exactly the same spot, no two can see exactly the same bow.

681. Sometimes two distinct bows are visible, one within the other. The inner one, which is called the Primary Bow, is the brighter of the two. The outer one is called the Secondary Bow; the rays that form it undergo one more reflection within the drop than those that form the primary bow, and are therefore fainter. In the primary bow, the arrangement of the colors is the same as in the solar spectrum; in the secondary bow, this order is reversed.

682. Whenever the air is filled with drops, and the sun shines on them at a certain angle, rainbows are formed, which are visible to an observer in a proper position. Hence they are often seen in the spray of water-falls and fountains.

683. Bows are sometimes similarly formed by moon-light, but they are faint and rarely seen. When so formed, they are called Lunar Rainbows.

684. HALOES.—Haloes are luminous or colored circles seen around the sun and moon under certain conditions of the atmosphere. They are more frequently seen around the moon, because the sun's light is so intense that they are lost in its superior brightness. Haloes arise from the refraction and dispersion of light by small crystals of ice floating in the higher regions of the atmosphere.

## Vision.

685. THE EYE.—The eye is the organ with which we see. Nothing more strikingly displays the wisdom of the

form of the rainbow? 680. Explain the principle on which the rainbow is formed. 681. When two bows are formed, what is each called, and which is the brighter? In what order are the colors arranged in the rainbow? 682. By what besides rain may bows be produced? 683. What are Lunar Rainbows? What is said of them? 684. What are Haloes? Where are they most frequently seen? How are haloes pro-

Creator than the nice adaptation of this wonderful instrument to the purposes for which it is designed.

686. *Parts of the Eye.*—The human eye is a spheroid, about an inch in diameter, resting in a cavity below the forehead, capable of being moved upward, downward, or sidewise, by muscles attached to it behind. It consists of ten parts :—

| | |
|---|---|
| 1. The Cornea. | 6. The Vitreous Humor. |
| 2. The Iris. | 7. The Ret'-i-na. |
| 3. The Pupil. | 8. The Choroid Coat. |
| 4. The Aqueous Humor. | 9. The Sclerotic Coat. |
| 5. The Crystalline Lens. | 10. The Optic Nerve. |

687. When we look at an eye as set in the head (see Figure 253), we see but three of these parts: the Cornea (*g*); the Iris (*i*); and the Pupil (*b*). The Cornea is a transparent coat, covering the whole front of the eye, and more convex than the rest of the ball. The Iris is the circular membrane in the middle of the cornea, according to the color of which we say that the eye is blue or black, hazel or gray. The Pupil is a circular opening in the iris, through which light passes into the interior of the eye. Fig. 254 represents a section of the eye. A A A is the cornea. I I is the iris, and the opening in the centre is the pupil. In the following description reference is made to this Figure.

Fig. 253.

Fig. 254.

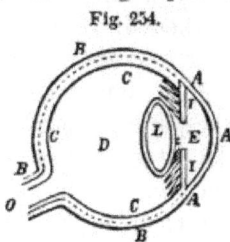

On passing through the cornea, a ray of light enters the narrow apartment E, between the cornea on one side and the iris and crystalline lens on the other. This is filled with a transparent liquid resembling water, and called the Aqueous Humor. Traversing this, the ray next enters a transparent body, L, called from its shape the Crystalline Lens. Behind this is the Vitreous Humor, D, a trans-

duced? 685. What is the eye? 686. Describe the eye. Of how many parts does it consist? Name them. 687. Which of these parts do we see when we look at an eye as set in the head? What is the Cornea? What is the Iris? What is the Pupil? With the aid of Fig. 254, name and describe the various parts of the eye. By what is

parent fluid which fills the greater part of the globe of the eye. This humor is enclosed within the Retina, C C C, a delicate fibrous membrane resembling net-work, formed by the expansion of the optic nerve, on which every image seen by the eye is formed. The Optic Nerve, O, passes through the back of the eye to the brain, and conveys to that organ the impressions made on the retina.

The retina is surrounded by another coat called the Choroid, represented in the Figure by a dotted line. The choroid coat is lined on its inner surface with black coloring matter, to prevent any reflection of light from the interior of the eye. Outside of all is the Sclerotic Coat, B B B, a strong membrane, to which the muscles that move the eye are attached. It envelopes the whole ball except the portion in front covered by the cornea, which fits into it just as the crystal of a watch fits into the case.

688. *Uses of the Different Parts.*—The outer coats of the eye protect the delicate parts within. The cornea reflects some of the light that falls on it, and this gives the eye its brilliancy. It transmits the greater part, however, and unites with the aqueous humor, the crystalline lens, and the vitreous humor, in bringing the incident rays to a focus and forming an image on the retina.

The iris intuitively regulates the supply of light admitted into the eye, contracting and thus enlarging the pupil in a faint light, expanding and thus diminishing it in a strong one. These changes are not instantly made. Hence, when we pass from a bright light into a room partially darkened, we can hardly discern anything till the pupil enlarges, so that more rays are admitted. When we go from a dark room into a bright light, the eye is pained, because the pupil, which had expanded to the utmost to accommodate itself to the faint light, does not immediately contract, and more light is admitted than the sensitive membrane can endure.

The pupils of cats, tigers, and animals generally that prowl at night for prey, are capable of being expanded to such a degree as to admit one hundred times as much light as when they are most contracted. They can therefore see as well by night as by day. The owl's pupil is exceedingly large;

the retina surrounded? With what is the choroid coat lined? What is outside of all? What are attached to the sclerotic coat? 688. What is the use of the outer coats of the eye? Of the cornea? Which parts unite with the cornea in bringing incident rays to a focus? What is the use of the iris? Give some familiar proofs that the iris accommodates itself to the intensity of the light. What is said of the pupil

In the day-time, even when contracted to the utmost, it admits so much light that the bird is nearly blinded, and has to remain stupidly on its roost.

689. DEFECTS OF VISION.—In a perfect eye, the rays that enter are brought to a focus on the retina, and an image is there formed. If the rays are not brought to a focus by the time they reach the retina, or come to a focus before reaching it, no impression is made on the optic nerve or communicated to the brain, and consequently no image is seen.

Hence arise two defects of vision. When the cornea is too convex, distant objects form images in front of the retina, and are not seen; only such objects as are very near the eye are visible, and hence persons with this defect of vision are called *near-sighted*. When, on the contrary, the cornea is not convex enough, the rays are not brought to a focus by the time they reach the retina, and no image is seen. The eyes of old people generally labor under this defect, in consequence of the waste of a portion of the vitreous and the aqueous humor, so that the crystalline lens and the cornea fall in. This falling in is just what the near-sighted person needs; accordingly it is often found that those who are near-sighted in youth see perfectly well when they grow old.

690. The two defects of vision mentioned above are remedied by the use of spectacles, which consist of lenses of different shapes placed in frames before the eyes. A near-sighted person uses glasses just concave enough to nullify the too great convexity of his eye. An old person uses glasses with sufficient convexity to make up the deficiency of his eye in that respect.

691. Spectacles were first used about the end of the thirteenth century. It is supposed that the world is indebted to Roger Bacon for their invention. Before that time all near-sighted and most aged persons had to remain in a state of comparative blindness.

692. Though all other parts of the eye be perfect, if the optic nerve does not perform its functions, blindness is the result. Images are formed on the retina, but there is no communication with the brain, and no impression

of beasts that prowl at night? What is said of the owl's pupil? 689. Where are images formed in a perfect eye? What will prevent an image from being seen? Describe the two defects of vision arising from images' not being formed on the retina. 690. How are these two defects of vision remedied? What sort of glasses does a near-sighted person use? An old person? 691. When were spectacles first used? By whom are they supposed to have been invented? 692. If the optic nerve does not

Is produced. For amaurosis, or paralysis of the optic nerve, there is no remedy.

693. IMAGES FORMED ON THE RETINA.—Images are formed on the retina, just as in a dark room, by light admitted through an aperture (see Fig. 235). In the latter case, as we have already, seen, the image is inverted, and it follows that images formed on the retina must be inverted also. Why then do we see them in their natural position? This question it is hard to answer. The explanation commonly given is this :—That we see all things inverted, and have always done so ; but, inasmuch as we know by experience that they are erect, the mind of itself, insensibly to us, corrects the delusion that the inversion would otherwise produce. We have no means of comparison; we see nothing erect, to serve as a standard and prove the general inversion.

694. Another question is sometimes asked :—Since we have two eyes, and two images are formed, one on each retina, why do we not see two images of every object? The answer is, because both eyes are inclined to any given object at nearly the same angle. The images produced on the retinas are very nearly the same. The impressions transmitted to the brain by the two branches of the optic nerve are identical and simultaneous, and but one perception is the result. If we press on one of our eyes, so as to incline it towards an object at a different angle from the other, we see two images. Drunken men often see double, because they lose control of the muscles of the eye, and do not direct both eyes towards a given object at the same angle.

695. VISUAL ANGLE.—The visual angle is the angle formed by two lines drawn from the eye to the extremities

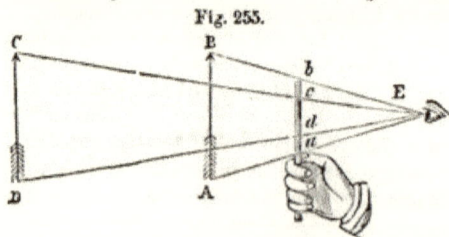

Fig. 255.

of a given object. In Fig. 255, the visual angle of the arrow B A is B E A; that of the arrow C D is C E D.

A given object

perform its functions, what is the consequence? 693. What kind of an image is formed on the retina, and why? Since an inverted image is formed on the retina, why do we see objects in an erect position? 694. Since we have two eyes, why do we not see two images of every object? How may we make two images visible? Why do drunken men often see double? 695. What is the Visual Angle? Show the

looks large or small, according to the visual angle that it forms. Two equal arrows held up before the eye at different distances, as in Fig. 255, form different visual angles, and therefore seem to be of different size. If we measure their apparent lengths with an interposed rod, we shall find the nearer one to measure the distance *a b*, the farther one only about half as much, *c d*. A small object placed near the eye may form as great a visual angle as a very large distant object, and may therefore entirely hide the latter when interposed between it and the eye.

Accordingly, the nearer an object is brought to the eye, the larger it appears to be, and the further it is removed the smaller it looks. When the visual angle is less than $\frac{1}{200}$ of a degree, an object becomes invisible. A bird flying from us grows smaller and smaller, till its visual angle diminishes so that it can no longer be seen, and we say that *it has gone out of sight*.

696. In the case of familiar objects, experience prevents us from being misled by their apparent size. Insensibly to ourselves, we make allowance for their distance, of which we judge by the distinctness of their outline and by intervening objects. A man at work on a lofty steeple may not look more than two feet high, yet we are in no danger of mistaking him for a dwarf. A distant tree seems to be no higher than a bush; but, if we see a horse feeding beneath it, we intuitively compare the two, and arrive at a correct idea of the tree's size.

A white object can be distinguished at a greater distance than one of any other color, and is visible twice as far when the sun shines directly on it as when simply illumined by ordinary light. An object is brought out most distinctly by a back-ground which contrasts strikingly with it in color. Dark-colored eyes, for the most part, see farther than light ones; and those who are in the habit of looking at remote objects, like sailors, can discern minute bodies at distances which render them invisible to ordinary sight.

697. ADAPTATION OF THE EYE.—One of the most remarkable properties of the eye is its power of adapting itself to different intensities of light and different distances. The pupil, by expanding and contracting, regulates in a measure the supply of light; still, the difference of intensity in the light admitted to the eye under different

visual angles of the arrows in Fig. 255. On what does the apparent size of an object depend? Illustrate this with the Figure. When does an object become invisible? When is a bird said to go *out of sight?* 696. In the case of familiar objects, what prevents us from being misled as to their size? Give some familiar examples. What color must an object be, to be distinguished at the greatest distance? How is an object most distinctly brought out? What is said of dark-colored eyes? 697. What is

12

circumstances is very great. We can read by the light of the moon and by that of the sun; yet the latter is 800,000 times as intense as the former.

698. Again, the eye adapts itself to different distances. If we look at a remote object through a telescope, we have to pull out the tube to a certain length, according to the distance, before we can see it to advantage. No such artificial adjustment is necessary with the eye. We look successively at objects 1, 5, 10, and 20 feet off; and in each case the eye instantly adapts itself to the distance, and we see without an effort.

699. An object may move with such velocity that we can not see it, as is the case with a cannon-ball. This is because the image formed on the retina does not remain sufficiently long to produce an impression. When an image is once formed, it remains from one-sixth to one-third of a second after the object has disappeared. Hence a burning stick whirled rapidly round seems to form a circle of fire, and a meteor or a flash of lightning, instead of appearing in a succession of luminous points, produces a continuous train of light in the heavens.

### Optical Instruments.

700. Several of the more important optical instruments remain to be described. They are for the most part combinations of the different lenses and mirrors already mentioned.

701. THE CAMERA OBSCURA.—We have seen that, when rays from an object brilliantly illuminated are admitted through an aperture into a dark room, an inverted image is formed. This image is apt to be indistinct. We may give it a sharper outline by placing a double convex lens in the aperture, and receiving the image on a white ground

one of the most remarkable properties of the eye? Give an example of the difference of intensity in the light admitted to the eye. 698. Show how the eye adapts itself to different distances. 699. Why is it that an object moving with very great velocity is not seen? When an image is once formed, how long does it remain after the object has disappeared? Give examples. 700. Of what are optical instruments for the most part combinations? 701. What is meant by the Camera Obscura? How

at its focus. Such an arrangement is called the Camera Obscura, or *dark chamber*.

For practical purposes, the camera obscura must be portable. A close box, painted black on the inside, is therefore substituted for the darkened room. This instrument enables the draughtsman to sketch material objects or natural scenery with great ease and accuracy, and is indispensable to the daguerreotypist and photographer.

702. *Draughtsman's Camera.*—Fig. 256 represents the camera as used by draughtsmen. To be conveniently traced, the image must be thrown on a horizontal surface, and this is effected by making the opening in the top of the box and receiving the rays on a mirror, A, inclined at an angle of forty-five degrees. From this mirror they are reflected to a meniscus, B, which crosses the aperture, and are by it refracted to the horizontal surface, C D, where, on white paper placed to receive it, is formed a distinct image, which can be readily traced with a pencil. The upper part of the draughtsman's person is admitted through an opening in the side of the box, over which a dark curtain must be drawn, so as to exclude all light except what enters from above.

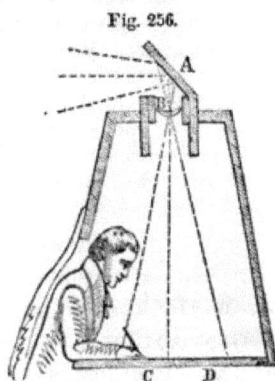

Fig. 256.

DRAUGHTSMAN'S CAMERA.

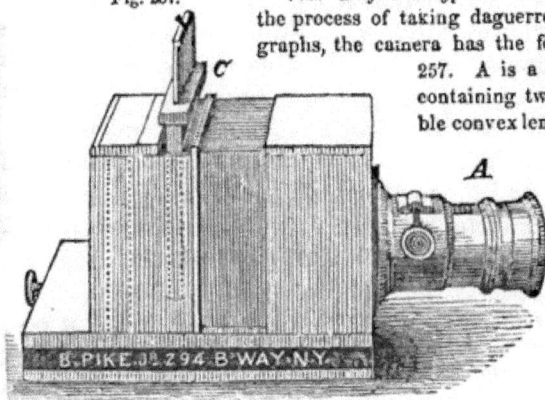

Fig. 257.

DAGUERREOTYPIST'S CAMERA.

703. *Daguerreotypist's Camera.*—As used in the process of taking daguerreotypes and photographs, the camera has the form shown in Fig. 257. A is a brass sliding tube, containing two achromatic double convex lenses, which is drawn out far enough to bring the focus at the right spot. The image is received on a piece of ground glass, fitted into a frame, which slides in a groove in the back of the camera. When a daguerre-

is the camera made portable? By whom is the camera used? 702. Describe the draughtsman's camera. 703. Describe the daguerreotypist's camera. How is the plate

otype is to be taken, the ground glass is withdrawn, and another frame, C, containing a prepared plate, carefully shielded from the light, is introduced in its place. A door in front of C is then raised, and the image formed by the lenses is thus allowed to fall on the plate.

The plate is of copper, covered on one side with a thin sheet of silver, which is rendered sensitive by exposure to the vapor of iodine. The rays transmitted through the camera, by that property inherent in them which we have called *actinism*, in a few seconds produce a chemical effect on the sensitive surface, and the plate is then removed to a dark room. No change is visible on its surface; but, as soon as it is exposed to the vapor of mercury, the picture begins to appear and soon becomes distinct. It is produced by the adhesion of small globules of mercury to those parts of the plate that have been affected by light, to the exclusion of the rest; and this adhesion is owing to some chemical change in the parts so affected. After being washed in a weak solution of hyposulphite of soda, and then in water, the plate is allowed to dry, and the image is fixed.

The photographic process is similar, except that the image is received on paper rendered sensitive by different preparations, instead of on a metallic plate.

704. THE MICROSCOPE.—The Microscope is an instrument which enables us to see objects too small to be discerned by the naked eye. This is the case with objects whose visual angle is less than $\frac{1}{360}$ of one degree; the microscope enables us to see them by increasing their visual angle.

Microscopes are divided into two classes, Single and Compound. A Single Microscope is one through which the object is viewed directly. With the Compound Microscope a magnified image of the object is viewed, instead of the object itself.

705. *The Single Microscope.*—The single microscope consists of a double convex lens (or sometimes more than one), through which we look at the object to be magnified. The principle on which it operates is shown in Fig. 258.

The arrow *b c* would be seen by the naked eye under the visual angle *b* A *c*. When the lens *m* is interposed, the rays are so refracted as to form the visual angle D A E, and the arrow appears to be of the size D E, much larger than it really is. Sometimes an exceedingly minute object becomes visible when brought very near the eye, but in that position the rays enter the eye with such divergency that a confused image is produced. The microscope corrects this excessive divergency, and presents a clear and magnified image.

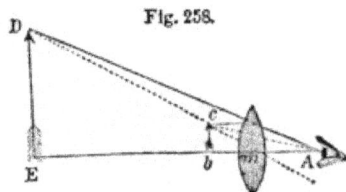

706. *The Compound Microscope.*—The compound microscope is a combination of two, three, or four convex lenses, through which we view a magnified image of an object instead of the object itself. The lenses are fixed in tubes moving one within the other, and suitable apparatus is provided for adjusting them, for holding the object under examination, and throwing on it a strong light. When but two lenses are employed, they are arranged as represented in Fig. 259.

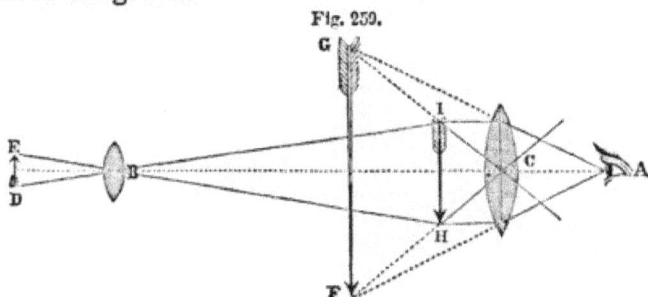

D E is the object, and B, the lens nearest to it, is called the *object-glass*. C, the lens nearest the eye, is called the *eye-glass*. A magnified image of the arrow is formed at H I by the lens B. This image is viewed through the lens C, and is thus still further magnified, being seen under an increased visual angle at F G. If the magnifying power of B is 20, and that of C 4, the image seen will be 80 times the size of life.

707. *Solar and Oxy-hydrogen Microscopes.*—These microscopes are used for throwing magnified images on a white screen in a darkened room.

In the case of the Solar Microscope, an aperture is made in one of the shutters. Outside of this a mirror is placed, in the sun, at such an angle as to reflect the rays that fall on it through a horizontal tube towards the object to be magnified. They first fall on a convex lens, and then on a second, which brings them to a focus on the object, and thus illuminates it brilliantly. Another lens, at the opposite extremity of the instrument, produces the magnifying effect. A screen, from ten to twenty feet off, receives the image, which increases in size with the distance. If the screen is too far removed, the image becomes faint; but so powerful is the light concentrated on the object that a very great magnifying effect may be produced without any lack of distinctness.

In the Oxy-hydrogen Microscope, the principle is the same, but the brilliant light produced by burning lime in a current of oxygen and hydrogen is substituted for the rays of the sun. Accordingly, with this instrument, the aperture in the shutter and the mirror on the outside are unnecessary. Fig. 260 shows the operation of the oxy-hydrogen microscope.

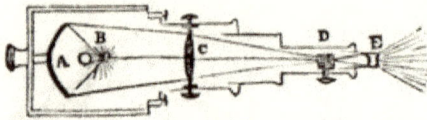

Fig. 260.

B represents an intense white light produced by the burning of a cylinder of lime in a current of oxygen and hydrogen combined. This light falls on the reflector A, by which it is thrown back on the double convex lens C, and this brings it to a focus on the object D. E is an achromatic lens, which throws a magnified image on the screen.

708. The microscope introduces us to new worlds, of the very existence of which we would otherwise have been ignorant. It reveals to us, in every drop of water in which vegetable matter has been infused, swarming myriads of moving creatures,—miniature eels, infinitesimal lobsters, ravenous monsters with distended jaws preying on their feebler fellows,—all endowed with the organs of life, and so minute that their little drop is to them a world nearly as large as ours to us. It shows us the feeding apparatus of the flea magnified to frightful dimensions, and his body arrayed in a panoply of shining and curiously jointed scales, studded at intervals with long spikes. The mould on decaying fruit it magnifies into bushes with branches and leaves,

displaying all the regularity and beauty of the vegetable creation. It discloses to us many striking facts connected with physiology and chemistry. It shows us the imperfection of the finest works of art, when compared with those of nature. The edge of the sharpest razor, viewed through a microscope, is full of notches; the point of a needle is blunt, and its surface is covered with inequalities. The magnified sting of a bee, on the other hand, is perfectly smooth, regular, and pointed. The finest thread of cotton, linen, or silk, is rough and jagged : whereas in the filament of a spider's web not the slightest irregularity can be detected.—In a word, the revelations of the microscope are in the highest degree wonderful and interesting ; and, to whatever we direct it, we always find abundant matter to reward our labor and stimulate us to further researches.

709. THE MAGIC LANTERN.—The Magic Lantern is an instrument for throwing on a screen magnified images of transparent objects. It operates on the same principle as the oxy-hydrogen microscope, but for its illuminating power has an ordinary lamp instead of the intense light produced by burning lime.

Fig. 261.

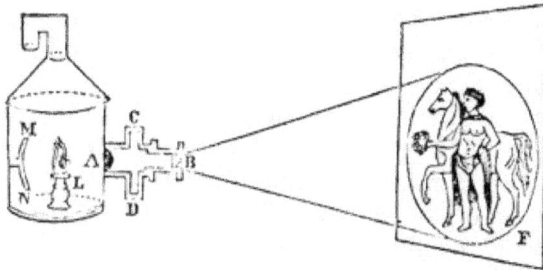

THE MAGIC LANTERN.

Fig. 261 represents the magic lantern. L is the lamp. M N is the reflector, which throws the light on the lens A. This lens brings it to a focus on the picture, which is painted on a glass slider and introduced into the opening C D. The lens B receives the rays from the slider, and throws a magnified image on the screen F.

710. *Phantasmagoria.*—When a powerful light is used, and the tube containing the magnifying lens or lenses is capable of being drawn out or pushed in, so as to bring them at different distances from the object, we have what is called a Phantasmagoria Lantern.

To exhibit the Phantasmagoria, a transparent screen is suspended, on one side of which is the exhibitor with his lantern, on the other the spectators. Having brought the lantern close to the screen and drawn out the tube till the image (which will be quite small) is perfect, the exhibitor walks slowly back. He thus gradually increases the size of the image, while he preserves its distinctness by pushing in the tube as he recedes. The effect on the spectators is startling. The room being dark, they can not see the screen, but only the illuminated image, which, as it grows larger, appears to be moving towards them; even those who are familiar with the instrument can hardly disabuse their minds of this impression. When the exhibitor approaches the screen and pulls out the tube, the image becomes smaller and appears to recede.

711. *Dissolving Views.*—Dissolving Views, in which one picture appears to melt into another, are produced by two magic lanterns, inclined so as to throw their images on the same spot. An opaque shade is made to revolve in front of the instruments, in such a way as gradually to intercept the rays from one and uncover the tube of the other. The first picture fades, and a new one takes its place, becoming more and more distinct as the other disappears.

712. THE TELESCOPE.—The Telescope is an instrument .for viewing distant objects. It appears to have been invented by Metius, a native of Holland, in 1608. The following year, Galileo, hearing of the new instrument, constructed one for himself, and was the first to make a practical use of the invention. To the Telescope, Astronomy is indebted for the important advances it has made during the last two centuries.

Telescopes are of two kinds, Refracting and Reflecting. In the former, which were the first constructed, lenses are used; in the latter, polished metallic mirrors.

713. *Refracting Telescopes.*—The simplest form of the telescope is that devised by Galileo. It is a tube containing a convex object-glass and a concave eye-glass. By the former parallel pencils are made to converge towards a focus, where they would form an inverted image; but be-

fore reaching the focus they fall on the concave lens, and
have their convergency so far corrected that an object is
distinctly seen by an eye at the extremity of the tube. The
Opera-glass consists of two Galilean Telescopes combined.
The night-glass used by sailors is on the same plan.

In the instrument called the Astronomical Telescope, both object-glass and
eye-glass are convex. The former produces an inverted image at its focus;
the latter, which is so placed that its focus falls at the same spot, refracts the
rays diverging from this image, and thus renders it visible to the eye. The
inversion of the image is of no consequence in observing the heavenly bodies;
but, when objects on the earth are viewed, we want an erect image, and therefore in the Terrestrial Telescope two additional lenses are introduced to correct the inversion.

714. *Reflecting Telescopes.*—In Reflecting Telescopes,
a speculum, or mirror, takes the place of the object-glass.
These instruments appear in several different forms. The
principle on which Herschel's is constructed, will be understood from Fig. 262.

The mirror S S is
placed at the farthest
extremity of the tube,
inclined so as to make
the rays that fall upon
it converge towards the
side of the tube in which

Fig. 262.

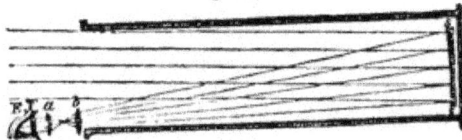

the eye-piece *a b* is fixed to receive them. The observer at E, with his back
towards the heavenly body, looks through the eye-piece, and sees the reflected image. His position is such as not to prevent the rays from entering the
open end of the tube. The advantage gained with this instrument depends
in a great measure on the size of the mirror; for all the rays that fall on it
are concentrated and transmitted to the eye.

715. The largest telescope ever constructed was made by the Earl of Rosse.
The great mirror is six feet in diameter, and weighs four tons. The tube, at
the bottom of which it is placed, is of wood hooped with iron. It is fifty-two
feet long and seven feet across. It is computed that with this instrument
250,000 times as much light from a heavenly body is collected and transmitted to the eye as ordinarily reaches it.

713. Describe the Galilean Telescope. Of what does the Opera-glass consist? Describe the Astronomical Telescope. How does the Terrestrial Telescope differ from
the Astronomical? 714. In reflecting telescopes, what takes the place of the object-
glass? With Fig. 262, explain the principle on which Herschel's Telescope operates.
On what does the advantage gained with this instrument depend? 715. Describe the
telescope of the Earl of Rosse. How great is the advantage gained with it?

1. (*See* § 594.) How long does it take a ray from the moon to reach the earth, the moon's distance being 240,000 miles?
2. The planet Jupiter is 496,000,000 miles from the sun. How long does it take a ray of light from the sun to reach the planet?
3. A ray of light from the sun is about 12,326 seconds longer in reaching the newly discovered planet Neptune than in reaching Jupiter. About how many miles farther from the sun is Neptune than Jupiter?
4. (*See* § 595.) A holds his book 1 foot, and B holds his 3 feet, from a certain candle. How much more light does A receive than B?
5. The planet Uranus is twice as far from the sun as the planet Saturn. How does the light received at Saturn compare in intensity with that received at Uranus?
6. (*See* § 650.) How many times is the ordinary heat of the sun increased by a burning glass with an area of 10 square inches, the focus of which has an area of $\frac{1}{10}$ of a square inch?
7. A convex lens has a focus $\frac{1}{8}$ of a square inch in area, and increases the heat of ordinary sun-light 200 times; what is the area of the lens?

---

# CHAPTER XV.

## ACOUSTICS.

716. ACOUSTICS is the science that treats of sound.

717. NATURE AND ORIGIN OF SOUND.—Sound is an impression made on the organs of hearing by the vibrations of elastic bodies, transmitted through the air or some other medium. These vibrations may be compared to the minute waves which ripple the surface of a pond when a stone is thrown in,—spreading out from a centre, but growing smaller and smaller as they recede, till finally they are no longer perceptible. They are produced by percussion, or any shock which gives an impulse to the particles of the sounding body. There is no sound that can not be traced to mechanical action.

718. Bodies whose vibrations produce clear and regular

sounds are called Sonorous. Bell-metal, glass, the head of a drum, are sonorous.

719. That sound is produced by vibrations is proved in various ways. A person standing near a piano-forte or an organ, when it is played, feels a tremulous motion in the floor of the apartment, as well as in the instrument itself if he touches it. We perceive the same tremor in a bell when in the act of being rung. In like manner, if we strike a tumbler so as to produce a sound, and then touch the top, we feel an internal agitation; and, when the vibrations are stopped, as they are by contact with the finger, the sound ceases with them. If we put water in a glass and produce a sound by rubbing the top with the finger, the liquid is agitated, and its motion continues until the sound dies away.—Place some fine sand on a square piece of glass, and, holding it firmly with a pair of pincers, draw a violin-bow along the edge. The sand is put in motion, and finally settles on those parts of the glass that have the least vibratory movement.—If a tuning-fork be struck and applied to the surface of mercury, minute undulations may be observed in the metal.

That these vibrations are communicated to the air and by it transmitted to the ear, also admits of easy proof. The rapid passage of a heavy cart or stage shakes the walls of a house. The discharge of artillery sometimes breaks windows. These effects are due to the vibrations suddenly produced in the air. If there is no air or other medium to transmit the vibrations to the ear, no sound is heard. We have already seen (§ 439) that a bell rung in an exhausted receiver can hardly be heard; if the air could be entirely removed, it would be wholly inaudible. Sound, therefore, does not leap from point to point, but is transmitted by vibrations communicated from one particle to another.

720. All sonorous bodies are elastic, but all elastic bodies are not sonorous.

Soft bodies are generally non-elastic, and consequently not sonorous. This is the case with cotton, for example, which yields little or no sound when struck by a hammer. It is on this account that music loses much of its effect in rooms with tapestried walls or curtained windows. Hence, also, a speaker finds it more difficult to make himself heard in a crowded room than in one that is empty.

721. TRANSMISSION OF SOUND.—All the sounds that or-

---

716. What is Acoustics? 717. What is Sound? How are sound-waves produced? To what is every sound traceable? 718. What bodies are called Sonorous? Give examples. 719. How is it proved by familiar experiments that sound is produced by vibrations? If a tuning-fork be struck and applied to the surface of mercury, what may be observed? How is it proved that these vibrations are communicated to the air and by it transmitted to the ear? 720. What property belongs to all sonorous bodies? What bodies are, for the most part, not sonorous? Give examples. What follows from the fact that soft bodies are not sonorous? 721. By what are the sounds

dinarily reach our ears are transmitted to them by the air. Any material substance, however, that connects our organs of hearing with a vibrating body, may transmit the vibrations in the same way. Thus, with our heads immersed in water, we can hear a sound produced under the surface at a considerable distance. Here water is the transmitting medium.

722. Liquids are better conductors of sound than aëriform bodies, and solids than liquids.

Persons in boats can converse with each other at a great distance, because water is a good conductor of sound. When the ear is applied to one end of a long stick of timber, the scratch of a pin at the other end can be distinctly heard, owing to the conducting power of the wood. An approaching locomotive can be heard at a great distance by placing one's ear on the rails. The American Indians knew by experience the facility with which solids transmit sounds, and were in the habit of applying their ears to the earth when they suspected the approach of an enemy, or wanted a more distinct impression of any sound that attracted their attention.

723. The denser air is, the more readily it transmits sounds. On the tops of high mountains, where, as we have already learned, the atmosphere is rare, the human voice can be heard only a few rods off, and the report of a musket sounds no louder than the snapping of a whip at the level of the sea. On the other hand, the air in a diving-bell let down to the bottom of the sea, which is condensed by the upward pressure of the water, transmits sound so freely that those who descend can hardly speak to each other above their breath; conversation in an ordinary tone would pain the ear.—Frosty air is a much better conductor of sound than warm air. In the polar regions, conversation has been carried on by two persons a mile apart.

Still air of uniform density transmits sounds more freely than air which is agitated by variable currents or contains strata of different density. This is one reason why sounds are more distinctly heard by night than by day. Falling rain or snow interferes with the vibrations, and tends to make sounds less distinct; so, likewise, do contrary winds.

724. If the air were perfectly still and of uniform density, sound transmitted through it *would decrease in loudness as the square of the distance from the vibrating body*

---

we ordinarily hear, transmitted? What else may transmit sound-waves in the same way? 722. How do solid, liquid, and aëriform bodies compare, as conductors of sound? Give a proof of the conducting power of water. State some facts illustrating the facility with which solids conduct sound. 723. How do rare and dense air compare, as conductors of sound? Give examples. How does cold air compare with warm in conducting power? Under what circumstances does air transmit sound most freely? What is the effect of falling rain or snow? 724. If the air were perfectly still

*increased.* The report of a cannon, for instance, would seem only one-fourth as loud at a distance of 200 feet as at a distance of 100 feet.

725. VELOCITY OF SOUND.—Under ordinary circumstances, *sound is transmitted through air with a velocity of* 1,120 *feet in a second,* which is at the rate of a mile in about 4⅔ seconds.

All sounds, whether loud or faint, high or low, are transmitted by a given medium with equal rapidity. Were it not so, there would be no such thing as harmony in musical performances, for the notes of the different instruments would reach the ear at different intervals.

Sound, it will be observed, travels much more slowly than light. The latter moves 192,000 miles while the former is going only 1,120 feet. The difference in their velocities is perceptible even at short distances. If we look at a man splitting wood a few rods off, we see the axe descend on the log some time before we hear the noise of the blow. So, the report of a cannon is not heard till after the flash is seen,—the interval being long or short according to its distance.

726. When the sound is accompanied with a flash, knowing the relative velocity of sound and light, we can calculate very nearly the distance from which it comes. We have only to notice the number of seconds that elapse after the flash is seen before the sound is heard, and multiplying this by 1,120, we get the distance in feet. The time which it takes the light to traverse the given distance and reach the eye, is so small that it does not enter into the calculation. For example, if a clap of thunder is heard 3 seconds after the accompanying flash is seen, the cloud from which they proceed is 3 times 1,120 (or 3,360) feet distant. The sooner the report follows the flash, the nearer the cloud.

727. Water transmits sound 4½ times as rapidly as air ; iron, 10 times ; and different kinds of wood, from 11 to 17 times.

Place the ear at one end of a very long stick of timber, and let some one strike the other end with a hammer. The wood conducts the sound to the ear so much more quickly than the air that the blow is heard twice. So,

and of uniform density, what would be the law for the loudness of a sound heard at different distances? Give an example. 725. What is the velocity of sound ? How is the velocity of sound affected by its loudness and pitch? What proof have we of this? How does the velocity of sound compare with that of light? Give some familiar instances showing their difference of velocity. 726. When the sound is accompanied with a flash, how may we calculate the distance from which it comes? Give an example. 727. With what velocity does water transmit sound, as compared with

when a bell at the end of a long iron tube is struck, two sounds are heard at the opposite extremity,—the first conducted by the iron, the second by the air within it.

728. DISTANCE TO WHICH SOUND IS TRANSMITTED.—So many changes are constantly taking place in the atmosphere, in its temperature, moisture, density, and the velocity and direction of its currents, that no universal law can be laid down as to the distance at which sound is audible. The human voice, when raised to its highest pitch and loudest tones, may be heard at the distance of an eighth of a mile ; the report of a musket, at 5 miles.

Through the water, or in the atmosphere directly over it, sounds are transmitted to a great distance. The ringing of a bell under water has been heard across the whole breadth of Lake Geneva, not less than nine miles. The "all's well" of the sentinel at Gibraltar has been distinguished twelve miles off, and naval engagements have been heard at a distance of 200 miles. An eruption of the volcano of St. Vincent has been heard at Demerara, 340 miles off,—the greatest distance on record to which sound has been transmitted by the atmosphere.

729. ACOUSTIC TUBES.—It is their dispersion in the surrounding air that makes sounds finally inaudible. Hence, when they are confined within tubes, they are carried to a much greater distance. The slightest whisper has been heard through an iron pipe 3,120 feet (more than half a mile) in length.

This fact has been turned to account in several ways. The voice is conveyed by speaking-tubes from one part of a building to another, frequently to a considerable distance and by a circuitous route. The Stethoscope, an instrument for examining the lungs and other internal organs, is an application of the same principle. It is a hollow cylinder of wood with a funnel-shaped extremity, which is placed on the organ to be examined while the ear is applied to the other end. The sounds produced by the vital action within are thus conveyed to the ear, and enable the experienced examiner to judge whether the organ is in a healthy state.

---

air? Iron? Wood? What experiments prove that solids conduct sound more rapidly than air? 728. What makes it impossible to lay down a universal law as to the distance at which sound is audible? How far may the human voice be heard? The report of a musket? What instances are mentioned showing the great distance to which sound is transmitted by water? What is the greatest distance on record to which sound has been transmitted by the atmosphere? 729. What makes sounds finally inaudible? How may this difficulty be in a measure removed? How far has a faint whisper been heard through a tube? How has this principle been turned to

730. *The Speaking-trumpet.*—Even if the tube is short, the more intense pulsation excited in a column of confined air makes a given sound audible at a much greater distance than if it is at once diffused in the atmosphere. This is proved by the Speaking-trumpet, an instrument used by seamen and others who wish to give additional power to their voices. The narrowness of the tube prevents the easy flow of the air which the voice sets in vibration. The organs of articulation, therefore, operate on it with concentrated force, as they do on condensed air ; and, consequently, when the vibrations escape from the tube, they are propelled to a greater distance. A loud voice with a speaking-trumpet 20 feet long, can be heard at a distance of three miles. No one can use the speaking-trumpet long without being exhausted, which shows that an unusual effort has to be made with the voice.

731. INTERFERENCE OF SOUND.—Two sets of vibrations of equal intensity, meeting in such a way that the depressions of one correspond with the elevations of the other, *interfere*, or neutralize each other, and an interval of silence is the result.

Cause a tuning-fork to vibrate and hold it over a cylindrical glass vessel. Vibrations will soon be communicated to the glass, and a musical note will be heard. Place a similar glass vessel at right angles to the first and opposite the tuning-fork, and the note previously heard will cease. Withdraw it, and the note is again heard. The vibrations of the first vessel produce the sound, but are neutralized by those of the second.

732. REFLECTION OF SOUND.—Vibrations striking a plane surface are reflected from it (like light and heat) in such a way as to make *the angle of reflection equal to the angle of incidence.*

733. *Echoes.*—When a sound is heard a second time by reflection, after a certain interval, an Echo is said to be produced. A sound is sometimes repeated more than once,

account? What instrument is constructed on this principle? Describe the Stethoscope, and its operation. 730. By whom is the Speaking-trumpet used? Explain the principle on which it operates. How far has a loud voice been heard with a speaking-trumpet? 731. What is meant by the Interference of sound, and how is it caused? Give an example. 732. What is the law for the reflection of sound? 733. What is an

according to the number of reflecting surfaces on which it strikes. An echo near Milan repeats a single syllable thirty times.

To be distinctly heard, the echo must not reach the ear till one-ninth of a second after the original sound has ceased. Otherwise they will run together and form one continuous sound. Hence, the reflecting surface must be a certain distance from where the original sound is produced. The farther it is off, the longer the reflected sounds will be in reaching the observer's ear, and the more syllables will be repeated. At Woodstock, England, there is an echo which repeats from 17 to 20 syllables; in this case the reflecting surface is distant about 2,300 feet. In mountainous regions echoes are quite common. There are several remarkable ones among the Alps; and the mountaineers contrive to sing one of their national songs in such time that the echo forms an agreeable accompaniment.

In ordinary rooms no echo is perceived, because the distance of the walls is so small that the reflected sound is mingled with the original one; but in large halls, unless the principles of Acoustics are regarded, an unpleasant echo follows the speaker's words and makes them confused and indistinct.

734. *Ear-trumpets.*—Ear-trumpets, used by deaf persons, concentrate and reflect to the interior membrane of the ear, vibrations that strike it, and thus render audible sounds that could not otherwise be heard. The principle on which they operate will be understood from Fig. 263.

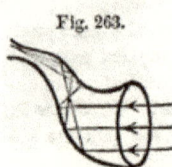

Fig. 263.

THE EAR-TRUMPET.

The sounds enter the large end, and are united by successive reflections at the small end, which is applied to the ear. The outer part of the ear is itself of such a shape as to collect the sound-waves that strike it and reflect them to the membrane within. To enable them to hear more distinctly, we often see people putting up their hands behind their ears, so as to form a concave reflecting surface; in which case, the hand acts somewhat on the principle of the ear-trumpet. Instinct teaches animals to prick up their ears when they want to catch a sound more clearly.

Shells of a certain shape reflect from their inner surface the vibrations that strike it from the external air, and hence the peculiar sound that is heard when they are applied to the ear.

Echo? In what case may a sound be repeated more than once? How often does an echo near Milan repeat a syllable? What is essential to the distinctness of an echo? On what does the number of syllables repeated depend? Give an account of the echo at Woodstock, England. Where are echoes quite common? What is said of those in the Alps? Why is there no echo in ordinary rooms? 734. How is it that Ear-trumpets render audible sounds that could not otherwise be heard? What is said of the outer part of the ear? How is the hand made to act on the principle of a speaking-trumpet? Why do animals prick up their ears? Explain the roaring of

735. *Whispering Galleries.*—Sound reflected from curved surfaces follows the same law as light and heat. Let two large concave brass mirrors be placed opposite to each other, as shown in Fig. 213; the ticking of a watch, or the faintest whisper in the focus of one is distinctly heard, after two reflections, at the focus of the other, though inaudible at any other point. Two persons with their backs to each other can thus carry on a conversation, while those between them are not aware that anything is being said.

An apartment in which such a reflection is produced by the walls is called a Whispering Gallery. An oval form is the best for such a gallery, because there are two points within, to either of which all the vibrations produced at the other are reflected at the same instant from every point of the surrounding walls. The dome of St. Paul's Church, London, and that of the Capitol at Washington, are examples of fine whispering galleries.

One of the most remarkable structures of this kind in ancient times was "the ear of Dionysius", a dungeon so called from the tyrant of Syracuse, by whom it was constructed. The walls and roof were so arranged that every sound from within was reflected and conveyed to a neighboring apartment, where the tyrant could ensconce himself and hear even the whispers of his unsuspecting victims.

736. Musical Sounds.—Musical Sounds are produced by regular vibrations, uniform in their duration and intensity.

737. *Loudness, Pitch, and Quality.*—In connection with musical sounds, three things must be considered; their Loudness, their Pitch, and their Quality.

The Loudness of a musical sound depends on the extent of the vibrations producing it. The greater the vibrations, the louder is the sound.

The Pitch of a musical sound depends on the rapidity

shells. 735. What law does sound reflected from curved surfaces follow? Illustrate this law in the case of sounds reflected from two concave mirrors. What is a Whispering Gallery? What is the best form for such a gallery, and why? What buildings contain whispering galleries? Give an account of "the ear of Dionysius". 736. How are Musical Sounds produced? 737. What three things must be considered in connection with musical sounds? On what does the Loudness of a musical sound

of the vibrations producing it.   The more rapid the vibrations, the higher is the pitch.

The slowest vibrations that produce audible musical sounds follow each
other at the rate of 8 in a second, and a very low note is the result.   As the
vibrations become more rapid the pitch rises, till they recur at the rate of
24,000 in a second, when a very high note is produced.   Beyond this the vibrations last so short a time that they no longer affect an ordinary ear, and
no musical sound is heard.

The Quality of a musical sound depends on the nature
of the vibrating body.   The human voice, the piano, and
the flute, may all produce a note of precisely the same
loudness and pitch, and yet we readily distinguish them
apart.   The difference lies in their Quality.

738. All musical sounds are produced by the regular
vibrations either of solids or confined air.   This gives rise
to a division of musical instruments into two classes :—
Stringed Instruments, like the violin ; and Wind Instruments, like the flute.

739. STRINGED INSTRUMENTS.—The strings used in musical instruments are made of metal or cat-gut.   They are
fastened at each end, and are set in vibration with the finger, as in the case of the harp,—or by the stroke of a hammer, as in the piano,—or by drawing across them an instrument made for the purpose, like the bow of a violin.

740. To produce notes of different pitch, two strings
must vibrate with different degrees of rapidity.   That they
may do so, one must be longer than the other, or thicker,
or stretched more tightly.

The longer a string is, with a given thickness and tension, the more
slowly it vibrates and the graver its tone.—The thicker a string is, with
a given length and tension, the more slowly it vibrates and the graver its
tone.—The more tightly a string is stretched, with a given length and thickness, the more rapidly it vibrates and the more acute its tone.

---

depend?   On what, its Pitch ?   How rapidly do the vibrations that produce the lowest audible musical sounds follow each other ?   How rapidly, those that produce the
highest notes ?   On what does the Quality of a musical sound depend ?   Give an example of difference in quality.   738. By what are all musical sounds produced ?   How
are musical instruments, then, divided ?   739. Of what are the strings used in musical instruments made ?   How are they set in vibration ?   740. How are two strings
made to produce notes of different pitch ?   State the three laws relating to the length,

Stringed instruments are tuned,—that is, brought to their proper pitch,—by turning pegs to which the strings are attached. Changes in the condition of the atmosphere affect the length and consequently the tone of the strings.

741. The music of the Æolian Harp is produced by the action of currents of air on strings which are stretched between two small uprights two or three feet apart. The most pleasing combinations of sounds sometimes proceed from this simple instrument, commencing with a strain, soft and low, as if wafted to the ear from a distance, then swelling as if it were coming nearer, while other notes break forth, mingling with the first with indescribable sweetness.

742. In the case of the drum, musical sounds are produced by the vibrations of a tense membrane acting on the same principle as strings.

743. WIND INSTRUMENTS.—In wind instruments, such as the flute, the trumpet, &c., musical sounds are produced by the vibrations of air confined within tubes. In tubes of equal diameter, the pitch of the note differs according to the length of the vibrating column ; the shorter the column, the higher or sharper the note.

There are two ways of producing notes of different pitch with the same instrument :—1. By joining tubes of different length and diameter, as in the organ. 2. By having but one tube and providing apertures in it at different intervals, by uncovering which the air is allowed to escape, and the internal vibrations are stopped at any desired point. This is the arrangement in the flute.

A wind and a stringed instrument produce notes of the same pitch when the column of air contained within the former vibrates with the same rapidity as the string which produces the note of the latter.

744. The tubes of wind instruments may be open at both ends, or closed at both ends, or open at one end and closed at the other. In the last case, the note produced is twice as low as in either of the other cases, the length of the tubes being the same.

745. Musical notes are produced with wind instruments by blowing into one end, by causing a current of air to enter an aperture, or by making

thickness, and tension of strings. How are stringed instruments tuned? What causes them to get out of tune? 741. How are the sounds of the Æolian Harp produced? Describe the music of this instrument. 742. How are musical sounds produced in the case of the drum? 743. How are musical sounds produced in wind instruments? On what does the pitch of the note depend? How many ways are there of producing notes of different pitch with the same wind instrument? Mention them. When do a wind and a stringed instrument produce notes of the same pitch? 744. What is said respecting the openings of the tubes of wind instruments? 745. What three modes of producing musical notes with wind instruments are men-

such a current act on thin plates of metal or wood properly arranged within.

746. A jet of hydrogen gas, ignited and made to pass through a glass tube about an inch in diameter, produces sweet musical sounds, which may be made soft or loud at pleasure by raising or lowering the tube. These sounds are caused by vibrations excited in the confined air by the burning hydrogen.

747. *The Organ.*—The grandest and most complicated of wind instruments is the organ. It combines the tones of almost every other wind instrument, in such a way that they may be used singly or together at the pleasure of the performer. An organ in Switzerland has tones so closely resembling those of the human voice, that visitors who hear it imagine they are listening to a full choir of singers. The great organ at Haarlem, in Holland, which is the most celebrated one in the world, has no less than 5,000 *pipes*, as the tubes of the organ are technically called.

The water-organ, or *hydraulicon*, was known more than two hundred years before the Christian era. Its invention is attributed to Ctesibius, the barber of Alexandria, already mentioned as the inventor of the lifting-pump. Wind-organs appear to have been little known until the eighth century after Christ, though perhaps invented some time before. We read that an instrument of this kind was sent to King Pepin, of France, in the year 757, by the Greek Emperor, Constantine.

748. THE GAMUT.—Notes are said to be *in unison* when the vibrations that produce them are performed in equal times.

Two notes, one of which is produced by twice as many vibrations as the other, are called Octaves. In passing from a note to its octave, there are several intermediate sounds, produced by intermediate numbers of vibrations, each of which the ear recognizes as a distinct note. These notes are distinguished by different names, as shown below. Assuming the number of vibrations producing the first to be 1, the relative number of vibrations producing

tioned? 746. How may musical notes be produced with a jet of hydrogen gas? 747. What is the grandest of wind instruments? What are combined in the organ? What is said of an organ in Switzerland? How many pipes has the great Haarlem organ? How long ago was the water-organ known? By whom was it invented? When do wind-organs appear to have first become known? 748. When are notes said to be *in unison?* What is meant by Octaves? Between a note and its octave,

the other notes will be expressed by the fractions respectively placed below them, the number of the eighth note being, as already stated, double that of its octave.

| Names of the notes, | C | D | E | F | G | A | B | C |
|---|---|---|---|---|---|---|---|---|
| or, | do | re | mi | fa | sol | la | si | do |
| Pronounced, | *do* | *ra* | *me* | *fah* | *sole* | *lah* | *se* | *do* |
| No. of vibrations, | 1 | $\frac{9}{8}$ | $\frac{5}{4}$ | $\frac{4}{3}$ | $\frac{3}{2}$ | $\frac{5}{3}$ | $\frac{15}{8}$ | 2 |

These eight notes constitute the Gamut, or Diatonic Scale. The notes of the next higher octave bear the same relations to each other, but are produced by vibrations performed in half the time, and therefore twice as numerous in each case. The notes of the next lower octave again bear the same relations to each other, but their vibrations take twice the time, and are therefore only half as numerous. In other words, a given note of any octave is produced by vibrations twice as rapid as the same note of the next octave below, and only half as rapid as the same note of the next octave above.

749. HARMONY.—Some notes, reaching the ear simultaneously, produce an agreeable impression in consequence of their vibrations' frequently coinciding, and constitute what is called *concord*. Other notes, whose vibrations rarely coincide, impress the ear unpleasantly and produce *discord*. A combination of concordant musical sounds is called a Chord. An agreeable succession of musical sounds constitutes Melody. A succession of chords constitutes Harmony.

The most agreeable concord is that of the octave; next, the fifth; then, the fourth; and then, the third. Thus, in the scale given above, concord is produced when C is sounded with its octave C, and with the notes G, F, and E.

750. THE HUMAN VOICE.—The sounds of the human voice, whether used in speaking or singing, are produced by the vibrations of two membranes stretched across a tube, which connects the mouth with the lungs. This tube is the wind-pipe; and the upper part of it, which consists

what occur? Name the notes by letters. Give their other names. Assuming the number of vibrations that produce C to be 1, mention the relative numbers that produce the other notes. What do these eight notes constitute? What relation do the notes of the next higher octave bear to these? The notes of the next lower octave? 749. What is meant by Concord? By Discord? What is a Chord? What is Melody? What is Harmony? Which is the most agreeable concord? Which next? Which next? 750. How are the sounds of the human voice produced? Describe

of cartilage, is called the Larynx. The larynx is flattened at the top, and terminates in two membranes, which nearly close the passage, leaving between them a narrow opening, known as the Glottis. These two membranes are called the Vocal Chords, and it is by their vibration, caused by the passage of the air breathed out from the lungs, that the sounds of the voice are produced. Small muscles enable us to stretch the vocal chords more or less tightly at pleasure, and also to enlarge or diminish the opening between them. By these means we produce notes of different pitch. To produce a change of note, we have only to make a difference of $\frac{1}{1200}$ of an inch in the length of the vocal chords.

Fig. 264.

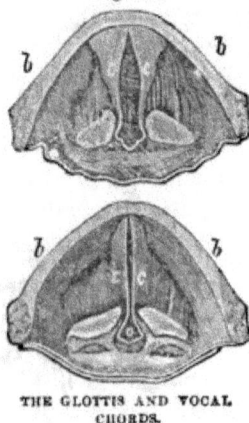

THE GLOTTIS AND VOCAL CHORDS.

Fig. 264 represents the glottis under different circumstances. The upper plate shows it at rest: b, b, represents the top of the larynx, and c, c, the vocal chords, relaxed so that the breath passing through the opening makes no sound. The lower plate shows the glottis in the act of emitting a musical sound, the chords being now tightly stretched, and made to vibrate by the air breathed out between them. o is a passage leading into the wind-pipe, which remains open, however close to each other the chords may be brought.

751. The vocal chords are shorter in boys and women than in men; hence the voices of the former are sharper or higher than those of the latter. When boys reach the age of 14 or 15, the vocal chords rapidly enlarge, and the voice is said to change.—The more forcibly the air is expelled from the lungs through the wind-pipe and larynx, the louder is the voice.

752. His surprising flexibility of voice enables man to imitate almost exactly, not only the cries of birds and beasts, but also the sounds of various musical instruments. This was shown by the performances of a band of twelve Germans a short time since in the principal cities. Each imitated a different instrument with his voice, and so accurately, that those who heard

the Larynx and the Glottis. What are the membranes stretched across the top of the larynx called? How do we produce notes of different pitch? How great a difference in the length of the vocal chords produces a change of note? Point out the different parts in Fig. 264. 751. Why are the voices of men deeper than those of boys and women? What causes the voices of boys to change? On what does the loudness of the voice depend? 752. What is said of the flexibility of the human

them could hardly believe they were not listening to an instrumental concert.

753. *Ventriloquism.*—Some persons have the faculty of uttering sounds and words without moving their lips. When, besides this, they can throw their voice into any object (as the expression is), or make it seem to come from a distance, they are called Ventriloquists. By practice ventriloquists attain to wonderful power over their voices.

Amusing exhibitions of ventriloquism are often given, in which the performer imitates to perfection the buzzing of bees, the grunting of pigs, the spitting of cats, the chirping of crickets, the drawing of corks, the gurgling of liquids, the moaning of the wind, the puffing of a locomotive, the cry of a young infant, conversation between different parties represented as approaching or receding, in different parts of the room, under tables, &c.—It is supposed that the priests of the ancient oracles practised ventriloquism, and thus made their responses appear to come from shrines, statues, &c.

754. *Stammering.*—Stammering is a defect in speech caused by the organs' not performing their respective parts in regular succession. A convulsive nervous action interferes with their operation.

755. The difficulty in the case of deaf mutes does not lie in any imperfection of the organs, but proceeds simply from their deafness. Having never heard their own voices or those of others, they are utterly unable to appreciate sounds or adjust the organs properly for their articulation.

756. VOICES OF THE INFERIOR ANIMALS.—Man alone has the power of articulation. The inferior animals utter cries of different kinds, according to the conformation of the larynx and the nasal cavities connected with it. Some of the cat's tones very closely resemble those of the human voice.

The sounds of insects are produced in various ways,—by the rapid vibration of their wings, the rubbing of their minute horns against each other, the striking of their organs on the bodies around them, &c.

757. THE HUMAN EAR.—The human ear consists of three distinct parts; the outer ear, the drum, and the in-

voice? What instance of its remarkable flexibility is given? 753. What is Ventriloquism? Describe some of the feats of ventriloquists. What use is supposed to have been made of ventriloquism in ancient times? 754. What is the cause of Stammering? 755. Why are deaf mutes unable to use their voices? 756. What is said of the tones of the inferior animals? How are the sounds of insects produced?

ner ear. These parts and their connections are represented in Fig. 265.

Fig. 265.

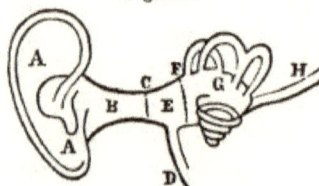

THE HUMAN EAR.

A A is the *outer ear*, which acts on the principle of the ear-trumpet, collecting the sound-waves and reflecting them along the pipe B to the membrane C, called the *membrane of the tympanum.* E is the *tympanum* or *drum*, bounded by the membrane C on the one side, and the membrane F on the other, and filled with air, which it receives from the tube D, communicating with the mouth. G, the *inner ear*, contains a number of ducts, and is filled with a liquid in which the acoustic nerve floats.

The sound-waves transmitted from the outer air cause the membrane C to vibrate. C excites vibrations in the air confined in the drum, and this in turn causes F to vibrate. The liquid in the inner ear receives the vibrations from the membrane F, and transmits them to the acoustic nerve, by which they are conveyed to the brain, and the sensation of hearing is produced. When a person takes cold, the tube which connects the drum with the mouth is apt to be obstructed, and temporary deafness is the consequence.

### EXAMPLES FOR PRACTICE.

1. (*See* § 724.) If the air were perfectly still and uniform in density, how would the report of a musket heard by a person 50 feet off compare in loudness with the same report heard at a distance of 250 feet?
2. A cannon is heard a quarter of a mile off with a certain degree of loudness. How far must a person be removed, to hear it with only $\frac{1}{100}$ of its former distinctness?
3. (*See* § 725.) How far does sound travel through air in 10 seconds? In 20 seconds? In one minute?
4. How much faster does the sound produced by the discharge of a cannon travel, than that produced by the snapping of a whip?
5. (*See* § 726.) I see the flash of a cannon two seconds before I hear its report. How far is it off?
6. A clap of thunder does not reach the ear till four seconds after the accompanying flash is visible. How far off is the thunder-cloud?
7. A thunder-cloud is distant about one mile. How many seconds will elapse between the flash and the clap?
8. (*See* § 727.) About how many feet will sound travel through water in 10 seconds? Through iron? Through wood?

---

757. Name the parts of which the human ear consists. With the aid of Fig. 265, point out the different parts, and show the operation of the organ. Why is temporary deafness produced by a cold?

# CHAPTER XVI.

## ELECTRICITY.

758. If a dry glass tube or a stick of sealing-wax be rubbed with a piece of flannel, and then held a short distance above some shreds of cotton, they will be instantly attracted to it, and after adhering to its surface for an instant again thrown off. A peculiar odor is perceived; and the face, when brought near the glass or wax, feels as if a cobweb were in contact with it. If the tube or sealing-wax be presented to a metallic body in a dark room, a spark, accompanied by a sharp cracking sound, will be seen darting from it to the metal.

The property thus developed by friction is called Electricity. The body in which it is developed is called an Electric, and is said to be *excited* or *electrified*. The attraction exerted by the excited electric over light bodies is called Electrical Attraction. The substance by whose friction the electric is excited is known as the Rubber.

759. ELECTRICITY AS KNOWN TO THE ANCIENTS.—The term *electricity* is derived from the Greek word *electron*, amber, the property in question having been first observed in that substance.

Thales, one of the seven wise men of Greece, who flourished 600 years B. C., is said to have discovered electricity in amber; Theophrastus and Pliny, at a later date, speak of the attraction of amber for leaves and straws. Both Pliny and Aristotle were acquainted with the electrical properties of the torpedo; and we are informed that a freedman of the Emperor Tiberius cured himself of gout by the use of its shocks. Yet the ancients appear to have known nothing more than a few isolated facts connected with the subject; and as a science Electricity had no existence till the commencement of the seventeenth century.

---

758. If a glass tube or a stick of sealing-wax be rubbed with flannel, what phenomena will be observed? Name and define the terms used in connection with this experiment. 759. What is the derivation of the term *electricity?* What allusions are made to this property by ancient authors? When did electricity originate as a sci-

13

760. SOURCES OF ELECTRICITY.—Electricity is developed—1. By friction. 2. By chemical action. 3. By magnetism. 4. By heat.

### Electricity developed by Friction.

761. Friction is one of the commonest sources of electrical excitement. Every one has noticed how his hair crackles under the comb in frosty weather. The same sound is heard on stroking the back of a cat, and if the room is dark sparks may be drawn from its fur.

A striking example of the exciting power of friction is often afforded in factories. The endless bands by their friction on the wheels develop electricity in great abundance, sometimes yielding sparks at a distance of two or three feet. In the carding-rooms of cotton mills, fibres of cotton are kept dancing to and fro by alternate attractions and repulsions, so that steam has to be let in from time to time to dissipate the electric fluid.

762. ELECTRICAL ATTRACTION AND REPULSION.—We have already noticed the alternate attraction and repulsion of shreds of paper, cotton, and similar substances by excited electrics. These phenomena may be further exhibited with the apparatus represented in Fig. 266, which consists of a pith ball suspended from a pillar by a long silken thread.

Fig. 266.

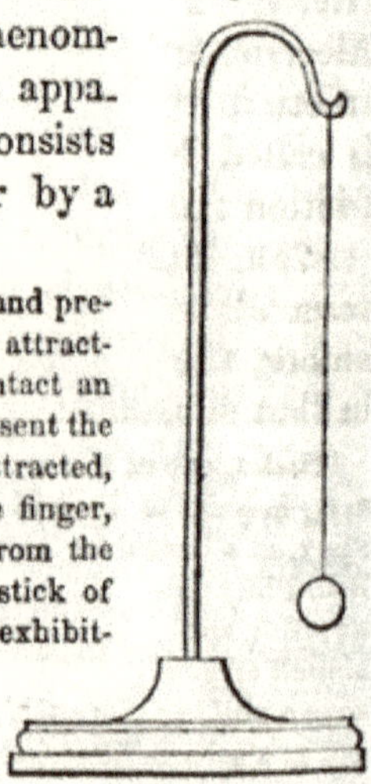

*Experiment* 1.—Rub a glass tube with flannel, and present it to the pith ball; the latter will be instantly attracted to the tube. After they have remained in contact an instant, the ball will be thrown off. If we now present the tube a second time, the ball, instead of being attracted, will be repelled. After touching the ball with the finger, to deprive it of the electricity it has received from the tube, repeat the experiment with an excited stick of sealing-wax, and the same phenomena will be exhibited,—that is, the ball will at first be attracted, but on the second application of the wax will be repelled. We find, then,—1. That both the

ence? 760. By what is electricity developed? 761. What familiar instances are mentioned of the production of electricity by friction? What striking example is afforded in factories? 762. What does Fig. 266 represent? What may it be used to illustrate? Describe the first experiment with the pith ball. What two facts are glass and the sealing-wax attract the ball before they have communicated to it any of their own electricity. 2. That, after so doing, they both repel the ball.

*Experiment 2.*—Suspend two pith balls from a pillar by silk threads, and present to them an electrified glass tube or piece of sealing-wax. They will both be attracted; but, on withdrawing the electric, instead of hanging vertically, they will repel each other, as shown in Fig. 267.

Fig. 267.

*Experiment 3.*—Excite the glass tube, present it to the ball represented in Fig. 266, withdraw it after a second or two, and then present the excited sealing-wax. The ball, instead of being repelled, is now attracted. Reverse the experiment by presenting first the excited wax and then the glass, and the latter in like manner will be found to attract the ball.

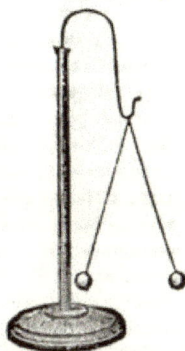

763. From these experiments it has been inferred that there are two kinds of electrical excitement: that produced by glass, which has been called Vitreous or Positive Electricity; and that produced by sealing-wax, which has been called Resinous or Negative Electricity. We may lay down the general law that *substances charged with opposite electricities attract each other, while those charged with like electricities repel each other.*

764. NATURE OF ELECTRICITY.—What electricity is,—whether it is an imponderable material substance, or consists in the vibrations of some subtile medium, or is simply a condition of matter,—we are unable to say. It was formerly supposed to be an exceedingly subtile fluid pervading all things, and for convenience' sake the expression *electric fluid* is still retained. The leading theories respecting the nature of electricity are Du Fay's, Franklin's, and Faraday's.

*Du Fay's Theory.*—Du Fay, a French philosopher, held that there are two distinct electric fluids (named by him Vitreous and Resinous), each of which attracts the other, but exhibits repulsion among its own particles. That in their natural state these fluids pervade all bodies in equal quantities, and

shown by this experiment? Describe the second experiment. The third experiment. 763. What has been inferred from these experiments? What general law may be laid down? 764. What is said of the nature of electricity? What was it formerly supposed to be? By what names are the leading theories respecting the nature of electricity distinguished? Give the substance of Du Fay's theory. Of

combining nullify each other; it is only when this quiescent compound fluid is decomposed by friction, or any other agency, that electrical phenomena are exhibited.

*Franklin's Theory.*—Dr. Franklin, whose views were once generally received by scientific men, believed that there is but one electric fluid, of which every body in its natural state possesses a certain quantity. That no evidences of the existence of this fluid are observed as long as a body retains its natural quantity; but, when it has either more or less than this, it exhibits certain phenomena and is said to be *electrified*. When overcharged, a body exhibits the phenomena displayed by glass when excited by flannel, and to such an electrical condition Franklin applied the term Positive; when deprived of its proper share, its phenomena are the same as those of excited resinous substances, and such an electrical state he called Negative. When communication is established between a positive and a negative body, the former shares its superfluous electricity with the latter, till equilibrium is established between them. Du Fay made the difference between the two electricities to consist in quality; Franklin, in quantity.

*Faraday's Theory.*—Faraday, an eminent English authority, regards electricity as simply a condition of matter. According to his theory, an electrified body is not pervaded by any fluid at all, but simply endowed with a certain property which under other circumstances it does not possess.

765. Why the electricity of one body when excited is positive and that of another negative, we can not tell. There is no law by which it can be determined, before experiment, what kind of electricity a body will exhibit. Indeed, the same body exhibits different kinds when rubbed by different substances. Thus, polished glass is positively electrified, when excited with flannel, but negatively when rubbed on the back of a cat. Rough glass is negatively electrified when rubbed with flannel, but positively when excited by dry oiled silk.

766. Electricity is confined to the surface of an excited body; it does not extend to the interior. A hollow ball may therefore contain just as much electricity as a solid ball of the same size.

767. Positive electricity is never produced without negative, or negative without positive.

When a glass tube is excited, the rubber is negatively electrified; and positively, when sealing-wax is excited. This may be shown by applying the rubber to a pith ball charged with the electricity which it has excited either in glass or sealing-wax. The ball is invariably attracted, which shows that the electricity of the rubber is opposite to that of the electric it has excited.

768. ELECTRICS AND NON-ELECTRICS.—All bodies can be electrified, but not with equal facility. Those that are easily excited, are called Electrics; those that it is hard to excite, Non-electrics. The metals generally are non-electrics.

769. CONDUCTION OF ELECTRICITY.—If we touch the two pith balls represented in Fig. 267 as repelling each other (because charged with the same electricity) with a glass rod, they will continue to repel each other; but, if we touch them with a metallic rod, they will fall and hang vertically. This is because glass does not draw off their electricity, while metal does. Some substances, therefore, conduct electricity, while others do not.

Substances that transmit electricity freely are called Conductors; those that do not, Non-conductors.

As a general thing, the non-electrics are conductors, and the electrics non-conductors. Some of the chief conductors are the metals (silver and copper ranking among the best), charcoal, water, snow, living animals, flame, smoke, and steam. Among the principal non-conductors are gutta percha, shellac, amber, the resins, sulphur, glass, transparent gems, silk, wool, hair, feathers, dry paper, leather, baked wood, air, and gases generally.

Good conductors, when brought in contact with excited bodies, at once draw off their electricity, and transmit it to all parts of their own surface, however extended. Bad conductors, on the other hand, receive electricity slowly, and diffuse it over their own surfaces no less slowly. A good conductor connected with the earth or a body of water, does not for an instant retain electricity communicated to it, but merely serves as a highway for its passage to either of those media.

770. *Insulators.*—The best non-conductors are called

one kind of electricity always accompanied? How may this be shown? 768. What are Electrics? Non-electrics? To which of these classes do the metals belong? 769. How may it be shown that there is a difference in the conducting power of different substances? What is a Conductor of electricity? A Non-conductor? To which of these two classes do the electrics generally belong? To which, the non-electrics? Mention some of the chief conductors. Some of the principal non-conductors. Show the difference between good conductors and bad conductors, when brought in contact with excited bodies. What is said of good conductors connected

Insulators, because they insulate electrified bodies,—that is, cut off their communication with such objects as would withdraw their electricity. The air is an insulator; were it not, no substance could remain electrified for an instant. When insulated, an excited body retains the electricity communicated to it, and is said to be *charged.* The pith ball in the experiment described in § 758 was insulated by the silk thread. Had it been suspended by a wire, the metal, being a good conductor, would have withdrawn the electricity from the ball as fast as it was received, and none of the phenomena that followed would have been exhibited.

Even when insulated, excited bodies will in time part with their electricity. This is because no insulation can be perfect.—Air, when imbued with moisture, acquires conducting power; and hence in damp weather it is impossible to keep an electric excited for any length of time. Well insulated bodies, slightly excited, may be kept several months in a dry atmosphere without any perceptible loss of electricity.

771. Path of an Electric Current.—An electric current always follows the best conductor, and of two equally good it takes the shorter.

772. Velocity of Electricity.—Various experiments have been made to determine the velocity of electricity. Their results show that electricity travels from 11,000 to 288,000 miles in a second, according to its intensity and the nature of the conductor along which it passes. In the case of the velocity last mentioned, which far exceeds that of light, and is so great as to be absolutely inconceivable, the conductor was copper wire.

773. Electrical Machines.—The Electrical Machine is an apparatus for developing large quantities of electricity by the friction of a rubber on a glass surface. Two kinds of electrical machines are in use, known as the Cylinder

with the earth or a body of water? 770. What is meant by Insulators? Why are they so called? Give an example of an insulator. When is an excited body said to be *charged?* Give an example. How is it shown that no insulation is perfect? Show the difference in conducting power between dry and damp air. 771. What path is always taken by an electric current? 772. How great is the velocity of electricity? 773. What is the Electrical Machine? How many kinds of electrical machines are

and the Plate Machine,—a glass cylinder being used in the former, and a circular plate of glass in the latter.

774. Experiments in electricity were originally performed with a glass tube rubbed with fur or flannel. Otto Guericke, the inventor of the air-pump, was the first to contrive a machine for developing the fluid more abundantly. It consisted of a globe of sulphur, turned with a winch, and submitted to the friction of the hand. Newton substituted a glass globe for the sulphur. About the middle of the eighteenth century, two further improvements were made,—the use of a rubber instead of the hand, and the addition of a metallic conductor.

775. *The Cylinder Machine.*—In the cylinder machine, represented in Fig. 268, electricity is developed by the friction of a rubber upon a glass cylinder, usually from 8 to 12 inches in diameter, supported between two uprights of well-dried wood, and made to revolve by a couple of wheels, as shown in the Figure, or (as is now generally preferred) by a simple winch attached to one end of the cylinder.

Fig. 268.

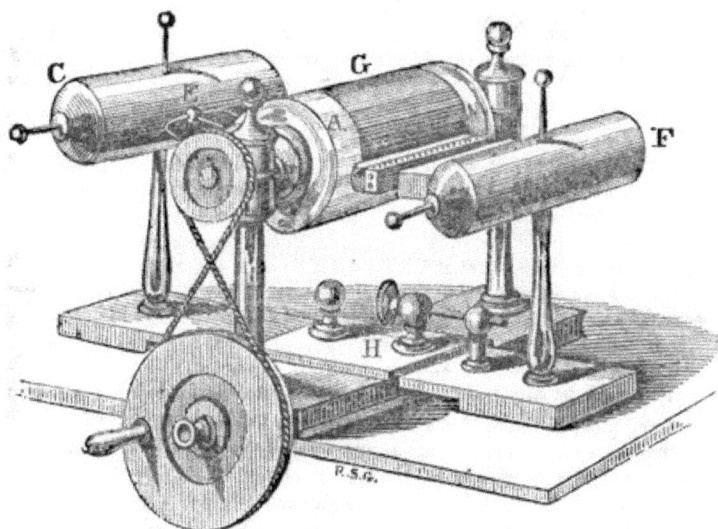

THE CYLINDER ELECTRICAL MACHINE.

in use? What constitutes the difference between them? 774. With what were experiments in electricity originally performed? Who first contrived an electrical machine? Describe Guericke's apparatus. What improvement did Newton make? What improvements were made about the middle of the eighteenth century?

A is the cylinder. The rubber, B, is a leather cushion stuffed with horse hair, and set on a spring which makes it press equally against the cylinder in all parts of its revolution. The intensity of its pressure is regulated by a sliding base-board, H, which can be moved by a screw towards or from the cylinder. Connected with the back of the rubber is the *negative conductor*, F, a hollow metallic cylinder, with round ends, insulated by a glass pillar. On the opposite side is a similar metallic cylinder, C, insulated in the same way, and called the *prime conductor*. Attached to this is a rod bearing a row of metallic points, E, like the teeth of a rake, projecting towards the cylinder and reaching to within a short distance of it. Several holes of different size are made in the upper surface of the prime conductor, to admit of the introduction of different pieces of apparatus used in experimenting. To prevent the electricity from escaping in the air before it reaches the prime conductor, a flap of black silk, G (which is a non-conductor), extends from the upper edge of the rubber, across the top of the cylinder, to within an inch of the metallic points.

776. *Operation.*—When the machine is to be used, its parts must be perfectly clean and dry. The rubber is rendered more efficient by spreading on it a thin coat of an amalgam of zinc, tin, and mercury, mixed with lard. The screw must be adjusted so that the rubber may press with moderate force on the glass, and the prime conductor so placed as to bring the metallic points about an eighth of an inch from the cylinder. If positive electricity is required, the negative conductor must be connected with the earth by a metallic chain. This done, the handle is turned. The electricity naturally present in the rubber is thus decomposed, and its positive part follows the revolving glass. On its reaching the metallic points, the neutral electricity naturally present in the prime conductor is decomposed; its negative element is attracted by the positive fluid of the cylinder, and rushes over the metallic points to unite with it, while its positive portion is repelled to the opposite surface of the conductor. The negative fluid received from the prime conductor neutralizes the positive fluid of the cylinder; but on reaching the rubber (which has meanwhile received a supply from the earth through the conducting chain) the process is repeated. The prime conductor does not, therefore, receive any positive electricity from the cylinder, but is rendered strongly positive by having its own negative fluid withdrawn.

If negative electricity is wanted, the chain connecting the machine with the earth must be attached to the prime conductor instead of the negative conductor, and the required electricity can then be drawn from the latter.

Water being a good conductor, if the air is damp the electricity is dissipated almost as soon as it is developed. This may be prevented by placing under the cylinder a small box containing a bar of red-hot iron. The radiation of heat from the bar keeps the atmosphere around the machine dry.

777. When the machine is working, present your knuckle to the prime conductor; a spark, accompanied by a sharp cracking sound, darts to your hand, producing a pricking sensation. This is called the Electric Spark. Any conductor will draw off a spark; but let a non-conductor, such as a piece of glass, be presented, and no spark will be received.

778. *The Plate Machine.*—In the Plate Machine, a circular plate of glass is used instead of a cylinder. The greatest electrical effects have been produced with these machines. Plates six and seven feet in diameter have been employed, with such power that a spark from their immense conductors is nearly sufficient to fell a man to the earth. The most powerful machine in the world, made in Boston, for the University of Mississippi, combines two plates, each six feet in diameter.

Fig. 269 represents the plate machine in one of its most convenient and efficient forms. A A is the plate, supported on an axis between two uprights and turned by the handle D. The plate is pressed by two pair of elastic rubbers, fastened on the inside of the uprights.

Fig. 269.

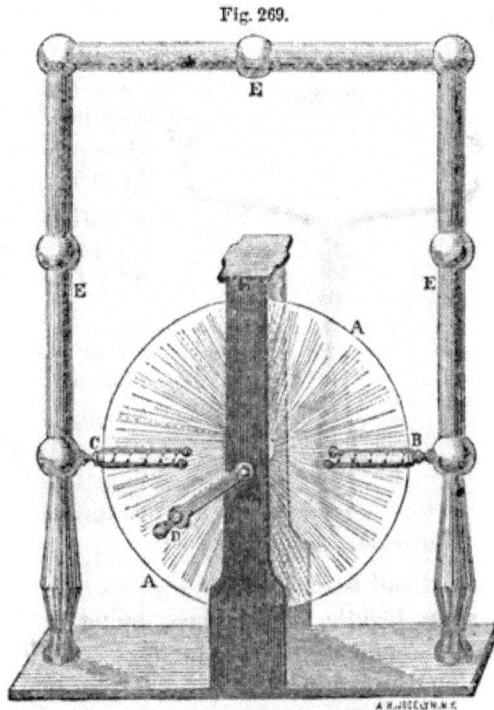

THE PLATE ELECTRICAL MACHINE.

E E E is the conductor, which consists of three long brass tubes joined at right angles, with large balls at intervals. Opposite the centre of the plate, two brass arms, B, C, provided with rows of teeth, extend on each side from the upright conductor. The plate being made to revolve by means of the handle D, the same results follow as in the case of the cylinder machine.

779. THE INSULATING STOOL.—The Insulating Stool consists of a platform of well-baked wood, supported on glass legs covered with varnish. A person on the stool, brought in connection with the prime conductor of a machine by holding in his hand a chain proceeding from it, may be charged with positive electricity. Sparks may be drawn from his person, and his hair, if fine and dry, will stand on end. If he holds in his hand a silver spoon full of alcohol, another person not on the stool may set the spirits on fire by simply presenting his finger to it, and thus producing a spark. The insulating stool is used when electricity is medically applied.

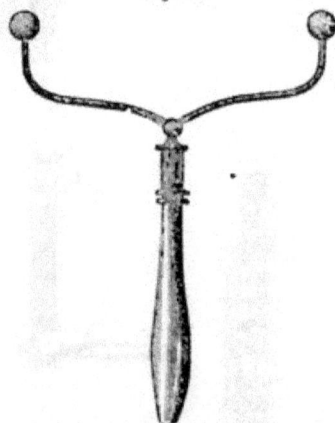

Fig. 270.

THE JOINTED DISCHARGER.

780. THE DISCHARGER.—The Jointed Discharger, Fig. 270, is an instrument with which an operator can discharge a conductor without having any of the electricity pass through his person. It consists of a couple of curved brass rods, terminating in balls at one end and at the other jointed and fixed in a socket, by which they are attached to a glass handle. The glass, being a non-conductor, cuts off communication with the operator's hand.

The Universal Discharger, represented in Fig. 271, is an instrument for passing a charge of electricity through any substance. Two wires, mounted on insulating pillars, are connected respectively with the positive and the negative conductor of a machine. The substance to be operated on is placed on a stand between two balls at the extremities of these wires, and thus made a part of the electric circuit traversed by the fluid when a discharge takes place.

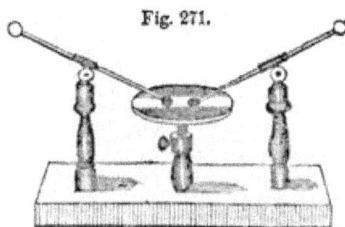

Fig. 271.

THE UNIVERSAL DISCHARGER.

781. THE LEYDEN JAR, OR VIAL.—The Leyden [li'-den] Jar is a glass vessel used for accumulating electricity. It is so called from having been first used at Leyden, Holland, in the year 1745.

Fig. 272.

LEYDEN JAR.

The ordinary Leyden jar (Fig. 272) consists of a glass vessel, coated inside and outside with tin-foil, to within about three inches of its mouth. It is closed with a dry varnished cork, through which passes a wire, terminating above in a brass knob, and below in a chain, which touches the inner coating. If the knob of such a jar be held within half an inch of the prime conductor when a machine is working, a succession of sparks will pass to the knob. In a short time they cease, and the jar is then said to be *charged*. The inside (being connected with the knob) is charged with positive, and the outside with negative electricity, which are prevented from uniting by the non-conducting glass between them.

If a person now grasp the outside of the jar with one hand, and touch the knob with the other, he will experience the peculiar sensation called "the electric shock", in his arms, and if the jar is large, through his chest. If, on the other hand, he apply one ball of the jointed discharger to the outer coat and the other to the knob, the jar will be discharged without his feeling anything, because his communication with the jar is cut off by the glass handle. A body through which a charge is to be sent must form part of the circuit between the inner and outer coating of the jar, so that a union of the positive and the negative fluid can not take place without passing through it.—So much electricity is sometimes accumulated in a jar that a discharge takes place through the glass, making a hole in it and rendering the jar useless.

Describe it and its mode of operation. 781. What is the Leyden Jar? Why is it so called? Of what does the ordinary Leyden jar consist? How is the jar charged? With what kind of electricity is the inside charged? The outside? How may the electric shock be taken? How may the jar be discharged without the operator's taking a shock? What is essential in order that a charge may be sent through a body?

Any number of persons may take a shock at once. Having joined hands so as to form a circle, let the person at one end take hold of a chain connected with the outside of a jar, while the one at the other end touches the knob with a piece of wire. The painful sensation experienced when a shock is taken, is caused by the obstructions which those parts of the body that are imperfect conductors present to the free passage of the electric fluid.

782. An interesting incident is related in connection with the experiments that led to the invention of the Leyden jar. Prof. Muschenbroeck, of Leyden, observing that excited electrics soon lose their electricity in the air, determined to see whether he could not collect and insulate the fluid in a vessel of non-conducting glass, so that it might be kept locked up, as it were, ready for use. Accordingly, he introduced a wire from a prime conductor into a bottle filled with water. After the machine had been working some time, an attendant, holding the bottle in one hand, attempted to withdraw the wire with the other, when he of course received a shock,—so unexpected and so unlike anything he had ever felt before, that it filled him with consternation. Muschenbroeck himself subsequently took a similar shock, which he described in a letter to a French philosopher. He says that he felt himself struck in his arms, shoulders, and breast, so that he lost his breath, and it was two days before he recovered from the effects of the blow and the fright. He would not, he adds, take a second shock for the whole kingdom of France.

783. THE ELECTRICAL BATTERY.—When a very heavy charge is required, a number of jars, coated in the usual way, are placed in a box lined with tin-foil, which forms a

Fig. 273.

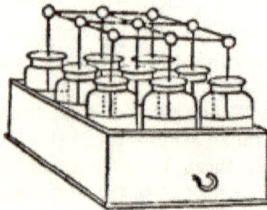

THE ELECTRICAL BATTERY.

communication between their outer coatings, while their knobs and consequently their inside coatings, are connected in the manner represented in Fig. 273. From its powerful effects, such a combination is called an Electrical Battery. By bringing one of the knobs in connection with a prime conductor all the jars may be charged as readily as one, care being taken to connect the outer coatings with the earth. The battery may be discharged in the same way as a single jar, but the operator must not let the charge pass through his person. The shock of a powerful battery will kill a man and fell an ox; even moderate discharges prove fatal to birds and the smaller animals.

784. EXPERIMENTS WITH THE ELECTRICAL MACHINE.— With the electrical machine and different pieces of apparatus that accompany it, a variety of experiments may be performed.

785. *Electrical Bells.*—This apparatus (Fig. 274) illustrates electrical attraction and repulsion. Two bells are suspended from a frame, with a brass clapper between them. One of these bells having been placed in connection with the prime conductor and the other with the ground, the machine is worked; when the former becomes charged with positive and the latter with negative electricity. The clapper is attracted to the positive bell, strikes it, becomes itself charged by the contact, and is repelled till it strikes the negative bell. Its positive electricity is there drawn off, and it falls back, to be again attracted and repelled. The clapper is thus made to strike the bells alternately.

Fig. 274.

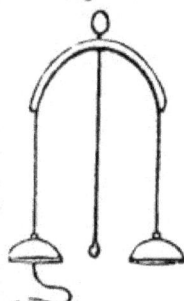

ELECTRICAL BELLS.

786. *The Electrical See-saw.*—The Electrical See-saw (Fig. 275) operates on the same principle. A brass beam, with a light figure on each end, is suspended on an insulating pillar, in such a way as to allow its extremities to move freely up and down. Two brass balls are supported at opposite sides of the stand, not far from the ends of the beam,— the one on a glass pillar, the other on a metallic rod. The insulated ball is connected with the inner coating of a Leyden jar, and the other with its outer coating. No sooner is the jar charged than the figure near the insulated ball is successively attracted and repelled, and this causes the beam to teeter. In the same way motion may be communicated to a figure swinging, a floating swan, an insect suspended in the air, &c.

Fig. 275.

ELECTRICAL SEE-SAW.

787. *Dancing Images.*—On a metallic plate supported by some conducting

Fig. 276.

DANCING IMAGES.

substance, place several light figures of pith or paper, and three or four inches above them suspend another plate from the prime conductor. As soon as the machine is worked, the figures will rise and dance up and down from one plate to another in a ludicrous manner, as shown in Fig. 276. If the lower plate is insulated, when they return to it after having been drawn up, the surplus positive electricity can not escape, and the dance ceases.

Fig. 277.

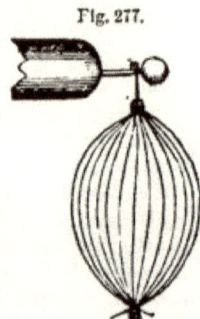

DIVERGING THREADS.

788. *Diverging Threads.* —Figure 277 represents twenty fine linen threads, eight or ten inches long, tied together at each end. Attach them to a prime conductor, and on working the machine, being all filled with electricity of the same kind, they will repel each other and assume an oval form.

Fig. 278.

THE ELECTRIFIED HEAD.

789. *The Electrified Head.*—On the same principle a head of hair is made to stand grotesquely on end, as shown in Fig. 278, by fixing the wire to which it is attached in one of the holes of a prime conductor. The hairs are charged with electricity of the same kind, and are therefore in a state of mutual repulsion. Draw off the fluid by presenting a knife-blade, and they at once fall.

Fig. 279.

ELECTRIC PAIL.

790. *The Electrical Pail.*—Suspend from the prime conductor by a chain a pail with a small hole in the bottom, and fill it with water. Before the machine is worked, the water falls from the hole drop by drop; but, as soon as the water is charged with electricity, it flows out in a stream, which in the dark seems to be of fire. This is owing to the repulsion excited in the particles of water by charging them with the same electricity.

791. *The Aurora Tube.*—This apparatus shows the phenomena

Dancing Images. Why do the images cease to move if the lower plate is insulated? 788. What does Fig. 277 represent? What takes place when these threads are attached to a prime conductor? 789. Describe the experiment with the Head of Hair.

produced when electricity passes through a vacuum. It is a glass tube, from two to three feet long, surmounted by a brass ball. This ball is supported on a wire, which passes into the tube through its air-tight top, and terminates a short distance below in a point. Inside of the tube, near the bottom, is another brass ball supported on a wire. The lower part of the tube is arranged so that it can be fitted to the plate of an air-pump, and is commanded by a stop-cock. Having thoroughly dried and warmed the tube, exhaust it by means of an air-pump; then, in a dark room, bring the upper ball in communication with a prime conductor. As soon as the machine is worked, the whole length of the tube is filled with a continuous stream of violet light; which, on a small scale, strikingly resembles the Aurora Borealis, or Northern Lights. This is a luminous appearance often visible in the north on clear and frosty nights, and peculiarly vivid in high latitudes. It is supposed that the Northern Lights are produced by the passage of currents of electricity through strata of highly rarefied air.

792. *Luminous Words.*—When the continuity of a conductor is broken, a spark darts from one part of it to another. Taking advantage of this fact, we may perform a variety of experiments, which in a dark room have a striking effect.

Fig. 281.

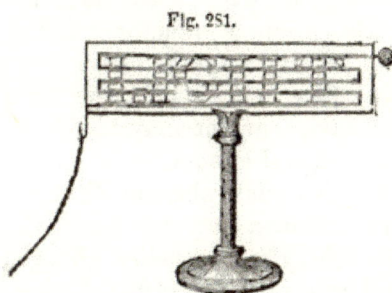

On a piece of glass paste some strips of tin-foil, with portions cut out so that the spaces may form letters, as shown in Fig. 281. Connect the first piece of foil with the prime conductor, and the last with the ground. When the machine is worked, sparks will pass between the different divisions of the foil, and the letters consequently appear like

Fig. 280.

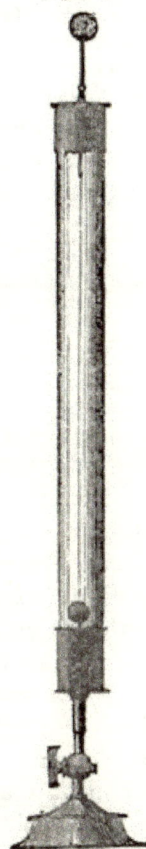

AURORA TUBE.

How may the hairs be made to fall? 790. Describe the experiment with the Electrical Pail. What causes the water to flow more rapidly when the machine is worked? 791. What is shown with the Aurora Tube? Of what does it consist? Describe the experiment with it. By what is it supposed that the Northern Lights are produced? 792. What takes place when the continuity of a conductor is broken? By taking ad-

characters of fire.—Serpentine and spiral lines of light, and other beautiful appearances may be produced, by arranging spangles on glass in the desired form about one-tenth of an inch apart, and subjecting them to the action of the machine.

793. *The Electrical Pistol.*—The electric spark may be made to explode a mixture of hydrogen and common air. In this experiment the Electrical Pistol (Fig. 282) is employed.

Fig. 282.

THE ELECTRICAL PISTOL.

The barrel of the pistol is of brass. Where the trigger is usually found, is a short ivory tube, which insulates a wire passing nearly across the barrel, and terminating on the outside in a ball. Hold the mouth of the pistol over a stream of hydrogen gas, and when enough has entered, close it with a cork. On passing a spark through the barrel from the extremity of the wire to the opposite surface, a loud report will be produced, and the cork will be discharged with considerable force.

794. MECHANICAL EFFECTS OF THE PASSAGE OF ELECTRICITY.—A pointed conductor receives and parts with the electric fluid much more readily than one with a spherical surface. Hence, in electrical machines, points connected with the prime conductor are brought near the excited glass, while the prime conductor itself is cylindrical.

Fix a pointed rod on the prime conductor, and a silent discharge will take place from it as long as the machine is worked. In this case, the prime conductor can not accumulate enough electricity to give a spark. In a dark room, the fluid is seen issuing from the point in the form of a luminous brush. The electric current may be felt if the hand is brought near the rod, and is sometimes strong enough to blow out a candle. No such phenomena occur near the surface of the conductor or a ball attached to it. The point parts with its electricity more readily, charges the air in contact with it, and repels it when charged, as in the case of the pith ball,—thus causing a constant current from the point.

vantage of this fact, what beautiful experiments may be performed? 793. For what is the Electrical Pistol used? Describe this instrument, and the experiment performed with it. 794. Why, in electrical machines, are metallic points connected with the prime conductor brought near the excited glass? Why is the prime conductor itself cylindrical? With what experiments is the silent discharge from points illus-

795. *The Phosphorus Cup.*—An interesting experiment, showing the passage of an electric current, may be performed with the apparatus represented in Fig. 283, known as the Phosphorus Cup. Two brass cups insulated on glass pillars are placed at the same height, about two inches apart, with a lighted candle midway between them. The cups, being each provided with a piece of phosphorus, are connected one with the prime conductor, and

Fig. 283.

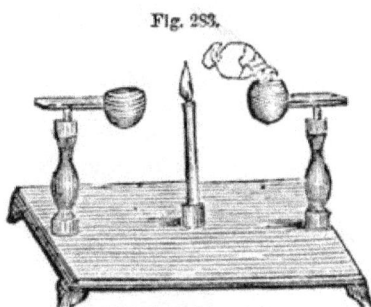

THE PHOSPHORUS CUP.

the other with the negative conductor, of a powerful machine. When the machine is worked, the flame sets in the direction of the negative cup, towards which it is carried by the current of positive fluid from the opposite cup. The phosphorus in the negative cup is soon set on fire by the heat thus produced, whereas at the positive cup there is no increase of temperature, and the phosphorus in it remains unignited. By reversing the connections with the machine, the opposite results may be produced, the flame being always carried towards the cup connected with the negative conductor.

796. When the electric fluid passes off from a pointed conductor, the reaction may be made to turn a wheel, and thus set delicate machinery in motion. To exhibit the effects of this reaction, different pieces of apparatus have been constructed, among which is the Electrical Flyer.

Fig. 284.

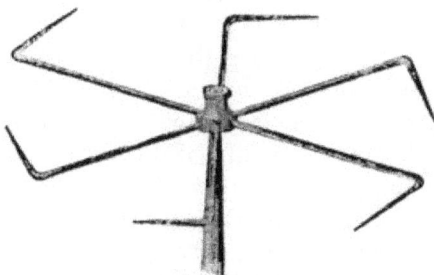

THE ELECTRICAL FLYER.

*The Electrical Flyer.*— The Electrical Flyer consists of a number of brass wires branching out from a common centre, having their ends bent at right angles in the same direction. Poise the flyer on a wire inserted in the prime conductor, and work the machine. A stream of fluid issues from each point, and the flyer is made to revolve in the opposite direction by the reaction of the air. When the room is darkened, the

trated? Explain how a lighted candle is blown out by an electric current. 795. What does Fig. 283 represent? Describe this apparatus, and the experiment performed with it. Towards which cup is the flame always carried? 796. How may delicate machinery be set in motion? How is this reaction shown? Describe the Electrical

points become luminous, and a circle of fire seems to be formed as they revolve.

On the same principle, horsemen (mounted on the ends of the flyer) may be made to move in a circle; wheels may be turned, the sails of a windmill set in motion, and a light body made to roll up an inclined plane.

797. *The Thunder House.*—The power of electricity, as a mechanical agent, may be further illustrated with an ingenious apparatus known as the Thunder House.

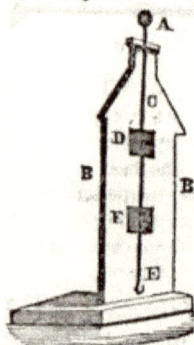

Fig. 285.

THE THUNDER HOUSE.

The Thunder House consists of a piece of baked mahogany, B B, shaped like the gable of a house, and attached to a stand. Down the centre runs a wire, C, terminating above in a ball, A. Several square pieces, D, F, about one-fourth of an inch thick, are cut out of the gable, and placed loosely in the holes from which they are cut. Across each square passes a wire in such a direction that by inserting the squares one way we have an uninterrupted line from C to E; but putting them in crosswise, we break the continuity of the conductor at D and F. Connect the end of the wire, E, with the outside of a Leyden jar; and, having inserted the square so that the conducting line may be unbroken, pass a charge through the wire by connecting the ball A with the inside of the jar. A report will be heard, but neither of the loose pieces will be displaced. Now let one of the pieces remain in the same position, and place the other crosswise; then, on passing a powerful charge through the wire, the former will remain undisturbed, while the latter will be thrown out of the gable by the mechanical action of the fluid in leaping over the break.

798. Among the mechanical effects of an electric discharge may be mentioned the perforation of thin nonconducting substances, such as a card or a piece of paper. Glass one-twelfth of an inch thick may be pierced by a discharge from a powerful battery.

799. THE ELECTRIC SPARK.—The color of the electric spark varies according to the medium through which it passes. In ordinary air and oxygen, it is bluish white; in rarefied air, violet, in nitrogen, a purplish blue; in hydrogen, crimson; in carbonic acid and chlorine, green.

Flyer. What is the effect of darkening the room? To what may motion be communicated on the principle of the flyer? 797. What apparatus further illustrates the mechanical power of electricity? Describe the Thunder House, and the experiment performed with it. 798. What other mechanical effect of an electric discharge is mentioned? 799. What does the color of the electric spark depend on? What

The length and intensity of the spark depend on the electrical intensity of the body from which it proceeds. Sparks may be taken from the prime conductor of a powerful machine at a distance of more than two feet. In a given machine, the positive conductor yields much more powerful sparks than the negative.

800. *Ignition by the Electric Spark.*—Inflammable substances may be set on fire by the electric spark, as is shown by several experiments.

Stand on the insulating stool, touch the prime conductor with one hand, and from the other transmit a spark to a burner from which a current of gas is issuing,—the gas will be ignited. In houses thoroughly dried by furnace heat, persons, by simply running over the carpet, have been sufficiently charged with electricity to light gas with a spark from the finger.—Present a candle just extinguished, with its wick still glowing, to a prime conductor, so that a spark may pass through the snuff to the candle, and it will be relighted.—A person on an insulating stool charged with electricity may set fire to a cup of ether by presenting to it an icicle, through which the spark is transmitted.—With a suitable apparatus, a fine wire may be melted by sending through it a charge from a powerful battery.

801. *The Electrical Fire House.*
—Rosin may be ignited with the apparatus known as the Electrical Fire House (Fig. 286). Brass wires, insulated by being enclosed in glass tubes, enter the opposite sides of the house, and terminate on the inside in two knobs, B, C, a short distance apart. These knobs are loosely covered with tow and sprinkled with powdered rosin. When a charge is passed from A to D, the rosin is ignited, and the flame seen through the windows gives the house the appearance of being on fire.

Fig. 286.

THE ELECTRICAL FIRE HOUSE.

802. *Apparatus for firing Gunpowder.*—This apparatus consists of two

is its color in ordinary air and oxygen? In rarefied air? In nitrogen? In hydrogen? In carbonic acid and chlorine? What do the length and intensity of the spark depend on? At what distance have sparks been taken from a powerful machine? How do the sparks from the positive conductor compare with those from the negative? 800. What is the effect of the electric spark on inflammable substances? Prove this with several experiments. What is the effect of sending a powerful charge through a fine wire? 801. Describe the Electrical Fire House, and the ex-

Fig. 287.

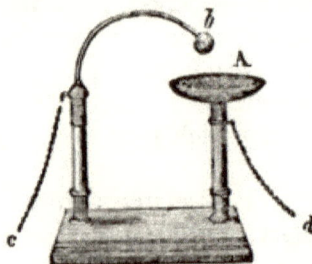

insulating glass pillars fixed in a stand, to one of which is attached a wire terminating in a ball, to the other a wooden cup for holding the powder. The chains *c, d*, being connected respectively with the inner and outer surface of a Leyden jar, a spark is made to pass from *b* to A, which ignites the powder.

803. THE ELECTROPHORUS. —Small quantities of electricity may be accumulated with a simple apparatus known as the Electrophorus, which to a certain extent answers as a substitute for the electrical machine.

The electrophorus consists of a cake of a resinous mixture 8 or 10 inches in diameter, and a somewhat smaller plate of metal with a rounded edge and a glass handle, by which it may be raised without drawing off the electricity. Excite the resinous mixture with fur, and placing on it the metallic plate, touch the upper surface of the latter for an instant to let its negative electricity escape. Then raise the metallic plate by the insulated handle, and on presenting a conductor a spark will be given. Place the metallic plate again upon the rosin, and on raising it another spark may be withdrawn. A Leyden jar may thus be slowly charged. Left on the rosin, the metallic plate will remain charged for a long time, and may be conveniently used as occasion requires in experimenting.

804. ELECTROSCOPES.—Electroscopes are instruments for detecting the presence of electricity, and determining whether it is positive or negative. They appear in various forms,—the simplest being the pith ball suspended by a silk thread, represented in Fig. 266. The attraction of the pith ball in its natural state by any substance presented to it, indicates the presence of electricity in the latter. When the pith ball is charged with positive electricity, its attraction by any substance indicates negative electricity in the latter, and its repulsion positive. When the pith ball is

periment performed with it. 802. Of what does the apparatus for firing gunpowder consist? 803. With what may small quantities of electricity be accumulated? Of what does the Electrophorus consist? How is it worked? 804. What are Electroscopes? What is the simplest form of the electroscope? How is the presence of electricity in any substance indicated? When the pith ball is positively charged, what does its attraction by any substance indicate? What, its repulsion? When the pith

charged with negative electricity, its attraction by any substance indicates positive electricity in the latter, its repulsion negative.

805. ELECTROMETERS.—Electrometers are instruments for measuring approximately the quantity of electricity in a given conductor or other body. Electrometers, more or less sensitive, are made in different forms; one of the simplest is the Quadrant Electrometer, shown in Fig. 288.

Fig. 288.

QUADRANT ELEC-
TROMETER.

A slender ivory rod, with a pith ball attached to its lower end, is suspended from a wooden pillar so as to swing freely like a pendulum. The pivot on which it turns is the centre of a semicircular scale attached to the pillar; and the whole apparatus terminates in a brass pin which may be inserted in the top of a prime conductor. The greater the quantity of electricity in the latter, the farther from the pillar the pith ball will swing,—and this distance is indicated by the scale.

806. ELECTRICAL INDUCTION.—An electrical atmosphere surrounds every excited body. An insulated conductor situated within this atmosphere becomes excited, and when thus affected is said to be electrified *by induction*. The phenomena of electrical induction are constantly exhibited.

Fig. 289.

INDUCTION APPARATUS.

807. Electrical induction is illustrated with the apparatus represented in Fig. 289. *c a d* is a brass cylinder with rounded ends, insulated on a glass support and furnished at one extremity with a pith ball electroscope, *f*. On bringing the end *d* within a few inches of a prime conductor, the pith balls, which before hung close together, instantly separate, indicating the presence of electricity. Since the cylinder is not in contact with the prime

ball is negatively charged, what does its attraction indicate? What, its repulsion? 805. What are Electrometers? What is one of the simplest forms called? Describe the Quadrant Electrometer, and its mode of operation. 806. By what is every excited body surrounded? When is a body said to be electrified *by induction?* 807. Describe the apparatus for illustrating electrical induction, and the experiments

conductor and receives no sparks from it, it is obviously electrified by induction. Its neutral and latent electricity is decomposed by the electrical atmosphere which surrounds the prime conductor: the negative portion is attracted towards $d$, and the positive repelled to $c$, where it charges the two balls, and thus causes them to separate. If the cylinder is removed from the neighborhood of the prime conductor, the pith balls immediately fall together; it is only when within the atmosphere of the prime conductor that they indicate any electrical excitement.

If the cylinder $c\,a\,d$, instead of being insulated, is connected with the earth, its positive electricity is driven off to the latter, while the negative portion is retained. If the cylinder is then removed, its communication with the earth being first cut off, it will remain excited with negative electricity.

808. ELECTRICITY FROM STEAM.—Electricity is developed during the escape of steam from an orifice. This fact was discovered in 1840 by a workman attending a steam-engine; who, happening to take hold of the safety-valve with one hand while the other was in a jet of steam escaping from a fissure, received an electric shock. The experiment was repeated, and it was found that a person with one hand in a jet of escaping steam could give a shock with the other to any one in contact with the boiler or the brick work supporting it. The electricity in question is produced by the friction of minute particles of water against the sides of the orifice.

As soon as this fact came to the knowledge of scientific men, an apparatus known as the Hydro-electric Machine was invented for the purpose of experiment. It consists of a steam boiler from three to six feet long, mounted on insulating pillars, with an arrangement for letting the steam escape in jets against a plate covered with metallic points, which acts like a prime conductor. This machine develops electricity in prodigious quantities, its power being equal to that of four large plate machines combined. It yields sparks twenty-two inches long, in such quick succession that they resemble a sheet of flame.

809. ATMOSPHERIC ELECTRICITY.—The atmosphere, besides the neutral and latent electricity which resides in it as in all other substances, contains more or less free elec-

performed with it? How may the cylinder be charged with negative electricity? 808. Under what circumstances is electricity produced by steam? State the circumstances attending this discovery. What was found when the experiment was repeated? How is the electricity in question produced? What instrument was invented for the sake of further experiment? Describe the Hydro-electric machine. To what is its power equal? What is said of its sparks? 809. What does the atmosphere

tricity, the quantity increasing with the distance from the earth's surface. This is proved by sending up arrows connected by a conducting metallic wire with a delicate electrometer. The higher the arrows rise, the more the electrometer is affected. An experimenter in England, by connecting a number of pointed conductors with an insulated wire a mile long and raised a hundred feet above the earth's surface, has collected enough electricity to charge a battery of fifty jars every three seconds.

810. *Origin.*—The free electricity in the atmosphere is due—1. To the *friction* of large masses of air of different densities on each other. 2. To the *condensation* of atmospheric vapors into a liquid form—a process which develops electricity in great abundance. 3. To the *chemical changes* involved in the growth of trees and plants. 4. To *evaporation*, particularly in the case of water filled with vegetable matter undergoing decomposition.

As these processes are not always going on with the same activity, it follows that the quantity of free electricity present in the atmosphere differs at different times and places.

811. *St. Elmo's Fire.*—When the atmosphere is very abundantly charged with electricity, its presence is indicated by various luminous phenomena. Hence the brilliant light called St. Elmo's Fire, which frequently appears at night on the tops of masts, the points of bayonets, and the tips of the ears of horses. It is simply the superabundant electricity of the atmosphere, attracted by a pointed conductor, into which it silently passes. Such phenomena are most common during thunder-storms, when as many as thirty have been seen in different parts of the same vessel. Sometimes they resemble sheets of flame, and extend three feet in length; at others they take the form of globes

contain? To what is the free electricity in the atmosphere proportioned? How is this proved? What has been done in this connection in England? 810. To what four processes is the free electricity in the atmosphere chiefly due? Why is the quantity of free electricity in the atmosphere different at different times? 811. When are luminous phenomena observed in the atmosphere? Describe the phenomenon known

of fire, attaching themselves to the yard-arms and mast-heads.

812. *Fire-balls.*—To electricity are also attributable the Fire-balls which are from time to time observed darting through the atmosphere, at heights of thirty miles and upwards, and with velocities of from five to thirty-three miles in a second. These balls sometimes vanish suddenly, leaving behind them a luminous track; at other times they explode into smaller balls or sparks; and at others again they are accompanied with showers of meteoric stones. Falling or shooting stars are the same phenomena on a smaller scale, and in lower regions of the atmosphere.

813. *Lightning and Thunder.*—The grandest of all the phenomena produced in the atmosphere by electricity is Lightning. Lightning is nothing more than the spark which accompanies the passage of the electric fluid from one cloud to another, or between a cloud and the earth. Thunder is the crackling sound produced at the same time by the sudden rush of air into the vacuum which the electric fluid, as it darts with inconceivable rapidity, leaves behind it. Flashes of lightning are sometimes several miles in extent; and, as the crackling sound is produced at every point of their course, it does not reach our ear all at the same instant. Hence the rolling or rumbling of thunder, which is in some cases prolonged by successive echoes from neighboring mountains or clouds.

814. That lightning and thunder are produced by an electric discharge, though previously suspected, was first experimentally proved in 1752, by Benjamin Franklin, whom the world recognizes alike great as a philosopher and a patriot.

Impressed with the conviction that lightning and the electric spark were identical, Franklin determined to test its truth by trying to collect electricity

as St. Elmo's Fire. At what time is it most common? What different forms does it assume? 812. What other phenomena are attributable to electricity? What becomes of these fire-balls? What are shooting stars? 813. What is the grandest of all the electrical phenomena of the atmosphere? What is Lightning? What is Thunder? How is the rolling of thunder accounted for? 814. By whom and when was it proved that lightning and thunder are produced by an electric discharge?

from the clouds during a thunder-storm. With this view he made arrangements for extending a wire to a great height from a steeple then in course of erection in Philadelphia. The work advanced but slowly; and while anxiously watching its progress one day, he observed a boy's kite far up in the air, and higher than he could hope to get his wire even when the steeple should be finished. It struck him at once that with this simple toy he could make the desired experiment, letting the string perform the part of the conducting wire. Accordingly, he made a cross of two strips of cedar, to the extremities of which he fastened the four corners of a silk handkerchief, using this as a covering that his kite might be able to withstand the rain and wind accompanying a thunder-shower. A sharp-pointed wire extended a foot from the top of the cross, to draw off the electricity from the clouds.

The kite thus constructed was raised by Franklin and his son in the first thunder-storm that occurred in June, 1752. Hempen twine was used, at the lower end of which a key was fastened for a prime conductor, while the whole was insulated by a silk ribbon fastened to a non-conductor sheltered from the wet. With intense anxiety the philosopher awaited the result. A cloud passed without any electrical indications, and he began to despair of success. Another came, and now to his indescribable joy he saw the loose fibres of the twine stand out every way and follow his finger as it passed to and fro. Presenting his knuckle to the key, he received a spark; and as soon as the twine was wet with rain, and its conducting power thus increased, the electricity was abundant. A Leyden jar was charged from the key, with which spirits were set on fire, and other experiments performed.—This discovery raised its author to the first rank among the philosophers of his day. His own feelings at the triumphant result of his experiment may be imagined. "Convinced of an immortal name, he felt he could have been content if that moment had been his last."

Franklin's experiment was repeated with success in various parts of Europe. There was no room left for doubting the identity of lightning with the electric spark. In later times this identity has been further confirmed by phenomena connected with the electric telegraph. Reports as loud as that of a pistol are often heard in telegraph offices during a storm, and to ensure the safety of the operators the wires have to be connected by conductors with the earth. Even in clear weather it is sometimes found difficult to fix the wires on the poles, in consequence of numbness produced in the hands by electricity conducted to them by the wires.

815. *Effects of Lightning.*—Lightning produces both mechanical and chemical effects. Its mechanical effects are very powerful. It crushes huge trees, rends off their branches, and sometimes tears their trunks into fragments.

Relate the incidents connected with Franklin's great discovery. What was the result of this experiment as regards the reputation of its author? As regards his own feelings? Where was the experiment repeated? How has the identity of lightning with the electric spark been since confirmed? 815. Mention some of the mechanical

14

When buildings are struck, large masses of masonry are displaced; a brick wall more than 12 feet long has been carried in one piece to a distance of 15 feet. These effects are analogous to the throwing out of the blocks of wood from the gable of the Thunder House, as described in § 797. It is only (as shown in that experiment) in the case of imperfect conductors,—that is, when obstructions are presented to the free passage of the electric fluid,—that these effects are produced.

Lightning is also a powerful chemical agent. It decomposes water and other substances into their elements. It sets fire to trees and houses, and melts metallic bodies. On the tops of mountains it is not unusual to see the surface of the hardest rocks perforated with deep cavities covered with a vitreous crust, owing to their having been struck with lightning.

816. *Lightning Rods.*—When a cloud becomes heavily charged with electricity, if another cloud in a different electrical state is near it, a discharge takes place between the two; in which case there is no danger. But sometimes there is no such adjacent cloud, and a flash of lightning darts from the charged cloud to the earth or sea: it is then said *to strike.* In such a case, the air being a bad conductor, the electric fluid in its descent follows any better conductor it can find, such as a house, a tree, the mast of a ship, a living animal, or a human being. Now, if the objects just mentioned were perfect conductors, the lightning would follow them to the earth without doing any injury; but they all offer some obstruction to its passage, and therefore all suffer more or less when struck.

The tallest objects, reaching nearest to the clouds, are the most likely to be struck. It is therefore imprudent to stand on the top of a hill or near a tree during a thunder-storm. In the house it is best at such a time not to sit near a damp wall, a bell wire, a gilded picture frame, or any metallic sub-

effects of lightning. Only in what case are these effects produced? State some of the chemical effects of lightning. 816. When does an electric discharge take place between two clouds? When, between a cloud and the earth? Why are houses, trees, &c., struck? Why do they suffer damage when struck? What objects are most likely to be struck? What positions is it imprudent to take during a thunder-

stance, as the electric fluid is sure to select the best conductor in its path to the earth if the house should be struck.

817. Having proved lightning to be an electric discharge, Franklin proceeded to devise means for preserving buildings from its effects. He thus became the inventor of the Lightning Rod, a simple contrivance which has been instrumental in saving life and property to an extent that can not be estimated.

The best material for a lightning rod is copper, but iron is cheaper and generally preferred. It must extend at least four feet above the building to be protected, and terminate above in one or more sharp points, which should be tipped with silver or platinum to keep them from rusting, and thus losing part of their conducting power. The rod should be continuous, and of such size that the fluid may follow it freely without danger of melting it,—say three-fourths of an inch across. It should be placed as close as possible to the wall and fixed securely to it. The lower end should be divided into two or more pointed branches, as shown at *a, a, a,* in Fig. 290. These branches should slant away from the building, and at least one of them should sink far enough into the ground to reach water or soil that is moist. If the build-

Fig. 290.

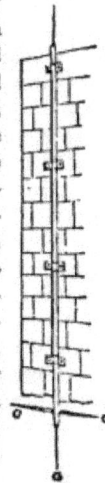

Fig. 291.

ing is large, and particularly if it has more than one point projecting upward, it should have several rods, either descending directly to the ground, like *c, d,* in Fig. 291, or connected together by a good conductor, and ultimately carried down like *c, f, g, h.*

818. The security afforded by lightning rods is twofold. In the first place, terminating in points, they generally draw off the electric fluid silently; and secondly, if a discharge takes place, the lightning in its descent will follow them rather than the inferior conductors to which they are attached, and finding a free passage through them will do no injury.—Lightning rods have not been found efficacious to a greater distance than forty feet. Within this limit, they protect a space around themselves equal to twice the height that

they project above the building; for example, a rod projecting five feet will protect every point of the surrounding surface within ten feet of itself.

819. ELECTRICAL FISH.—The torpedo, the Surinam eel, the si-lu'-rus electricus, and several other species of fish, have a peculiar organ with which they can give electric shocks, more, or less powerful according to their size. They use this organ for defending themselves against enemies, and for stunning and thus securing their prey. The power of giving shocks ceases with life; its too frequent exercise exhausts the fish and ultimately kills it. The shock of a torpedo fourteen inches long is borne with difficulty; and the Surinam eel has been found of such size that its shock proved immediately fatal.

The Surinam eel gives as many as twenty shocks a minute, yields the electric spark in the air, and charges a Leyden jar. Faraday computed that the average shock of one of these eels on which he experimented was equal to the discharge of a battery of fifteen jars, containing 3,500 square inches of glass, charged as heavily as possible.—The South American Indians catch these eels by driving a number of wild horses into a pond containing them. The eels, roused from their muddy retreats, vigorously defend themselves by pressing against the stomachs of the horses and repeatedly discharging their electrical battery. The poor beasts, panting from their struggles, with mane erect and haggard eyes expressing fright and anguish, seek to escape from their invisible foes, but are driven back by the Indians who surround the pond, armed with long reeds, and making terrible outcries. After several of the horses are stunned and drowned the eels become exhausted by their continued discharges, and are no longer objects of dread to the Indians. Slowly approaching the shore, they are captured with harpoons fastened to long cords; and to such a degree is their electrical power weakened that hardly any shock at all is received in drawing them ashore.

The silurus is a fish twenty feet long, found in the Nile and the Niger; its electrical apparatus lies immediately below the skin and extends round the whole body.

## Voltaic Electricity;

### OR, ELECTRICITY PRODUCED BY CHEMICAL ACTION.

820. Having considered electricity produced by friction, we proceed to treat of that developed by chemical

action. This branch of the subject is known as Galvanism.

821. GALVANI'S DISCOVERY AND THEORY.—The first discoverer in this department of science was he from whom it received its name, Galvani [*gal-vah'-ne*], Professor of Anatomy in the University of Bologna, Italy. The effects of atmospheric electricity on the animal frame had long engaged his attention. In the year 1790, having prepared the hind legs of some frogs suitably for experiment, and hung them on copper hooks till they should be needed, he observed to his surprise, on accidentally pressing the lower extremities against the iron railing of a balcony, that they were drawn up with a singular convulsive action. He found upon experiment that similar contortions were produced whenever copper and iron, connected with each other, were brought in contact, the one with the nerves of the thigh, the other with the muscles of the leg.

Galvani's experiment is often repeated at the present day. To perform it, separate the lower extremities of a frog from the rest of the body, skin them, and pushing back the muscles on either side of the back-bone, lay bare the lumbar nerves. Stretching out the legs in the position shown in Fig. 292, lay a thin curved rod of zinc under the nerves, and touch the muscles of the leg with a similar rod of copper. As long as the rods are kept apart, there is no movement in the legs; but the instant they are brought in contact, a violent convulsive motion takes place, the legs are drawn into the position shown by the dotted lines, and these contortions are repeated as often as the rods are separated and again brought together.

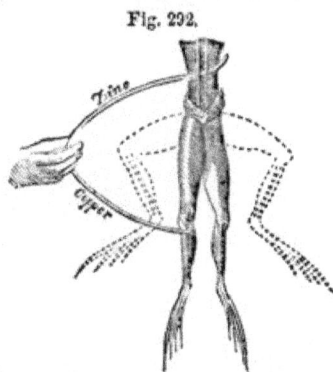

Fig. 292.

Galvani attributed this convulsive movement to a certain vital fluid which he supposed to reside in the nerves, and to pass to the muscles over the metallic conductors, in a manner similar to the passage of electricity between the inner and the outer coating of a

eels? What is said of the silurus? 820. What is Galvanism? 821. From whom did it receive its name? Give an account of Galvani's discovery. How may Galvani's experiment be repeated at the present day? When do the contortions take place? To what did Galvani attribute this convulsive movement? What did he call this

Leyden jar when it is discharged. He therefore called this supposed fluid Animal Electricity; but in compliment to its discoverer it soon became known as Galvanic Electricity, or the Galvanic Fluid.

822. VOLTA'S THEORY AND THE VOLTAIC PILE.—Prof. Volta, of Pavia, experimenting further on the subject, soon laid aside Galvani's theory of a " vital fluid ", and held that the effects in question were caused by the contact of the two dissimilar metals; that the legs of the frog had no agency in producing the galvanic excitement, but merely gave indications of its presence, like the pith ball electroscope in the case of ordinary electricity. To prove this, he combined the metals apart from all animal organizations; and advancing step by step, about the year 1800, he gave to the world his celebrated PILE, the appearance of which marked a new era in the history of electrical science.

Volta's "contact theory" was at one time generally received; but it is now known that the galvanic excitement is not produced by the mere contact of the metals, but by chemical action. A third element, such as the moisture of the hand, animal fluids, an acid, or some saline solution, must act chemically on one of the metals. It is believed that no chemical action ever takes place without the development of free electricity, though the quantity may be so small as to escape our senses.

823. Volta's Pile consisted of a number of circular plates of copper and zinc, and pieces of cloth moistened with a weak acid or saline solution, alternating as follows, the same order being observed throughout. At the base of the pile was a plate of copper, and on this a zinc plate, the two constituting *a pair*. On this pair was a piece of cloth moistened as above, then a second similar pair (the copper always below), then a piece of cloth, a third pair, and so on to the top of the pile. The whole was insulated on glass, and a wire was attached to each end. The wire connected with the zinc plate at the top of the pile yielded positive electricity; that connected with the copper plate at the base, negative. When the ends of these wires were brought

together or separated, a bright spark was produced. A
very fine platinum wire, half an inch long, stretched between
the ends of the wires, was made red hot. A person taking
one of these wires in each hand, received a succession of
shocks, like those from a Leyden jar, but slighter,—their
intensity depending on the number of plates. These effects
were produced as long as the arrangement and condition
of the plates remained unchanged.

Volta's pile, immediately connected as it was with the Galvanic Battery
(which has since superseded it), was one of those inventions to which science
is most largely indebted. It has immortalized its author, in honor of whom
this species of excitement produced by chemical action is now generally
called Voltaic Electricity.

824. FAMILIAR EXPERIMENTS.—The effects of voltaic
electricity may be illustrated with familiar experiments.

*Experiment* 1.—Place a piece of zinc under the tongue, and on the tongue
a silver coin. As long as the metals do not touch, nothing is perceived;
but as soon as they are brought in contact, the voltaic circuit is formed, a
thrilling sensation is felt in the tongue, a taste somewhat like copperas is
perceived, and, if the eyes are closed, a faint flash of light is seen. Here
electricity is developed by the chemical action of the saliva upon the zinc.

*Exp.* 2.—Lay a silver dollar on a sheet of zinc, and on the coin place a
living snail or leech. No sooner does the creature in moving about get
partly off the dollar and on the zinc, than it receives a shock and re-
coils. In this case it is the slime of the snail or leech that acts chemically on
the zinc.

825. GALVANIC BATTERIES.—Soon after inventing the
pile, Volta proposed another arrangement for the metallic
plates, identical in principle, but more convenient for use.
He discovered that electrical excitement was exhibited
whenever slips of copper and zinc were immersed in a vessel containing some diluted acid, if the circuit was completed by bringing the metals themselves, or wires connected with them, in contact above the vessel. Such an
arrangement is called a Simple Galvanic Circle; it is

scribe some of its effects. How long were these effects produced? What is said of
the invention of Volta's Pile? What is electricity produced by chemical action now
generally called? 824. What is the first experiment with which the effects of voltaic
electricity are familiarly illustrated? The second experiment? 825. Soon after inventing the pile, what discovery did Volta make? What is such an arrangement

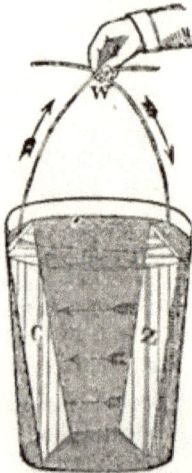

Fig. 293.

shown in Fig. 293. Combining a number of vessels similarly prepared, Volta made the first galvanic battery, known as the Couronne des Tasses [*koo-rone' da tahs*].

826. *The Couronne des Tasses*, or " crown of cups", represented in Fig. 294, consisted of any number of vessels, each containing a slip of copper and zinc, the copper of one vessel being

Fig. 294.

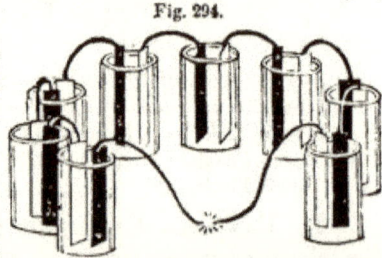

COURONNE DES TASSES.

SIMPLE GALVANIC CIRCLE.

connected by a conductor with the zinc of the next. To complete the circuit, wires attached to the extreme metallic slips of the series were brought together, when a spark and other electrical phenomena were produced.

827. *Trough Battery.*—Instead of the separate cups used by Volta, one long vessel divided into cells was subsequently employed. The zinc and copper plates, connected in pairs by a slip of metal, and arranged at such

Fig. 295.

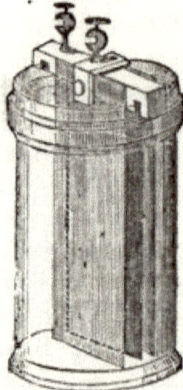

distances as to enclose a partition between the zinc and copper of each pair, were fastened to a common frame, so that they could all be immersed in acid and thus subjected to chemical action at the same time. This improved arrangement was known as the Trough Battery.

828. *Smee's Battery.*—Smee's Battery (see Fig. 295) has three metallic plates suspended, without touching each other, from a wooden frame. The middle plate is of silver coated with platinum. The outside ones are of amalgamated zinc,—that is, zinc coated with mercury. The whole are immersed in dilute sulphuric acid contained in an earthenware vessel. No action takes place till communication is established between the metals, when a bubbling immediately commences in the liquid, and voltaic electricity is produced. This battery, though not so powerful as those hereafter de-

SMEE'S BATTERY.

scribed, is economical, may be kept in operation for several days, and is much used in plating the inferior metals with gold and silver. With certain modifications it is also employed in working the magnetic telegraph.

829. In the batteries thus far described but one fluid was used, and two metals of such a nature that one was more readily acted on by the fluid than the other. Dilute sulphuric acid being used as the fluid, zinc (which it readily acts upon) was generally taken for one of the metals. Great improvements have been made on these single fluid batteries. With the exception of Smee's, they have been entirely superseded by instruments in which two fluids are employed, and which are not only more powerful, but also more regular and permanent in their action. The most important of these we proceed to describe.

830. *Daniell's Constant Battery.*—The two-fluid batteries are all modifications of Daniell's, which was invented in 1836. It consists of an outer cylinder of copper, within which is a cup of unglazed porcelain, of the shape represented in Fig. 296. Within this cup is a solid cylinder of amalgamated zinc. From both the zinc and the copper cylinder project brass cups (see Fig. 297) provided with screws for the insertion of wires ; the extremities of which, if there be but one cell, are called the Poles of the battery. If there be several cells, strips of metals inserted in these cups connect the zinc of one with the copper of the next, and wires for conducting the fluid are attached to the zinc of one of the extreme cells and the copper of the other. The porous cup is filled with dilute sulphuric acid. The copper cylinder is filled with the same fluid saturated with sulphate of copper ; and on a perforated shelf near its top (represented by the circular dotted lines in the figure) is placed some of the solid sulphate, that as fast as this substance is used up by the chemical action a fresh supply may be obtained, and the operation of the battery thus made constant.

Fig. 296.

Fig. 297.

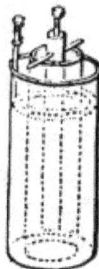

As soon as the poles are joined, a powerful action commences, which, instead of constantly diminishing as in the single fluid batteries, is maintained for hours without losing any of its efficiency. For ordinary use two dozen such cells are combined in a battery. One of the chief improvements in this apparatus is the introduction of the porous cup, which

---

used ? 829. In the batteries thus far described, what are employed for the purpose of producing chemical action ? Which is the most efficient of the single fluid batteries ? How do the single fluid batteries compare with those in which two fluids are used ? 830. By whom and when was the first two-fluid battery invented ? Describe Daniell's Constant Battery, and its mode of operation. What is one of the chief im-

14*

keeps the liquids apart, yet does not prevent the passage of voltaic currents.

831. *Grove's Battery.*—Grove's Battery is the most powerful one yet constructed. It operates on the same principle as Daniell's, but employs different metals and fluids, which render it more active. The porous cup contains a strip of platinum immersed in strong nitric acid, and is itself contained in

Fig. 298.

GROVE'S BATTERY.

a zinc cylinder filled with dilute sulphuric acid. The whole is set in a vessel of glass or earthenware. Fig. 298 shows one of Grove's batteries consisting of six cells, as arranged by Benjamin Pike, jr., of New York. The platinum of each cup is connected with the zinc of the next. At the extremities of the circuit, wires are attached respectively to the platinum of one cell and the zinc of the other, the former of which exhibits positive electricity and the latter negative.

Grove's battery is the best for performing the more striking experiments of galvanism, being nearly twenty times as powerful as a zinc and copper battery containing the same amount of metallic surface. Its superiority is owing to the absorption of the hydrogen evolved, the high conducting power of the fluids employed, and the ease with which nitric acid is decomposed.

832. *Bunsen's Battery.*—The cost of platinum renders Grove's apparatus expensive. Bunsen therefore devised a battery, in which plates of carbon acted on by nitric acid are substituted for platinum. In other respects it is like Grove's, but it is less efficient.

833. DRY PILES.—Feeble galvanic currents may be produced by compressing a great number of circular pieces of copper and zinc paper (sometimes called gold and silver paper), placed back to back, in a varnished glass tube, which they exactly fit. As in Volta's pile, the same order must be observed throughout. The electrical excitement produced by a Dry Pile (as such an apparatus is called) lasts a long time. Bells have been kept constantly ringing for eight years by the alternate attraction and repulsion of a clapper suspended between two such piles.

provements in this apparatus ?　831. Describe Grove's Battery. How does it compare in power with a zinc and copper battery ?　To what is its superiority owing ? 832. What is the objection to Grove's battery ?　To remove this, what modification did Bunsen propose ?　833. How are Dry Piles formed ?　What evidence is adduced

834. QUANTITY AND INTENSITY.—The quantity of voltaic electricity produced by a battery, depends on the size of the metallic plates employed; its intensity, on their number.

The difference between the quantity and the intensity of the electric fluid is analogous to the difference between the quantity of a solid dissolved in a given liquid and the strength of the solution. Into a hogshead of water throw a wine-glass full of salt, and into a tea-spoon full of water put as much salt as it will dissolve. The former solution will contain a greater quantity of salt than the latter, but it will be less strong.

835. THEORY OF THE GALVANIC BATTERY.—Let us now inquire how electricity is developed with the galvanic battery. Take, as an example, Volta's single fluid apparatus. When the zinc and copper plates are immersed in acidulated water, and connection is established between them, the water is decomposed into its elements, oxygen and hydrogen. The oxygen combines with the zinc, for which it has a strong affinity, and forms oxide of zinc; while the hydrogen appears about the copper in the form of minute bubbles. The zinc, in consequence of the chemical change produced in its surface, parts with its positive electricity to the liquid, and remains negatively electrified. The copper, not acted on by the liquid as the zinc is, attracts from it this same electricity, and becomes positively electrified. The acid mixed with the water tends to dissolve the oxide of zinc as fast as it is formed, and thus to keep a fresh surface of the metal exposed to the liquid.

836. The terminal wires of a battery, or, when no wires are attached, the plates from which they would proceed, are called its Poles. The pole connected with the metal most easily acted on by the fluid, always exhibits negative electricity; the other, positive. For *pole* some substitute the term *electrode*, meaning the path by which a voltaic current enters or leaves a body. The positive pole they

of the permanency of their action? 834. On what does the quantity of voltaic electricity produced by a battery depend? On what, its intensity? Illustrate the difference between the quantity and the intensity of the electric fluid. 835. Give the theory of the operation of the galvanic battery. 836. What is meant by the Poles of a battery? Which pole exhibits negative electricity? Which, positive? What term

call the Anode (ascending or entering path) ; the negative, the Cathode (descending or departing path). When the electrodes are brought in contact, the galvanic circuit is said to be *closed*. The two currents then meet and neutralize each other ; but, as fresh currents are all the time being produced, the action continues without interruption.

837. DIFFERENCE BETWEEN FRICTIONAL AND VOLTAIC ELECTRICITY.—Voltaic electricity and that developed by friction are the same in kind, but are characterized by certain points of difference.

1. The electricity developed by friction is far more intense; that produced by chemical action is far greater in quantity.

A simple galvanic circle (§ 825) develops as much electricity in three seconds as would be accumulated in a battery of Leyden jars by thirty turns of a powerful plate machine. Yet so weak is this voltaic electricity that a person receiving it through his system would hardly be aware of its passage, while the same quantity from the Leyden jars might prove fatal to life. It takes a galvanic battery of about fifty pair of plates (no matter what their size) to affect a delicate electroscope, and one of nearly a thousand pair to make pith balls diverge.

2. The voltaic fluid will not pass through an insulating medium, as the electric spark does. If the circuit is broken, all action at once ceases. It will pass thousands of miles over a conducting wire, but will not leap a break the fiftieth part of an inch.

3. The chemical effects of the voltaic fluid are incomparably greater than those of frictional electricity.

The galvanic battery produces the most intense heat, and readily decomposes compound substances; no such effects belong to the electrical machine. An ordinary galvanic battery will decompose a grain of water into oxygen and hydrogen. To do this with frictional electricity would require the power of an electrical plate having a surface of 32 acres,—which would be equivalent to a flash of lightning.

838. EFFECTS OF VOLTAIC ELECTRICITY.—Among the

is by some substituted for *pole?* What is the Anode? What is the Cathode? When is the galvanic circuit said to be *closed?* What then takes place? 837. What is the first point of difference between frictional and voltaic electricity? State some facts illustrating this difference. What is the second point of difference between frictional and voltaic electricity? The third point of difference? What facts are stated in the

effects of voltaic electricity on substances brought within the circuit, may be mentioned the following :—

839. *Decomposition.*—Compound substances may be decomposed into their elements with the galvanic battery; and it is a singular fact, that of the elements so obtained some always arrange themselves about the positive pole, and others about the negative. Thus, oxygen, chlorine, iodine, and the acids, invariably fly to the positive pole, when set free from any compound substance ; hydrogen, the oxides, and the alkalies, to the negative. As the elements must be in an opposite electrical state to the poles that attract them, we conclude that oxygen, chlorine, &c., are naturally negative,—and hydrogen, the oxides, and alkalies, positive. Every chemical compound seems to consist of a positive and a negative element, held together by electrical attraction.

The great discovery that water could be decomposed by voltaic electricity was made in 1800, immediately after the announcement of Volta's pile, by an experimenter, who observed that gas bubbles rose when the terminal wires were immersed in water. Several years later, Davy, after a long course of experiments, decomposed the earths and alkalies, which had before been universally regarded as simple substances, and thus brought to light a number of new metals, the existence of which had not even been suspected.

840. *The decomposition of water* is effected with the apparatus represented in Fig. 299. A large glass goblet has a frame fitted to its rim, from which are suspended two small receivers for the purpose of collecting the two gases evolved. As water consists of two parts of hydrogen to one of oxygen, one of the receivers should be twice as large as the other. Two holes in the bottom of the vessel, to which screw cups are attached, admit the electrodes from a battery, and terminate on the inside in strips of platinum, which enter the receiver. The vessel being filled with water and the battery set in operation, decomposition at once commences. Oxygen passes to the positive electrode

Fig. 299.

text to illustrate this difference? 839. What is the first effect of voltaic electricity? What singular fact is stated respecting the elements thus obtained? What elements go to the positive pole? What, to the negative? What is inferred from this fact? When and under what circumstances was it discovered that water could be decomposed by voltaic electricity? What great discovery was made by Davy? 840. Describe the mode of decomposing water with the galvanic battery. How is the process

(which should be inserted in the smaller receiver) and hydrogen to the negative. The identity of the gases may be proved by subsequently experimenting on them. As water is not a very good conductor of voltaic electricity, the process is facilitated by the addition of a little sulphuric acid.

Fig. 300.

841. *The decomposition of a neutral salt* may be performed with the apparatus represented in Fig. 300. A glass tube shaped like a V is fitted at each end with a cork and screw. Through these screws pass the wires from a battery, terminating inside in platinum strips. The tube having been filled with a solution of sulphate of soda or any other neutral salt, colored blue with tincture of violets, the battery is set in action. No sooner is a current passed from pole to pole through the liquid, than the latter is decomposed. The acid passes to the positive pole, and the alkali to the negative. This is shown by the change of color produced, the liquid becoming red around the positive wire and green around the negative. If the poles be transposed, the effects will be reversed.

842. The decomposing power of the galvanic battery is turned to practical account in the various processes of ELECTRO-METALLURGY. This is the art of depositing on any substance a coating of metal from a metallic solution decomposed by voltaic electricity. One of the branches of this art is Plating, which consists in covering the inferior metals with a thin coat of gold or silver. When the metal coating is not to adhere permanently to the surface on which it is deposited, but to form a copy of it and be removed, the process is called Electrotyping.

The different processes of Electro-metallurgy differ somewhat in their details and in the apparatus employed, but the principle involved is the same in all; viz., that any compound metallic solution is decomposed by the passage through it of a voltaic current; whereupon the pure metal is attracted to the negative pole, while the substance before combined with it goes to the positive. A medal, an engraving, or any conducting substance, has therefore only to be attached to the negative pole, and the metal in question will be deposited on it, the thickness of the coat de-

facilitated? 841. With what apparatus, and how, may a neutral salt be decomposed? 842. How is the decomposing power of the galvanic battery turned to practical account? What is Electro-metallurgy? In what does Plating consist? In what, Electrotyping? What is the principle involved in all the processes of electro-metallurgy? When any conducting substance is attached to the negative pole, what takes

pending on the length of time it is left to the action of the battery.

Reversed copies are thus obtained; the minutest indentations on the surface of the original being represented by elevations on the copy, and projections on the original by corresponding indentations in the copy. If an exact and not a reversed copy is wanted, a mould, taken from the original in wax or plaster, must be submitted to the above process.

This metallic deposit will take place only on a good conductor; if, therefore, the object to be copied is not such, it must be endowed with conducting power by dusting over it some fine plumbago. On the contrary, if there is any part of which a copy is not wanted, it may be covered with varnish which is a non-conductor.—That the copy may be readily removed from the original, the surface of the latter should be rubbed with oil or powdered plumbago.

843. The most convenient mode of electrotyping is as follows:—Fill a trough with a solution of sulphate of copper, and over its top extend two parallel rods of wood a short distance apart. Run the positive wire from a battery along one of these rods, and the negative along the other. From the negative wire suspend in the fluid the object to be copied, and from the positive one a piece of copper plate. Sulphate of copper is composed of sulphuric acid and copper. When the battery begins to operate, this fluid is decomposed; the copper is drawn to the negative pole and deposited on the object attached to it. The sulphuric acid goes to the copper plate, and combining with it forms sulphate of copper, thus providing fresh metallic solution as fast as the original supply is used up.

844. Much use is made of the electrotype process. It has to a certain extent taken the place of stereotyping in the preparation of plates from which books, charts, maps, &c., are printed. Copper plates being harder than those of type-metal, a far greater number of copies can be printed from them, and they are therefore preferable for works that are likely to have an extensive circulation. When the types are set, a mould of each page is taken in wax, brushed over with plumbago, and subjected to the above process till a thin deposit is formed, which is made of sufficient thickness to print from by backing it with type-metal. This book is printed from electrotype plates.

Engravings both on wood and copper are reproduced in the same way, their fine lines being brought out with exquisite perfection. The originals are put away, and the duplicates alone used in printing. By multiplying copies, which is done with little or no injury to the face of the original, any number of impressions can be obtained.—Fac-similes of delicate leaves, the wings of insects, and even daguerreotypes, may be made in a similar way.

---

place? What sort of copies are thus obtained? What must be done, to obtain fac-similes? On what alone will this metallic deposit take place? How may it be made to take place on a bad conductor? What precaution is necessary, to enable us to remove the copy from the original? 843. Describe the most convenient mode of electrotyping. 844. For what is the electrotype process used? In what case are copper plates preferable to those of type-metal? State the process gone through in pre-

845. *Protection of Metals.*—Voltaic electricity has been applied to the protection of metallic surfaces from corrosion. If a given metal is acted on by an acid or saline solution, we have only to immerse in the liquid some other metal more readily acted on by it, and close the circuit by connecting the two, when the chemical action on the former metal at once ceases and is transferred to the latter.

Davy proposed on this principle to protect the copper sheathing on the bottom of vessels from the action of sea-water. Strips of zinc were fastened at certain distances on the copper, and it was found that the latter metal was thus perfectly preserved from corrosion. No practical use, however, could be made of this proposed improvement; for shell-fish, sea-weed, &c., which had before been kept off by the poisonous properties of the corroded copper, now adhered to the bottom in such quantities as to make the vessel sail more slowly.

846. *Luminous and Heating Effects.*—When the galvanic circle is closed or broken,—that is, when the two terminal wires are brought in contact or separated,—a bright spark passes between them. With the proper apparatus, this spark may be intensified into the most brilliant light yet produced by art, known as the Electric Light, or the Voltaic Arch.

To produce the electric light, connect the poles of a powerful battery with the rods of a universal discharger (§ 780), and to the extremities of these rods fix charcoal points, or pieces of graphite pointed like a pencil. The battery being set in operation, the charcoal points are brought in contact, and then gradually withdrawn from each other a short distance, when the space between them is spanned by an arch of intensely bright light.

The voltaic arch is widest in the centre; its length varies with the power of the battery, ranging between three-fourths of an inch and four inches. No luminous appearance is produced unless the points first touch, no matter how close together they are brought, the air between being an insulator and

paring the plates. What else are reproduced by the electrotype process? 845. To what has voltaic electricity been applied? How may a metal acted on by a liquid in which it is immersed be protected from corrosion? What application of this principle was proposed by Davy? What was the result of the experiment? 846. What takes place when the galvanic circuit is closed or broken? Into what may this spark be intensified? How is the electric light produced? What is the shape of the arch.

breaking the circuit. In a vacuum, however, the arch may be formed without previous contact; and even in the air, if when the points are brought near each other a charge from a Leyden jar is passed from one to the other.

The electric light, like the electric spark, is entirely independent of combustion. None of the carbon is consumed, though a portion of it is mechanically carried over with a sort of hissing sound from the positive to the negative electrode, as is shown by the change of shape in the points when the experiment is over. The electric light may be produced in a vacuum and even under water, which shows that it is not the result of combustion.

The intensity of the electric light depends rather on the size of the metallic plates employed than on their number; that is, on the quantity of electricity developed more than its intensity. The arch produced with a powerful battery is about one-third as intense as that of the sun; while the Drummond light, which stands next to it among artificial lights in point of brilliancy, has only about $\frac{1}{150}$ of the sun's intensity. It has been proposed to use the electric light for illuminating the streets of cities; but the great expense of maintaining a sufficient voltaic current has thus far prevented its introduction for that purpose.

847. Heat, as well as light, is produced in the greatest intensity yet known to man by the galvanic battery. The hardest substances introduced within the voltaic arch, or brought between the electrodes of a powerful battery to close the circuit, are instantly ignited or fused. Platinum, which withstands the fiercest heat of the furnace, melts like wax in the flame of a candle. Quartz, the precious stones, the earths, the firmest and most refractory compounds, are fused in like manner. Thin leaves of metal subjected to the action of a battery burn with great brilliancy and beauty, yielding flames of different colors. Gold and zinc burn with a vivid white light, silver with an emerald green, copper and tin with a pale blue, lead with a brilliant purple, and steel watch-spring with the brightest scintillations.— The heat produced by a battery, like its light, depends on the size of the plates rather than their number.

The heating power of a galvanic battery may be shown by experiments with wires of different metals stretched between the electrodes. A wire so

and its length? What is essential to its production in the air? Is this necessary in a vacuum? How is it proved that the electric light is not the result of combustion? On what does the intensity of the electric light depend? How does its intensity compare with that of the sun and the Drummond light? For what has it been proposed to use the electric light? 847. What is said of the heat produced by the galvanic battery? State some of its effects. On what does the heat produced by a battery

placed instantly becomes hot; if not too long, red hot. By reducing its length, we may raise it to a white heat, and by shortening it still further we may fuse or ignite it. Experiments with different metallic wires of the same size and length, show that they are not all heated to the same degree by a given battery. The best conductors allow the current to pass with the least obstruction, and are therefore heated the least.

Platinum wire (which is one of the poorest metallic conductors and therefore most readily heated), immersed in a small quantity of water between the electrodes of a battery, causes the water to boil. Passed through phosphorus, ether, and alcohol, it ignites them. Gunpowder is exploded by contact with such a wire, a fact which is turned to account in the firing of blasts and submarine batteries. The platinum wire being carried through the powder and connected with the positive and negative electrodes, no matter how far off the battery may be, the moment the circuit is completed the platinum becomes red hot, and the explosion takes place. By thus simultaneously firing a number of charges of powder placed in deep holes at certain distances, 600,000 tons of rock have been instantly blown off from the face of a cliff, with an immense saving of labor, and with perfect safety on the part of the operator, who with his instrument was a fifth of a mile from the scene of the blast.

848. *Physiological Effects.*—The singular effects of the galvanic fluid on the nerves and muscles of animals, originally led, as we have seen, to the development of the science of Galvanism, and were carefully investigated in the earlier stages of its history. The more powerful instruments since invented have enabled experimenters to push their researches still further.

When we grasp the electrodes of a battery of fifty cups, one in each hand, we feel a peculiar twinge in the elbow and sometimes in the shoulder, as if the joints were being wrenched apart. This sensation continues as long as the electrodes are held in the hands, and when we first grasp them or let them go is sufficiently sudden and vivid to be called a *shock*. A number of persons may take the shock at once by joining their hands, which should be previously

depend? What is the effect of the galvanic battery on metallic wires? When wires of different metals are used, what is found? How is this explained? What experiments may be performed with a platinum wire fixed between the electrodes of a battery? Describe the process of firing a blast with such a wire. What instance is mentioned of the practical application of this process? 848. What originally led to the development of galvanism as a science? What sensation is experienced on grasping the electrodes of a battery? How may a number of persons take the shock?

moistened. A weak current passed through the eyes produces a faint flash ; passed through the ears, a roaring sound; and through the tongue, a metallic taste.

The effects of the galvanic battery on the animal system, unlike its luminous and heating effects, are found to depend on the number of plates employed rather than their size,—that is, on the intensity of the electricity produced, and not its quantity. A battery of several hundred pair of plates proves fatal to life. One of a hundred pair gives a shock that few would like to bear a second time, though, if the plates are small, it has no effect on wires stretched between the electrodes. Put the same amount of metallic surface in a few pair of very large plates, and such a battery will instantly fuse wires subjected to its action, while its shock will hardly be felt.

849. There seems to be a remarkable analogy between a voltaic current and the nervous energy. Experiment has shown that, if a nerve be divided, a galvanic current directed through the region in which it runs will in a measure supply its place. The part, which would otherwise be palsied from a want of nervous energy, may thus be restored to its usual action. If, for example, the nerves of the stomach are divided, digestion ceases; but it is resumed if the stomach is subjected to galvanic influence. Galvanism is therefore medically applied in asthma, paralysis, and other diseases arising from a prostration of the nervous system.

850. Among the most remarkable effects of voltaic electricity are the violent contortions it produces in bodies just deprived of life.

A few years ago, the body of a murderer hanged in Glasgow was subjected, about an hour and a quarter after his execution, to the action of a battery consisting of 270 pair of four-inch plates. One pole was applied to the spinal marrow at the nape of the neck, and the other to the sciatic nerve in the left hip, when the whole body was thrown into a violent tremor as if shivering with cold. On removing the wire from the sciatic nerve to a nerve in the heel, the leg was thrown out so violently as nearly to overturn one of

What is the effect of passing a weak current through the eyes ? Through the ears? Through the tongue ? On what do the effects of the galvanic battery on the animal system depend ? Compare the different effects of a given amount of metallic surface, when thrown into many small plates, and a few large ones. 849. To what does the voltaic current bear a remarkable analogy ? What has been shown by experiment ? Give an example. In what diseases is galvanism medically applied ? 850. What is one of the most remarkable effects of voltaic electricity ? Describe the experiments

the assistants, who tried in vain to prevent its extension. On directing a current to the principal muscle of respiration, the chest heaved and fell, and labored breathing commenced. When one of the poles was applied to a nerve under the eyebrow and the other to the heel, the most extraordinary grimaces were produced: "every muscle of the countenance was simultaneously thrown into fearful action; rage, horror, despair, anguish, and ghastly smiles, united their hideous expression in the murderer's face." Several spectators were so overcome by the sight that they had to leave the room, and one gentleman fainted. In the last experiment, the fore finger, which had previously been bent, was instantly extended, and shaking violently, with a convulsive movement of the whole arm, seemed to point to the persons present, some of whom thought that the body had really returned to life.

## Thermo-electricity;

### OR, ELECTRICITY DEVELOPED BY HEAT.

851. How produced.—If two strips of metals which differ in their conducting power, are soldered together at one end so as to form an acute angle with each other, and heat is applied at the place of junction, a current of electricity is produced, which may be carried off by any good conductor. Antimony and bismuth exhibit this phenomenon in its greatest perfection, and are generally used in performing the experiment. Electricity thus developed by heat is known as Thermo-electricity. Its properties are the same as those of frictional electricity.

Fig. 301.

THERMO-ELECTRIC BATTERIES.

852. Thermo-electric Batteries.—Thermo-electricity may be developed abundantly by combining a number of thin bars of antimony and bismuth, or platinum and iron. They may be arranged in either of the forms represented in Fig. 301, or may be laid flat one upon another, with pasteboard between to prevent them from touching except at their extremities. By heating the points of junction at one end, *a, a, a, a,* and cooling those

at the other, $b$, $b$, $b$, $b$, an electric current is produced, the intensity of which is equal to the sum of the intensities of the separate pairs. With a wire attached to the first bar of bismuth and another attached to the last bar of antimony, the thermo-electric current may be conducted wherever it is desired.

When thirty or forty such combinations are needed, thin metallic bars are used, connected alternately at their extremities, and arranged for convenience' sake in parallel piles of five or six each. Such a battery indicates changes of temperature at its junctions so minute that they can be detected in no other way,—even to the hundredth part of a degree of the thermometer. The heat radiated from the hand is sufficient to produce a slight electric current.

853. Electricity, besides being produced by friction, chemical action, and heat, is also developed under certain conditions by magnetism. When so produced, it is called Magneto-electricity. This branch of the subject can not be understood till we have treated of Magnetism, and will therefore be considered in the next chapter, which is devoted to that subject.

---

# CHAPTER XVII.

## MAGNETISM.

854. A MAGNET is a body which has the property of attracting iron and being attracted by it.

855. Magnetism is the science that treats of the laws, properties, and phenomena of magnets.

### Kinds of Magnets.

856. There are two kinds of magnets, Natural and Artificial.

thermo-electric battery formed? What is the usual arrangement when a large number of such combinations are needed? How minute changes of temperature are indicated with such a battery? 853. By what other agency is electricity also developed? What is it then called?

857. NATURAL MAGNETS.—The natural magnet, or loadstone, is an ore of iron, found in great quantities in different parts of the earth, which has the property of drawing to itself steel filings, needles, or small pieces of unmagnetic iron. Its texture is hard, and its color varies from reddish-brown to grey. Besides the loadstone, nickel, cobalt, and brass when hammered are found to have magnetic properties, though in an inferior degree.

858. The attraction of the loadstone for particles of iron appears to have been known to the Greeks, Chinese, and other nations in remote antiquity. It is distinctly alluded to by Homer and Aristotle. Pliny speaks of a chain of iron rings suspended one from another, the first of which was upheld by a loadstone. He tells us, also, that Ptolemy Philadelphus proposed to build a temple at Alexandria, the ceiling of which was to be of loadstone, that its attraction might hold an iron statue of his queen Ar-sin'-o-e suspended in the air. Death prevented Ptolemy from carrying out his design; but St. Augustine, at a later day, mentions a statue thus actually held in suspension in the temple of Ser'-a-pis, at Alexandria.—The magnet (*magnes* in Greek) is supposed to have received its name from Magnesia, a city of Asia Minor, near which it was first found.

859. *Poles.*—The attractive power of a natural magnet does not reside equally in all its parts, but is strongest at its extremities and diminishes towards the middle, where it is entirely wanting. This is shown by rolling a piece of loadstone in iron filings. They will be found to cluster about the ends, those that first adhere being endowed with the power of attracting others, till large tufts are formed, while the middle is left entirely bare.

The points at which the greatest attractive power is exhibited, are called the Poles of the magnet. The central part, where it is wanting, is called the Neutral Line.

If a piece of loadstone is broken, each portion becomes a perfect magnet, and has poles of its own.

---

854. What is a Magnet? 855. What is Magnetism? 856. How many kinds of magnets are there? Name them. 857. What is the natural magnet? What other metals have magnetic properties? 858. To whom and when was the attraction of loadstone for iron known? What ancient authors allude to it? Of what does Pliny speak? What use did Ptolemy Philadelphus propose to make of the loadstone? What is mentioned by St. Augustine? From what did the magnet receive its name? 859. What is shown by rolling a piece of loadstone in iron filings? What is meant by the Poles of the magnet? What is the Neutral Line? If a piece of loadstone is

860. *Power of Natural Magnets.*—When quite small, a natural magnet will sustain many times its own weight of iron. Sir Isaac Newton is said to have worn, in a ring, a piece of loadstone weighing three grains, which would lift 750 grains of iron. Their attractive power, however, does not increase with their size. Large pieces of loadstone never support more than five or six times their own weight, and rarely as much. The most powerful natural magnet known is capable of lifting 310 pounds.

861. *Armature.*—The power of a natural magnet is increased by applying vertically to its opposite polar surfaces thin strips of soft iron, projecting a little below, and bent, as shown in *a p*, *b n*, Fig. 302. The attractive force then centres in *p* and *n*, which become the new poles. This arrangement is called an Armature, and a magnet so prepared is said to be *armed*.

Fig. 302.

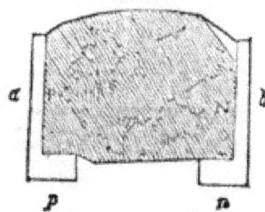

To keep the armature in its place, metallic bands, A B, C D (Fig. 303), are passed round the whole. A ring, R, is attached to the top for convenience of handling. The effect of the magnet is further increased by uniting its poles with a transverse piece of soft iron, K, called the Keeper. To this a hook is attached for suspending a scale-pan and weights.

Fig. 303.

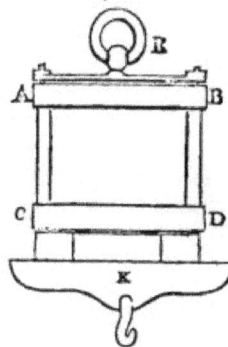

ARMED MAGNET.

862. ARTIFICIAL MAGNETS.—A piece of iron or steel brought in contact with a natural magnet or very near it, acquires its peculiar properties, and will itself attract steel-filings, needles, &c. Soft iron loses these properties on being withdrawn from the magnet; but a piece of steel retains them permanently, nor does the natural magnet from

which it receives them suffer any diminution of power in consequence.

A piece of iron or steel to which magnetic properties have been imparted in any way, is called an Artificial Magnet.

863. *Kinds of Artificial Magnets.*—There are several kinds of artificial magnets, called from their shape Bar Magnets, Horse-shoe Magnets, and Magnetic Needles. The first two are most powerful when formed of several similar pieces riveted together, in which case they are called Compound Magnets.

Fig. 304.

COMPOUND BAR MAGNET.

Fig. 305.

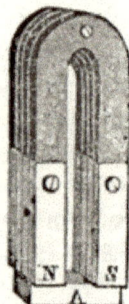

HORSE-SHOE MAGNET.

Fig. 304 represents a Compound Bar Magnet; Fig. 305, a Compound Horse-shoe Magnet. N, S, represent the poles. The Horse-shoe magnet has an armature, A, attached, which increases and preserves its power, and should always be kept on when the magnet is not in use.

Magnetic Needles are very light magnetic bars (see Fig. 306), poised at their centre on a pivot, on which they move freely either horizontally or up and down. In the former case, they are called Horizontal Needles; in the latter, Vertical or Dipping Needles.

Fig. 306.

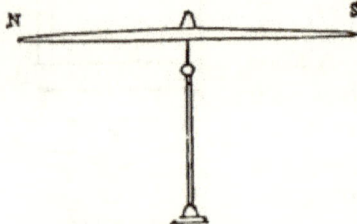

MAGNETIC NEEDLE.

864. Artificial magnets are more efficient and regular in their action than natural ones, and are therefore preferred for purposes of experiment. The horse-shoe is more

powerful than the bar magnet. A horse-shoe of one pound has been known to sustain 26½ pounds.

865. *Poles.*—The poles of an artificial magnet,—that is, the points in which the greatest attractive force resides,—are found to be about one-tenth of an inch from the extremities. In very long bar magnets, besides the two poles always situated near the extremities, two other poles, nearer the centre, are sometimes, though rarely, found.

866. The power of a magnet, whether natural or artificial, may be increased by daily adding a little to the weight which it will support. If, for instance, a given magnet just sustains two pounds of iron, by putting on a small additional weight every day, we may perhaps make it sustain three or even four pounds. If, on the other hand, we overload it, so that the armature falls off, the power of the magnet will be impaired. Any rough treatment, such as hammering the magnet, rubbing it violently, or letting it fall, has the same effect. Heat, also, diminishes the power of a magnet. Red heat destroys it altogether, even after the magnet has cooled.

867. Air is not essential to the action of a magnet; all its phenomena may be exhibited in a vacuum.

## Properties of the Magnet.

868. ATTRACTION.—As stated above, all magnets attract unmagnetic iron. They are also attracted by it.

Suspend a magnetic needle by a thread. Bring a piece of iron near either extremity, and the needle will be drawn towards it.

869. Magnetic attraction acts with undiminished power through any thin substance.

In the last experiment interpose a piece of glass or paste-board between the iron and the needle; the latter will be attracted none the less.

---

shoe compare with the bar magnet? 865. Where do the poles of an artificial magnet lie? What are sometimes found in very long bar magnets? 866. What is the effect of adding a little daily to the weight which a magnet supports? Give an example. What is the effect of overloading a magnet? Of treating it roughly? Of heating it? 867. Is air essential to the action of a magnet? Prove it. 868. What is the first property of magnets? What experiment shows the attraction of iron for a magnet? 869. What is the effect of interposing any thin substance? How may this fact be

15

838          MAGNETISM.

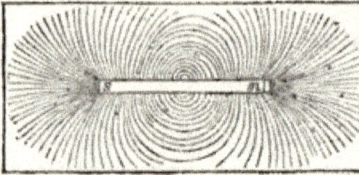

Fig. 307.

MAGNETIC CURVES.

Hold a piece of paper over a bar magnet, and dust on it some steel filings. Under the influence of the magnetic attraction transmitted through the paper, they will arrange themselves in regular lines, as shown in Fig. 307. These lines are called Magnetic Curves.—The superior attractive power of the poles is also shown by this experiment; for the filings are thickest directly over those points, the curves appearing to converge there from all directions.

Magnetic figures of any description may be formed on a steel plate by marking on it with one of the poles of a bar magnet, and then sprinkling iron filings on the surface. They will at once adhere to the lines which the magnet has traced. The result is the same if paper is laid on the steel surface before the bar is drawn over it, the magnetic influence being transmitted through the paper.

870. *Law.—Magnetic attraction decreases in intensity as the square of the distance from the magnet increases.*

If two similar substances are situated respectively 1 inch and 2 inches from a given magnet, the former will be attracted 4 times as strongly as the latter. This law corresponds with that of gravitation, light, and heat.

871. POLARITY.—A magnetic needle, left free to move, always points north and south, or nearly so. Often as it may be disturbed from its natural position, it invariably resumes it after a few vibrations. This property is called Magnetic or Directive Polarity.

It is to be observed in connection with magnetic polarity that the same extremity of the needle always points to the north, and the same extremity to the south. That which points north is called the North Pole; and that which points south, the South Pole. Turn the needle round till its north pole points south, and it will not rest till it has traversed a semicircle and got round again to the north.

872. If the poles of a bar or horse-shoe magnet be presented successively to the north pole of a magnetic needle,

illustrated? How are Magnetic Curves formed? What does this experiment show? How may magnetic figures be formed? What is the effect of interposing paper between the magnet and the steel surface? 870. What is the law of magnetic attraction? Give an example. 871. What is meant by Magnetic or Directive Polarity? What is to be observed in connection with magnetic polarity? What name is given

one of them will be found to attract it and the other to repel it. If the experiment be tried with a number of different needles, the same pole will always be found to attract, and the same to repel. This shows that the two poles of the magnet have different properties, which we indicate by giving them different names. The one that attracts the north pole of the needle we call the South Pole of the magnet, and the one that repels it, the North Pole.

873. *General Law.*—*Like poles of magnets repel each other, and unlike poles attract each other.* This law corresponds with that of electrical attraction and repulsion.

Balance a bar magnet with weights on a pair of scales. Beneath its positive pole bring the positive pole of another magnet, and the scale containing the bar will rise owing to the repulsion of the like poles. Substitute the negative pole, and the scale will descend owing to the attraction of the unlike poles.

874. Like poles neutralize each other's attraction for unmagnetic iron.

Immerse the positive poles of two magnets separately in iron filings. On withdrawing them, both will be covered with large tufts. Now bring them together, and the filings will immediately drop off from both. The result will be the same if the experiment be tried with the negative poles of two magnets. If the positive pole of one magnet and the negative of the other be used, the filings, instead of falling off, will join in a festoon between the two unlike poles.

875. *The Astatic Needle.*—The polarity of two needles of equal power may be neutralized by supporting them on the same pivot, one above the other, parallel and with unlike poles pointing in the same direction. An instrument so formed is called the Astatic Needle.

Fig. 308 represents an astatic needle. The north pole of the upper one points the same way as the south pole of the under one, and

Fig. 308.

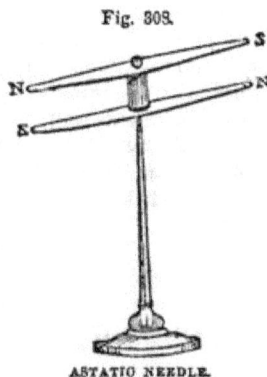

ASTATIC NEEDLE.

to the two poles of the needle? 872. How is it shown that the poles of a bar magnet have different properties? How are these poles distinguished? 873. What is the law of magnetic attraction and repulsion? Illustrate this law with an experiment. 874. What is the effect of like poles on each other's attraction? Show this experimentally. 875. How may the polarity of two needles of equal power be destroyed?

*vice versa.* The consequence is that the polarity of both is destroyed; the needles will remain in whatever direction they are placed.

876. When a magnet is divided, each portion becomes a perfect magnet in itself, and has its own poles, even though the parts in which the new poles lie exhibited no magnetic attraction at all before the division. Those extremities of the divided portions which lie towards the north pole of the original magnet will all be north poles, and the extremities towards its south pole will all be south poles.

877. *Magnetic Variation.*—In a given place, all magnetic needles point in the same direction. This direction is called the Magnetic Meridian.

In some parts of the earth the magnetic meridian runs due north and south; that is, a plane extended in the direction in which the needle stands would pass through the north and the south pole of the earth. The magnetic meridian would then correspond with the geographical meridian. In most places, however, the magnetic meridian deviates more or less from the geographical meridian. This deviation is called the Variation of the Needle, or Magnetic Variation.

The variation of the needle is different at different places on the earth's surface, and is constantly changing at the same place. Recorded observations in the old world show that for a series of years the needle kept varying more and more towards the west; till, having attained its western limit, it turned back towards the east, in which direction it is now moving. The cause of this periodical change and the law which regulates it are as yet unknown. At Washington City the variation is now between 2 and 3 degrees west; that is, the needle points between 2 and 3 degrees west of north. Every year it becomes somewhat greater, the annual rate of increase being about 3'.

Two irregular lines (which are constantly changing) may be traced on the earth's surface, one in each hemisphere, along which the needle points due north and south. They are called Lines of no Variation.

878. *Magnetic Dip.*—An ordinary steel needle, poised

Describe the Astatic Needle. 876. When a magnet is divided, what is said of each portion? Which extremities of the divided portions will be north poles, and which south? 877. What is the Magnetic Meridian? In some parts of the earth how does the magnetic meridian run? How, in others? What is meant by Magnetic Variation? What do recorded observations show? What is the present variation at Washington City, and how is it changing from year to year? What is meant by

on its centre of gravity so as to move freely up and down, remains in any position in which it may be placed ; if magnetized, in most parts of the earth it inclines more or less towards the horizon. This inclination is called the Dip of the Needle, or Magnetic Dip. It was discovered in 1576, by an optician of London.

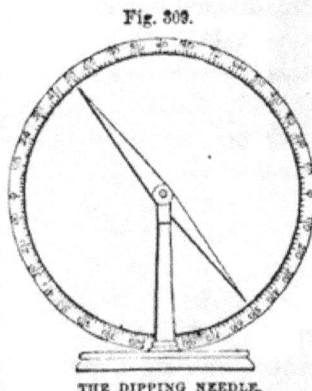

Fig. 309.

THE DIPPING NEEDLE.

With the Dipping Needle and graduated scale attached, represented in Fig. 309, the magnetic dip at any given place can be measured. Experiments with this instrument show that there are two points, one in the northern hemisphere (latitude 70), the other in the southern (lat. 75), in which the needle stands vertical, and the dip is therefore 90 degrees. That, on the contrary, there is a circle of points near the equator, at which the needle is parallel to the horizon, and the dip is 0; this line is called the Magnetic Equator. At different intermediate points the dip is different, increasing, though not regularly, as the distance from the magnetic equator increases. The dip, like the variation, keeps changing at a given place. In the latitude of New York it is now about 70 degrees, and is constantly decreasing.

879. *The Compass.*—The polarity of the magnetic needle, applied in the Compass, enables us to determine, at any place, a given direction or the bearing of a given object.

The Land or Surveyor's Compass is simply a magnetic needle set in a shallow case covered with glass, on the bottom of which is a circular card, having its circumference divided into 360 degrees. At a distance of one-fourth of the circumference apart stand the letters N, E, S, W, denoting the four *cardinal points*—North, East, South, West. As the needle is stationary, while the card moves, the order of the points is reversed ; that is, when we hold the instrument so as to have the point S next to us, E is on the left, and W on the right.

880. It is to the navigator, who relies entirely on it for guidance over the trackless ocean to his desired port, that the compass is most important. Arranged for his use, it is called the Mariner's Compass.

Fig. 310.

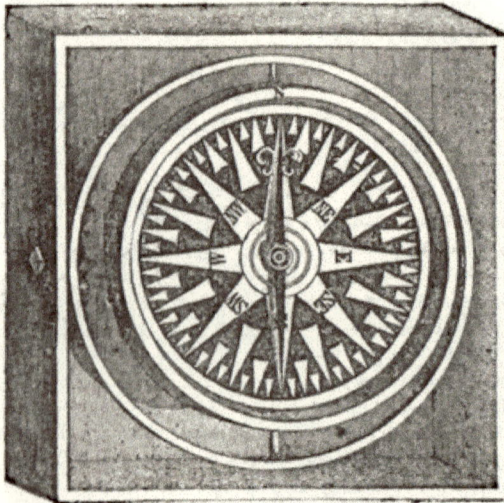

THE MARINER'S COMPASS.

In the mariner's compass, represented in Fig. 310, the circular card is attached to the needle and turns with it. The circumference of the card is divided into 32 equal parts, denoted by marks and sometimes subdivided into halves and quarters. These marks have names given to them, indicating the different directions, which are called Points of the Compass. Mentioning the points of the compass in their order is called *boxing the compass.*—The compass box is suspended within a larger box by means of two brass hoops, or *gimbals* as they are called, supported at opposite points on pivots, so that however the vessel may roll or pitch the needle may retain its horizontal position.

It is believed that the Chinese were the first to avail themselves of the magnet in navigation, many hundred years before the Christian era; and that from them various other eastern nations learned to use it for the same purpose. The compass of these early times was probably nothing more than a piece of loadstone mounted on a cork and allowed to float on water. The magnetic needle and the card attached to it were no doubt the inventions of Europeans, among whom a knowledge of the rude compass used in the East appears to have been introduced in the twelfth century after Christ. Flavio Gioia [*flah'-ve-o jo'-yah*], a Neapolitan who flourished about the year 1300,

the magnetic needle applied? Describe the Land Compass. 880. To whom is the compass most important? Describe the Mariner's Compass. What is meant by *boxing the compass?* How is the compass box suspended? Who are thought to have been the first to use the magnet in navigation? What did this ancient compass probably consist of? When did it first become known in Europe? What improvements were soon made? How did the name of Flavio Gioia become connected with

by some regarded as the inventor of the compass, probably merely improved its construction, or extended its use among the maritime nations of Europe.

No one can estimate how much the invention of the mariner's compass has contributed to the progress of the world. Relying on his little needle, which never betrays its trust, the mariner is no longer obliged to keep his bark within sight of land, and to direct his course by sun and star which clouds may obscure for days and nights together. He fearlessly ventures into unknown seas, explores the remotest regions, pursues his way under lowering skies and in utter darkness, well knowing whither he is sailing and how to steer when he wishes to retrace his course. This simple instrument has thus made the ocean a safe and frequented highway, extended the commerce and knowledge of the world, linked its most distant families in friendly intercourse, and brought whole continents virtually into being.

881. The compass needle, like all other magnetic needles, is subject to variation and dip.

Its variation seems to have been known two hundred years before the time of Columbus; but that this variation differs in different places was discovered by that navigator on his memorable voyage across the Atlantic in 1492. As he went westward, he observed that the variation increased from day to day. The fact was soon discovered by his crew, and filled them with consternation. It seemed 'as if the very laws of nature were changing, and they were entering a new world subject to mysterious influences'. It required all the ingenuity of Columbus to induce them to proceed; which he did by allaying their fears with an explanation of the phenomenon satisfactory to them, though it was far from satisfying himself.

As the compass needle must be perfectly horizontal, the dip is counterbalanced by loading the end that tends to rise with a small weight, which may be shifted to suit any latitude.

## Theory of Magnetism.

882. The theory of magnetism is analogous to that of electricity. An agent, which for convenience' sake we call the magnetic fluid, may be supposed to pervade all things. In its quiescent state it is a combination of two fluids, which may be distinguished as North or Positive, and South or Negative. When combined, these fluids neutralize each other, and no magnetic phenomena are exhibited; but

---

the compass? What is said of the effects which this simple instrument has wrought? 881. To what is the compass needle subject? How long ago was the variation of the needle known? What discovery did Columbus make respecting it? What was the effect of this discovery on his crew? How is the dip counterbalanced in the compass needle? 882. State the theory of magnetism. How are the phenomena exhib-

when through any agency they are separated, the substance
containing them displays magnetic properties, and is said
to be *magnetized.*

In loadstone, as found in nature, the two fluids do not
combine at all.  In soft iron and steel they are easily separated, but in the former re-combine as soon as the separating agency is withdrawn, while in the latter they remain
permanently apart.  In most substances they are united so
strongly as to be almost incapable of separation, and such
substances are magnetized with the greatest difficulty.

When the magnetic equilibrium is disturbed, and the two fluids are separated as described above, they seem to take up their abode in opposite sides
of the individual particles of the magnetized body, the positive fluid taking
the same side throughout, so that the positive pole of one particle is contiguous to the negative pole of the next.  Both fluids remain in the body, but
without combining; one is not wholly expelled, as in the case of the electric
fluid.  The opposite fluids nullify each other at the centre of the magnetized
body, but not at the extremities, which become their chief seats of action.
If a new extremity is formed by breaking a magnet, a new pole is formed,
opposite in kind to the one at the other end.  When a piece of iron or steel is
brought near the positive pole of a magnet, its neutral magnetic fluid is decomposed.  The negative portion is attracted by the positive pole towards
the end nearest it, which consequently becomes a negative pole; while the
positive element is repelled towards the other end, and forms there a positive pole.

883. Terrestrial Magnetism.—The polarity of the
needle is best explained by supposing the earth itself to be
a vast magnet.  At the magnetic equator, as at the centre
of a bar magnet, the two fluids neutralize each other, and
there are no magnetic phenomena.  Hence at this line there
is no dip.  The chief seats of magnetic energy are two
points which lie towards the geographical poles of the
earth, and which are called its Magnetic Poles.

That point of the earth which attracts the north or positive pole of the
needle, must be its south or negative magnetic pole.  It lies near Hudson's

ited by loadstone, soft iron, and steel, explained?  How is it with most substances?
When the two magnetic fluids are separated, where do they seem to take up their
abode?  Where do the two fluids nullify each other, and where not?  What follows
if a new extremity is formed by breaking a magnet?  What follows when a piece of
steel is brought near the positive pole of a magnet?  883. How is the polarity of the
needle explained?  Why is there no dip at the magnetic equator?  What is meant

Bay, in 70 degrees of north latitude, and was reached by Captain Ross during his Arctic expedition of 1829. At this point he found the dipping needle to stand vertical, with its north pole towards the earth. The north or positive magnetic pole of the earth has never been exactly reached, but is supposed to lie south of New Holland, in about 75° south latitude. The dipping needle would there also stand vertical, but with its south pole towards the earth. A point has been found near the region alluded to, in which the needle is very nearly vertical, the dip being 88²/₃ degrees.

The changes in the variation and dip appear to be in some way connected with the solar heat received by the earth.

884. Magnets draw small pieces of iron to themselves; but it must be remembered that the magnetic attraction of the earth only affects the direction, and does not tend to change the actual position. A magnetic needle mounted on a cork and placed on the surface of a pond, is made to point north and south by the earth's magnetic attraction, but is not drawn to the north side of the pond.

885. *Magnetic Intensity.*—A magnetic needle suspended by a delicate fibre, when turned from the direction in which it naturally rests, resumes it, but not immediately. The magnetic attraction of the earth brings it back, but its inertia carries it past the point, and thus a series of vibrations, like those of a pendulum, take place before it finally settles. The number of such vibrations occurring in a given time evidently depends on the intensity of the earth's magnetic attraction. Now this number (and consequently the intensity of terrestrial magnetism) is found to be different at different places, and at different times in the same place.

*The magnetic intensity varies according to the square of the number of vibrations made in a given time.* By applying this law, it is ascertained that the greatest magnetic intensity thus far found on the earth's surface is three times as great as the least. The magnetic intensity is found to be least in Southern Africa.

### Production of Artificial Magnets.

886. Artificial magnets should be made of well hardened steel, of fine grain and uniform structure, free from

by the Magnetic Poles? Where is the earth's south magnetic pole? By whom was it reached, and what was found there? Where is the earth's north magnetic pole? How near has it been reached? With what do the changes in variation and dip seem to be connected? 884. What alone is affected by the magnetic attraction of the earth? Give an illustration. 885. How is the intensity of the earth's magnetic attraction shown to be different at different places? What is the law for ascertaining

15*.

flaws, and having level and polished faces. The breadth of a bar magnet should be one-twentieth of its length, and its thickness about one-seventieth of its length. In a horse-shoe magnet, the distance between the poles ought not to be greater than the breadth of one of the sides.

887. Magnetism may be imparted to steel or iron in four different ways:—1. By induction. 2. By the sun's rays. 3. By contact with a magnet. 4. By electric currents.

888. INDUCTION, A SOURCE OF MAGNETISM.—A magnetic atmosphere surrounds every magnet. A piece of iron or steel brought within this atmosphere, even without touching the magnet, has its neutral fluid decomposed, and exhibits magnetic properties. It is then said to be magnetized *by induction.*

Present half a dozen bars of iron at different angles to the positive pole of a magnet, without letting them touch it. They will all be magnetized by induction, the ends towards the magnet becoming negative poles and the opposite ends positive.

Suspend two pieces of soft iron wire by threads, parallel to each other and on the same level. On bringing either pole of a magnet a short distance below them, they become magnetized by induction. Like poles are formed in their contiguous extremities, and consequently instead of hanging parallel as before, they repel each other and diverge.

Bring one end of an unmagnetized steel bar near the north pole of a magnetic needle, and the latter will be attracted to it. Now place the positive pole of a powerful magnet near the other end of the bar, and the needle will soon be repelled. This is because the bar becomes magnetized by induction. The end nearest the needle becomes a positive pole by which the positive pole of the latter is repelled.

889. The earth magnetizes by induction. A bar of soft iron placed in the direction of the dipping needle, acquires magnetic properties by the inductive influence of the earth acting as a magnet. A few blows with a hammer on the

the magnetic intensity? What is found by applying this law? Where is the magnetic intensity found to be least? 886. Of what should artificial magnets be made? What should be the comparative dimensions of a bar magnet? What is essential in a horse-shoe magnet? 887. Name the four ways in which magnetism may be imparted to a piece of steel or iron. 888. When is a piece of iron said to be magnetized *by induction?* Illustrate magnetic induction with an experiment. Describe the experiment with two pieces of soft iron wire. What other experiment proves that a bar may be magnetized by induction? 889. How is it proved that the earth magnetizes by induction? What experiment shows the inductive influence of the earth?

upper end, by causing the particles to vibrate, help them to receive the magnetic influence.

Hold a bar of soft iron horizontally with one end near the north pole of a magnetic needle. The iron, being unmagnetized, attracts the needle. Now hold the bar in the direction of the dipping needle, give it one or two blows with a hammer, and the north pole of the needle will be repelled,—showing that the bar is magnetized, and a north pole formed in its lower end, by the inductive influence of the earth.

Iron bars that have long stood in a vertical position, or in the direction of the dipping needle, often acquire magnetic properties in an inferior degree. The same may be said of iron bars raised to a red heat and allowed to cool in the positions above mentioned, as well as of augers, gimlets, &c., that have been much used. Iron wire is frequently made magnetic by twisting it till it breaks.—All these are instances of magnetism by induction.

890. THE SUN'S RAYS, A SOURCE OF MAGNETISM.—Sunlight constitutes a second source of magnetism. The violet rays of the solar spectrum, concentrated by lenses on steel needles, have been found to endow them with magnetic properties.

891. CONTACT WITH A MAGNET, A SOURCE OF MAGNETISM.—A third and more efficient mode of exciting magnetism in iron or steel is by bringing it in contact with a magnet. Till recently this was the way in which artificial magnets were almost exclusively produced.

There are several different ways of magnetizing by contact. The principal are as follows :—

892. *Magnetizing Needles.*—An ordinary sewing needle may be magnetized by simply touching one of its ends to either pole of a powerful magnet. The end in question becomes negative if touched to the positive pole, and positive if touched to the negative.

893. *Magnetizing Bars.*—Steel bars may be magnetized either by *single touch* or *double touch.* Single Touch consists in applying but one pole of a magnet to the bar, or one pole to one-half, and the opposite pole to the other.

Double Touch consists in applying both poles at the same time throughout the whole length of the bar.

894. *To magnetize a bar by single touch*, apply midway of its length one of the poles of a magnet, and draw it to either end. Return it through the air to the middle of the bar, and draw it again to the same end as before. Repeat this process several times, always using the same pole and drawing it in the same direction. Then place the other pole on the middle of the bar, and draw it to the opposite extremity, repeating the strokes as in the former case. This must be done on both sides of the bar.

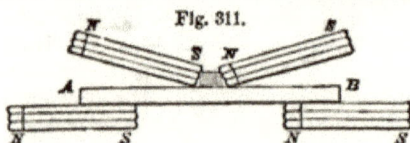

Fig. 311.

Another mode is represented in Fig. 311. The opposite poles of two magnets, kept about one-fourth of an inch apart by a piece of wood, are placed on the centre of the bar A B, so as to form angles of about 30 degrees with its surface. They are then slowly drawn in contrary directions from the middle to the extremities. This process is repeated several times, the magnets being raised when they reach the ends and replaced in the middle. The bar is then turned over, and the same thing done on the other side. The process is facilitated by resting the ends of the bar on the opposite poles of two other magnets, as shown in the figure.

895. *To magnetize a bar by double touch*, apply the opposite poles of two magnets as just described, only let them be perpendicular to the surface. Then, instead of drawing them to opposite extremities as before, move them together from the middle to one end, then through the air to the opposite extremity, and over the bar to the same end again, and so on—drawing them in the same direction over the bar, letting neither of the applied poles pass beyond its extremity, and finally stopping in the middle.

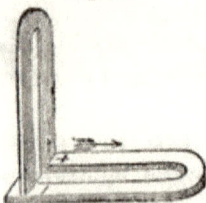

Fig. 312.

896. *Magnetizing Horse-shoe Bars.* —Horse-shoe magnets are produced by placing a piece of soft iron, as a keeper, across the ends of a steel bar bent in the proper form; and then, as shown in Fig. 312, applying perpendicularly to the extremities a horse-shoe magnet, whose arms are the same distance apart. Move it slowly to the bend, then carry it back through the air to the extremities, and draw it to the bend again. This must

be done about a dozen times; then, without removing the keeper, turn the bar over and do the same on the other side. The poles of the magnet produced will in this case be of the same character as those respectively brought in contact with them.

897. The best mode of magnetizing a horse-shoe bar is represented in Fig. 313. Lay the horse-shoe, A B, flat on a table, with its ends in contact with the poles of a horse-shoe magnet, N, S. Then place a piece of soft iron on these poles, and draw it slowly six or eight times towards the bend of the bar, in the direction of the arrow, raising it as often as it reaches the bend, and replacing it as at first. This process performed on both sides endows the horse-shoe with strong magnetic properties. The end which touches the positive pole of the horse-shoe magnet becomes negative, and the other positive.

Fig. 313.

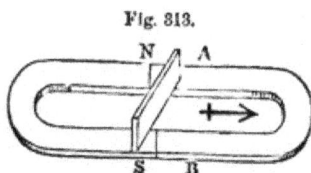

Two straight bars may be readily magnetized at once in the same way, by placing one extremity of each against the poles of the horse-shoe magnet, and connecting the opposite ends with a keeper.

898. ELECTRIC CURRENTS, A SOURCE OF MAGNETISM.— A bar of iron or steel is endowed with magnetic properties in the highest degree, by passing a current of voltaic electricity over a conductor placed in a certain position relatively to the bar. The details of this process belong to that branch of the science which is known as Electro-magnetism.

### Electro-magnetism.

899. Electro-magnetism treats of the phenomena and principles of magnetism excited by the passage of electric currents.

900. EFFECTS OF ELECTRIC CURRENTS ON THE MAGNETIC NEEDLE.—As a science, Electro-magnetism owes its origin to a discovery made in 1819 by Prof. Oersted, of Copenhagen. He found that a wire along which a voltaic current

produced? What will be the character of the poles in the magnet produced? 897. With Fig. 313, describe the best mode of magnetizing a horse-shoe bar. How may two straight bars be magnetized at once? 898. How is a bar of steel endowed with magnetic properties in the highest degree? 899. Of what does Electro-magnetism treat? 900. To what does electro-magnetism owe its origin? Give an account

was passing tended to turn the magnetic needle from its natural position to one perpendicular to the direction of the current. The conducting wire, of whatever metal it might be, was thus rendered magnetic by the electric current which it transmitted. It was subsequently found to attract iron filings; which, when the battery was in full action, clustered around it to the thickness of a quill, but gradually thinned off as the energy of the battery diminished, and left it entirely bare the moment the circuit was broken.

The direction in which the needle is turned depends on its position relatively to the wire, and the direction in which the current is passing. When the needle is on a different level from the wire, that is, directly above or below it, it retains its horizontal position; but its north pole is turned east or west, according to whether it is above or below the wire, and according to the direction in which the current moves. When the needle is on the same level with the wire, but on one side of it, it does not then swerve east or west; but its north pole is made either to dip or to rise, according to the side of the wire it is on and the direction in which the current moves. The following rule enables us always to determine the direction in which the needle will be turned :—

*Imagine yourself, with arms extended perpendicularly, lying along the conducting wire, with your head towards the point from which the current is coming, and your face turned towards the north pole of the needle ; then this north pole will be deflected in the direction of your right hand, whether it be up or down, east or west.*

The magnetic influence of the electric current is not therefore exerted in the plane of the conducting wire, but rather perpendicularly to that plane, so as to produce circular motion round the wire.

901. The deflection of the needle by an electric current may be shown with the apparatus represented in Fig. 314.

A brass wire is bent into rectangular form, and provided with a screw-cup at each extremity, P, N, for the reception of the wires from a galvanic battery, so that a current may be passed above and below a magnetic needle, N, S, suspended within the rectangle. The arms proceeding from P and N

of Oersted's discovery. How was it proved that the conducting wire was rendered magnetic by the electric current? On what does the direction in which the needle turns depend? How does it turn, when on a different level from the wire? How, when on the same level with the wire, but on one side of it? State the rule for determining the direction in which the needle will be turned? How is the magnetic influence of the electric current exerted? 901. Illustrate the deflection of the needle

are insulated from each other where they cross. No sooner is a positive current passed over the upper wire from north to south, than the needle is turned, its north pole deviating towards the east and its south pole to the west.

Here the *under* current, passing in the *opposite* direction to the upper one, tends to turn the

Fig. 814.

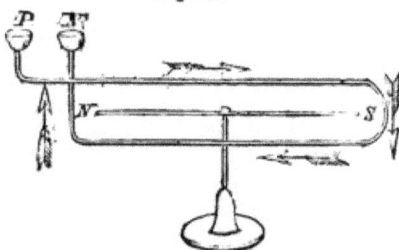

needle in the same direction; and the *deflecting force*, as it is called, is therefore twice as great as if the current passed in one direction only. If the wire be bent so as to make two rectangles about the needle, the deflecting force will be twice as great as when but one is formed; if five rectangles are made, as in Fig. 315, it will be five times as great, &c. In these cases, the wire must be covered with silk thread, or some other non-conductor, so as to insulate its arms from

Fig. 315.

each other, and oblige the current to traverse its whole length. It is on this principle that the Galvanometer is constructed.

902. *The Galvanometer.*—The Galvanometer is an instrument for measuring the force of galvanic currents by the deflection of the magnetic needle. It consists of a long wire bent into an oval or rectangular coil, the parts of which are prevented from touching by being wound with silk. The wire terminates in screw-cups, for convenience of connection with a galvanic battery. Within the coil a magnetic needle is delicately poised; and the instrument is placed so that the wire may have the same direction as the needle. They retain this direction till a galvanic current passes over the wire, when the needle is turned towards the east—more or less, according to the force of the current. A graduated scale fixed below the needle, with its circumference divided into degrees, measures the deflection, and consequently the quantity of electricity passing over the wire.

Fig. 316.

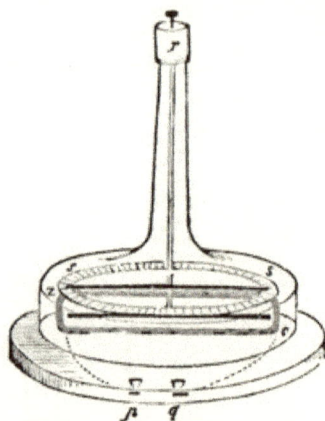

GALVANOMETER WITH ASTATIC
NEEDLE.

903. *Galvanometer with Astatic Needle.*—Instead of the ordinary needle, an astatic needle (see § 875) is sometimes used in the galvanometer. In this case, the needle, having its polarity neutralized, is more readily turned. The instrument is consequently more sensitive, indicating the presence of electric currents which would otherwise entirely escape detection.

Fig. 316 represents the Galvanometer with the Astatic Needle. The needles are suspended by two parallel silk threads from *r*, so that one of them may hang directly over the top of the coil *z c*, and the other below it. *p q* are the screw-cups terminating the wire which forms the coil, and *s s* is the graduated scale. The upper needle hangs above the coil; but as its poles point in opposite directions to those of the under one, it will tend to move in the same direction as the latter when galvanic action takes place.

904. CONNECTION BETWEEN ELECTRICITY AND MAGNETISM.—That there is an intimate connection between electricity and magnetism, was established by Oersted's experiment. It is further shown by the fact that compass-needles often have their poles reversed or their polarity weakened by lightning; that a spark has been drawn from a magnet; that a charge of electricity passed through a needle renders it magnetic; and that a bar may be permanently magnetized with an electric current more efficiently than in any other way.

These facts have led to the theory that magnetism is not an independent agent, but simply one of the forms assumed under certain circumstances by that subtile all-pervading agent which we call THE ELECTRIC FLUID. According to this theory, frictional electricity, voltaic electricity, thermo-electricity, magneto-electricity, and electromagnetism, are all one and the same thing, identical in kind, but differing in intensity, quantity, and properties, in consequence of the different modes in which they are developed.

905. ELECTRO-MAGNETIC ROTATION.—When a magnetic pole and a wire over which an electric current is passing are brought near each other, the pole tends to revolve round the wire, and the wire has a similar tendency to revolve round the magnet in a plane perpendicular to the direction of the current. With suitable apparatus, the following phenomena of electro-magnetic rotation may be exhibited :—

1. The conducting wire being fixed, the magnet will revolve about it.

2. The magnet being fixed, the conducting wire will revolve about it.

3. Both magnet and wire being left free to move, tney will revolve in the same direction round a common centre, each appearing to pursue and be pursued by the other.

4. The conducting wire being dispensed with, a magnet may be made to turn on its own axis by the passage of an electric current along half its length.

906. To show the revolution of a magnet about a conducting wire, Faraday used the apparatus represented in Fig. 317. A magnet, *n* S, is immersed in a vessel of mercury, with its north pole, *n*, a short distance above the liquid, and its south pole, S, connected by a silk thread with the conducting wire C, which passes through the bottom of the vessel. *a b* is another conducting wire, which enters the mercury from above. When *a b* is connected with the positive pole of a galvanic battery, and C *d* with the negative, a descending current of positive electricity passes along the conductor (the mercury completing the circuit), and the north pole, *n*, will revolve round the fixed wire, *a b*, in the direction of the hands of a watch. If, on the contrary, *a b* be connected with the negative

Fig. 317.

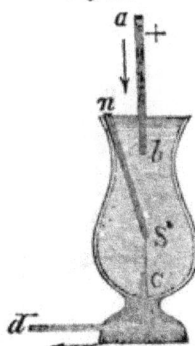

pole, and C $d$ with the positive, an ascending current will be formed, and the magnet will revolve in the opposite direction.

Mercury is used in this experiment, because, being a liquid, it allows the magnet to move through it, while at the same time, being a conductor, it completes the circuit, and carries off the magnetic influence from the south pole immersed in it. Were it not for this, the south pole, by its tendency to move in the opposite direction to the north, would keep the magnet stationary.

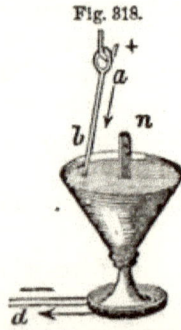

Fig. 318.

907. Fig. 318 illustrates the revolution of a conducting wire around a fixed magnet. Again we have a vessel of mercury, with a conducting wire, $d$, passing through its bottom, and another wire, $a b$, suspended from a hook directly over the magnet, entering the mercury from above. $n$ is the north pole of the fixed magnet. On connecting the hook and the wire $d$ with the poles of a galvanic battery, the wire will revolve round the magnet, the direction depending, as before, on whether the electric current is ascending or descending.

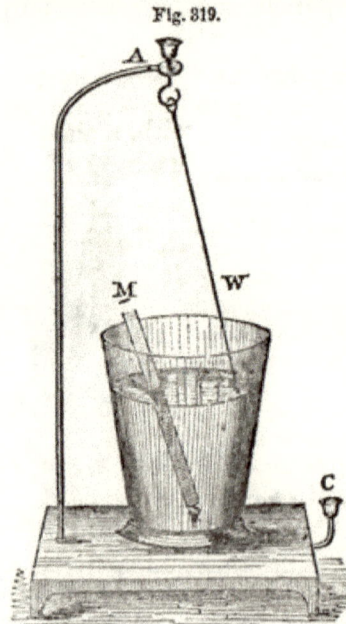

Fig. 319.

908. By ingeniously combining the two pieces of apparatus just described, we may exhibit the simultaneous revolution of both magnet and wire round a common centre. The magnet, M, is immersed in a vessel of mercury about half its length, that the current may affect only one pole. It is connected at the bottom with a conducting wire and screw-cup, C, in such a way as to allow it freedom of revolution. The wire, W, is suspended from a hook, so as to move freely. On transmitting a current, which is done by connecting A and C with the poles of a battery, both the magnet and the wire commence revolving in the same direction as if chasing one another

909. EFFECT OF ELECTRIC CURRENTS ON STEEL AND SOFT IRON.—The deflection of a magnetic needle by a wire

over which an electric current is passing, has been described in § 900. If a bar of soft iron is placed across such
a wire, it becomes a temporary magnet, as is shown by its
attracting iron filings. A bar of steel so placed is made a
permanent magnet.

910. *The Helix.*—The magnetizing power of the wire is
greatly increased, if, instead of touching the bar in but a
single point where they
cross, it is wound a number
of times spirally round the
latter, as shown in Fig. 320.

Fig. 320.

Such a coil of wire is called a Helix (plural, *hel'-i-ces*).

A helix may be familiarly made by winding some copper wire tightly
round a small bottle, and then drawing the bottle out. As the magnetizing
power of the helix increases with the number of times that the electric current passes round the bar, each turn of the wire is pushed close up to the
one before it; and, to increase the effect still further, several coils or layers
of wire may be formed, one on top of another. Direct communication between contiguous parts of the wire must be prevented by winding silk or
some other insu-
lating material
round it. When
the ends of the
wire are connect-
ed with the poles
of a galvanic bat-
tery, the current
is thus obliged to
pass through its
whole length. Fig.
321 represents a

Fig. 321.

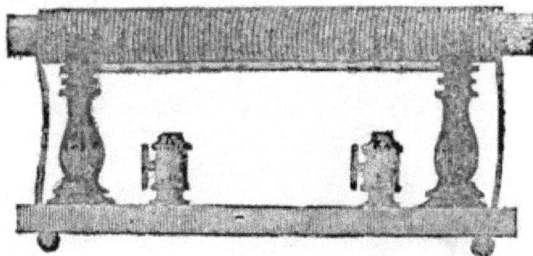

A HELIX.

helix mounted on a stand. An iron bar extending through the centre is seen
projecting at each end.

911. *Magnetizing Power of the Helix.*—A steel bar
introduced within a helix becomes permanently magnetized
the moment an electric current is passed over the wire. A
needle laid inside of it is sometimes so powerfully acted on

apparatus. 909. What is the effect of a wire over which a current is passing on a bar
of soft iron placed across it? On a bar of steel so placed? 910. How is the effect
greatly increased? What is such a coil of wire called? How may a helix bo made?
How is the effect of the helix increased? With what is the wire covered, and why?
What does Fig. 321 represent? 911. What is the effect of a helix on a steel bar in-

as to be lifted up and held suspended in the air in the middle of the helix. A bar of soft iron placed in the same position is endowed with strong magnetic properties for the time, but instantly loses them when removed, or when the current ceases to pass. To be magnetized, the bar must always be placed lengthwise of the helix, —that is, at right angles to the direction in which the current is passing.

Fig. 322.

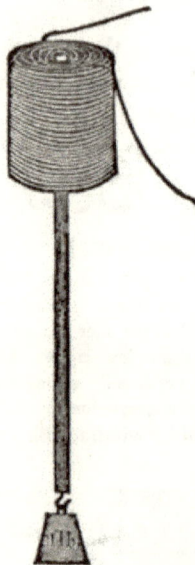

One of the most remarkable effects of the helix is the suspension in the air, without any visible support, of a heavy iron bar loaded with weights. A helix consisting of a very long wire, forming several coils one upon another, and charged by a powerful battery, is held in a vertical position, as shown in Fig. 322. An iron bar brought within the helix just at its base, will be lifted up half way into it, and held there in the centre of the hollow cylinder, without touching it, as long as the current continues to pass. If pulled down a little way, it immediately springs back to its former position. The moment the current ceases, the bar falls. With a powerful apparatus, a weight of eighty pounds has been thus kept suspended in the air.

Fig. 323.

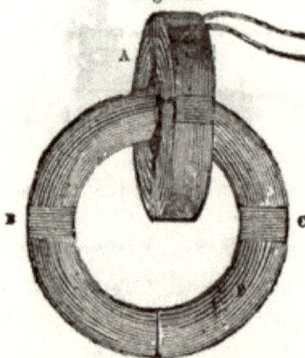

A no less interesting experiment, showing the power of the helix, may be performed with the apparatus represented in Fig. 323. The helix, A, is in the form of a ring. B, C, are two semi-circular pieces of soft iron, having their ends accurately fitted to each other. When B and C are brought together so as to form a circle, with one pair of their joined ends within the helix, they are endowed with so strong an attraction for each other that two men can hardly pull them apart.

912. *Electro-Magnets.*—An electro-magnet consists of a bar of soft iron within a helix.

It is strongly magnetic as long as a current passes over the wire, but loses its power the moment the current ceases.

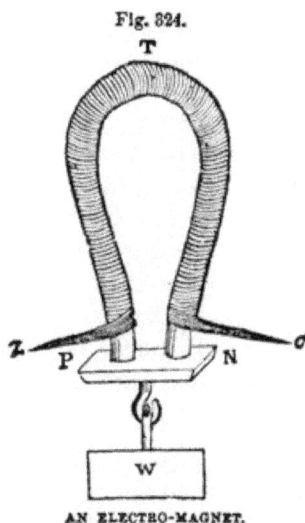

Fig. 324.

The most powerful electro-magnet is made by bending a bar of soft iron into the form of a horse-shoe, as shown in Fig. 324, and winding closely round it a large quantity of insulated copper wire so as to form a helix of several layers. The ends of the wire, Z, C, are connected with a powerful battery. A soft iron keeper, P N, connects the poles, having a hook beneath, to which weights may be attached. So strongly is this keeper attracted that an enormous force is required to separate it. An electro-magnet prepared as above has supported over 4,000 pounds.

AN ELECTRO-MAGNET.

913. Electro-magnets furnish us with the most efficient means of magnetizing an ordinary horse-shoe bar. The mode of using them for this purpose is shown in Fig. 325.

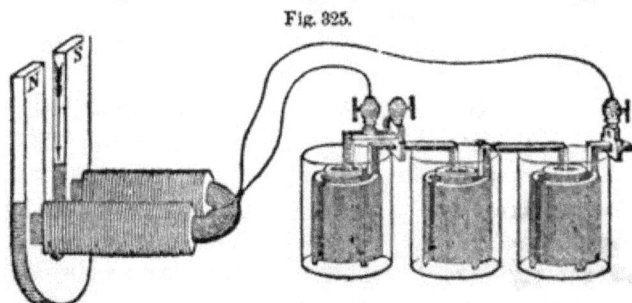

Fig. 325.

The electro-magnet is applied at the bend, one pole on each arm, and drawn towards the extremities, N, S. This is done several times on both sides, when the bar is rendered permanently magnetic. To deprive it of its magnetic power, reverse the process, by applying the poles of the electro-magnet to the ends N, S, and drawing them towards the bend.

914. ELECTRO-MAGNETISM, AS A MOTIVE POWER.—We have seen that an electro-magnet is instantly endowed with

In Fig. 323. 912. Of what does an electro-magnet consist? How is the most powerful electro-magnet made? How great a weight has been supported with such an electro-magnet? 913. What is the most efficient means of magnetizing a horse-shoe

great attractive power for iron on being connected with a galvanic battery, and as instantly divested of it when the connection is severed. It may thus be made to impart motion to an iron rod, and through it to various kinds of machinery. So strong at one time was the impression that the enormous attractive power of the electro-magnet could be advantageously used as a mechanical agent, that the United States government appropriated $20,000, and Russia $120,000, for experiments on the subject; and various machines were contrived in which it was used as a motive power. In none, however, thus far invented, has it been found to approach steam in efficiency or economy.

A boat 28 feet long with a dozen persons on board has been propelled against the current at the rate of three miles an hour by electro-magnetic action. A locomotive engine has also been driven from ten to twelve miles an hour. But this is the utmost that has been effected, and in both cases the cost of keeping the galvanic battery in operation was much greater than that of producing an equivalent quantity of steam. The difficulty appears to be twofold. First, the attractive power of the magnet rapidly diminishes as the distance from it increases. Secondly, electric currents opposite in direction to the primary one are excited in the moving machinery; which, increasing in power with its velocity, nullify much of the effect of the magnet. Until these difficulties are removed, electro-magnetism can not be advantageously used as a mechanical agent.

915. The Electro-magnetic Telegraph.—Although unavailable as a motive power, electro-magnetism has been turned to practical account in the Telegraph, one of the crowning triumphs of human ingenuity. For this great invention as at present perfected, which enables us, almost with the rapidity of thought, to communicate with distant points, over miles of intervening land or sea, the world is chiefly indebted to an American—Samuel F. B. Morse.

916. *Morse's Telegraph.*—The principles on which Morse's Telegraph operates are as follows :—

bar? Describe the process. 914. On what principle may an electro-magnet be made to impart motion to an iron rod? For what were appropriations made by the United States government and Russia? What has been effected with machinery moved by electro-magnetism? How does the expense compare with that of steam? What difficulties interfere with the usefulness of electro-magnetism as a motive power? 915. In what has electro-magnetism been turned to practical account? To whom is

1. An electro-magnet may be alternately endowed with and deprived of the property of attracting iron by connecting and disconnecting it with a galvanic battery.

2. The battery may be miles away from the magnet. If wires connect the two, the electric current will still be carried to the helix and produce the same effects.

3. A person stationed near the battery may complete and break the circuit at pleasure. As he does so, one end of a lever placed near the poles of the distant magnet will be attracted or released. When it is attracted, the other end of the lever, which is furnished with a point, is made to indent a strip of paper passed in front of it by machinery, with dots or dashes, according to the time that the operator by the battery keeps the circuit complete. If, now, different combinations of dots and dashes are agreed upon to represent certain letters, it is evident that a message can be communicated from the one point to the other.

Fig. 326 represents Morse's recording apparatus.

Fig. 826.

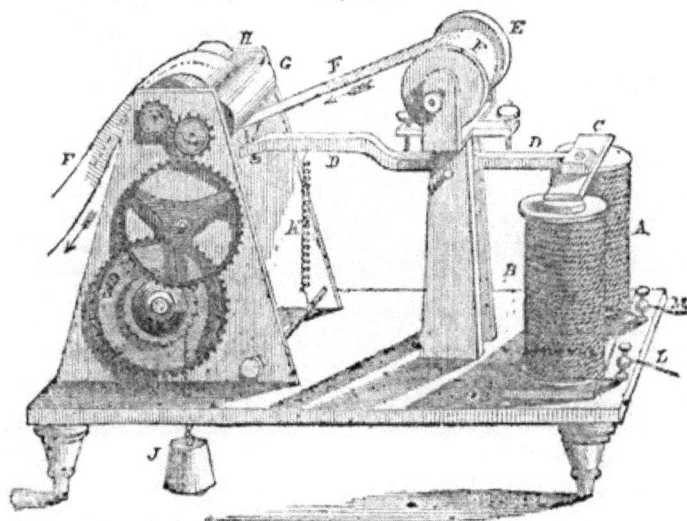

A B is the electro-magnet, connected with the distant battery by the wires L, M, which are raised on poles and insulated by glass supports. C is an armature of soft iron attached to one end of the lever D D, so as to rest about one-eighth of an inch above the poles of the magnet. The other end of the lever carries a point or style, I, which is raised as C is depressed. A strip of paper, F, F, rolled on the spool E, is made to pass in front of the style, between the two cylinders G, H, by means of wheel-work set in motion by the weight J when the current passes. K is a spring, to pull down the end of the lever bearing the style when the other end is released by the magnet. A striking apparatus was formerly connected with the machinery in such a way as to give warning to the attendant with the first motion of the lever; but it is now generally dispensed with, as the clicking sound produced by the lever is found to be sufficient for the purpose.

Instead of carrying both wires over poles from the electro-magnet to the battery, the earth is now generally made to form one-half the circuit. This is effected by carrying down the wire from the magnet, and connecting it with a metallic plate buried in the ground; a similar plate must be buried where the battery is stationed, and a wire from the latter connected with it. If this is done, but one wire need pass over the poles to complete the circuit.

917. The apparatus used by the operator where the battery is stationed, to complete and break the circuit, is called the Signal Key. It is represented in Fig. 327.

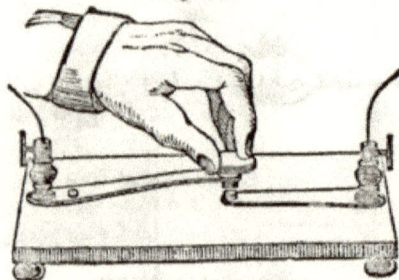

Fig. 327.

THE SIGNAL KEY.

By pressing on the knob, the screws in which the wires are fastened are connected, and the circuit is completed. On removing the hand, the knob springs up, the circuit is broken, and the current ceases. If the knob is kept pressed down, the paper at the other end is indented with a continuous line; but by tapping on it so as to form different combinations of dots and dashes, which stand for letters, and are understood at both ends of the line, a message is transmitted. According to Morse's system, the following combinations are used to represent the different letters and figures:—

LETTERS.

| | | |
|---|---|---|
| a - — | j — - — - | s - - - |
| b — - - - | k — - — | t — |
| c - - - | l —— | u - - — |
| d — - - | m — — | v - - - — |
| e - | n — - | w - — — |
| f - — - | o - - | x - — - - |
| g — — - | p - - - - - | y - - - - |
| h - - - - | q - - — - | z - - - - |
| i - - | r - - - | & - - - - |

FIGURES.

| | |
|---|---|
| 1 | - — —. |
| 2 | - - — - - |
| 3 | - - - — - |
| 4 | - - - - — |
| 5 | — — — |
| 6 | - - - - - - |
| 7 | — — - - |
| 8 | — - - - - |
| 9 | — - - — |
| 0 | —— |

To prevent confusion, a small space is left after each letter, a longer one between words, and a still longer one at the end of a sentence. The operators in telegraph offices become so familiar with this alphabet that they understand a message from the mere clicks of the lever, without looking at the paper on which it is recorded.

918. An electric current is transmitted by a wire to a great distance, but not with undiminished power. When, therefore, the stations are very far apart, the electro-magnet is charged too feebly to make the style indent the paper. In this case, the wire from the original battery is made to act on a very delicate armature, so as to complete the circuit of a second battery placed near the machine. This Relay Battery, as it is called, acts on the recording apparatus as described above, or transmits a fresh and vigorous current to another relay battery. In this way lines of any length may be formed.

As relay batteries do not interrupt the circuit, any number of them may be placed at intervals along a line. Each may work a recording apparatus of its own, and a given communication may thus be registered simultaneously at a multitude of different stations.

Relay batteries may be dispensed with by increasing the number of plates employed and distributing them in groups along the line. It has been computed that if a telegraph wire could be carried round the earth, 1200 of Grove's pint cups, distributed in equi-distant groups of fifties, would supply the galvanic power for the whole distance.

of operation. How are the different letters represented? 918. What difficulty is there when the current is transmitted to a great distance? How is this remedied? How does the Relay Battery act? How may a given message be registered simultaneously at different stations? What may be substituted for relay batteries? How many cups would supply the galvanic power for a telegraph round the earth?

919. *House's and Bain's Telegraph.*—Morse's apparatus, having been first introduced and being very simple and not likely to get out of order, is more used than any other, both in this country and in Europe. There are other ingenious systems, however, which are employed to some extent. Among these are House's Printing Telegraph and Bain's Electro-chemical Telegraph.

House's apparatus is one of the most wonderful achievements of inventive art. Making use of the electro-magnet in connection with ingenious and somewhat intricate machinery, it enables the operator, by playing on twenty-eight keys like those of a piano (representing the twenty-six letters and two punctuation points), to print ordinary letters on a strip of paper at the other end of the line at the rate of about two hundred a minute. The great advantages of House's system are that there is little or no liability to mistake in transmitting a message, and that the latter, being produced in Roman capitals, need not be transcribed, but may be sent just as it comes from the machine to the person for whom it is intended.

In Bain's Electro-chemical Telegraph no magnet is used. The point of the wire, which is stationary, constitutes the pen, and rests lightly on a metallic plate, which is made to revolve by machinery. On this plate is placed paper which has been previously moistened with some chemical preparation decomposable by voltaic electricity. When the connection is made by the distant operator, the current passes from the wire to the plate through the paper, and in passing decomposes the chemical compound with which the paper is impregnated. The result is a deep blue spot on the paper, which renders the dot or dash visible, just as the indentation does according to Morse's system. As even a feeble voltaic current has the power of decomposition, there is not the same necessity for relay batteries on Bain's line as on either of the others.

920. *Submarine Telegraphs.*—Submarine Telegraphs are telegraphs connecting points separated by water, in which the wire is submerged. The first successful telegraph of this kind was laid in 1851 across the English Channel, and connected Dover with the French coast. This was followed by several others; and in 1858, after several unsuccessful attempts, a telegraph cable nearly 2,000 miles in length was laid across the Atlantic Ocean, between Valen-

---

919. What other telegraph systems besides Morse's are in use? What is said of House's apparatus? What are its great advantages? What is the principle involved in Bain's Electro-chemical Telegraph? What advantage is there connected with this system? 920. What are Submarine Telegraphs? Where and when was the first submarine telegraph laid? In 1858 what great enterprise was carried through? De-

tia Bay, Ireland, and Trinity Bay on the coast of New-foundland. It consisted of a group of seven copper wires insulated and protected by a casing of gutta-percha, the whole surrounded by strands of iron wire, and sunk to the bottom of the ocean, at a depth nowhere exceeding 2½ miles.

Public interest was strongly excited in this great enterprise; but it has thus far been doomed to disappointment. After transmitting several mes-sages, the Atlantic Telegraph, for some unexplained reason, ceased to work, though signals have from time to time been received. There is little doubt, however, that the work is feasible, and that we shall soon have regular tele-graphic communication between the opposite sides of the Atlantic.

921. *History of the Telegraph.*—The fact that frictional electricity could be conveyed by wires to a great distance was known more than a hundred years ago. Franklin, in 1748, set fire to alcohol by means of a wire from an elec-trical machine carried across the Schuylkill River. The first attempt to transmit a communication by electricity, however, was made in 1774 by Le Sage [*luh sahzh*], a Frenchman, at Geneva.

Le Sage used twenty-four wires insulated in glass tubes buried in the earth, each of which represented a letter of the French alphabet. The wires were connected with an electrical machine in the order necessary to spell out the words, and electroscopes attached to them at the other end indicated this order by their successive divergence to an attendant stationed there.

922. Volta's discovery in 1800 furnished a far more effi-cient agent for telegraphic communication than frictional electricity, and was followed in a few years by a plan for an electro-chemical telegraph, requiring thirty-five wires, to represent the different letters and figures, and to act by the decomposition of water.

The great discovery of electro-magnetism in 1819 called forth many new suggestions,—among others, the use of the deflections of the needle as signals; but none of the plans proposed were practicable on a large scale. A more per-

scribe the Atlantic cable. What is said of the working of the Atlantic telegraph? 921. What fact relating to frictional electricity was known more than a hundred years ago? What experiment was performed by Franklin in 1748? Who made the first attempt to transmit a message by electricity? Describe the plan of Le Sage. 922. By what was the discovery of voltaic electricity followed? What suggestions were called

manent galvanic power was needed; and this was not supplied till 1836, when Daniell brought out his constant battery. The appearance of this battery and the improved electro-magnets prepared by Prof. Henry, was followed in 1837 by the invention of apparatus for transmitting and recording communications, by Samuel F. B. Morse, who had been experimenting on the subject for five years. Application was at once made to the Congress of the United States for aid to construct a line of sufficient length to test the invention; and after discouraging delays, in 1843, the sum of $30,000 was appropriated by that body, with which a line was established between Baltimore and Washington, a distance of forty miles. The enterprise was crowned with complete success; and the first news transmitted was the proceedings of the democratic convention of 1844, then sitting in Baltimore, by which James K. Polk was nominated for the presidency.

So manifold were the advantages of telegraphic communication, that immediately on the announcement of Morse's success companies were formed, and wires were soon seen threading the country in all directions. The various lines now in operation in the United States and British Provinces make a total of about 45,000 miles, on nine-tenths of which Morse's apparatus is used, House's and Bain's being chiefly employed on the remainder. With Morse's instruments about 9,000 letters may be transmitted in an hour. The construction of the line costs not far from $150 a mile.

The same year in which Morse perfected his invention (1837), plans for telegraphic communication based on the deflections of the needle were announced by Wheatstone in England, and Steinheil [stine'-hile], a German philosopher, to whom the discovery that the earth could be made to complete the circuit seems to be due. They are therefore sometimes mentioned as entitled to share with Morse the honor of his great invention. Their systems, however, were but modifications of what had been proposed some years before; though practicable, they could not compete in rapidity of operation with Morse's, and consequently never came into general use.

923. ELECTRO-MAGNETIC CLOCKS.—American ingenuity

forth by the discovery of electro-magnetism? By whom and when was the first perfect apparatus for transmitting and recording communications invented? What two improvements prepared the way for Morse's invention? How was Morse enabled to test his invention? What was the result? What was the first news transmitted? How many miles of telegraph are now in operation? On how much of this is Morse's apparatus used? What is the cost of constructing a telegraphic line? Who are sometimes mentioned as sharing with Morse the honor of inventing the telegraph?

has applied electro-magnetism to the determining of minute intervals of time and the regulation of clocks. The time of astronomical observations may thus be fixed with perfect precision to the tenth of a second.

The pendulum of a clock, for instance, is, by some mechanical contrivance, made by its vibrations to close and break a galvanic circuit. With Morse's apparatus, each vibration is indicated by a dot on a strip of paper passed in front of the style. If now an observer have a signal-key connected with the same circuit, by depressing it the instant a star passes one of the wires of his telescope, he permanently records its transit on the same paper by a dot intermediate between two vibration-dots, the exact time of which is known.

924. By the same agency a number of clocks may be made to keep uniform time.

This is effected by connecting any number of distant clocks, by means of wires, with one standard time-piece, which is itself connected with a galvanic battery,—so that the circuit may be closed and broken by all the pendulums simultaneously. Wheels connect the pendulums with the hands of the clocks, which are thus made to move with perfect uniformity. Some railroad companies use an arrangement of this kind to make the clocks at their different stations keep time together.

925. ELECTRO-MAGNETIC FIRE-ALARM.—The principle of the telegraph has been used for raising a simultaneous alarm of fire at a number of different stations connected with one principal station by wires. By completing and breaking the galvanic circuit, an attendant who is constantly on watch at the principal station, and receives his information by telegraphic signals from the district in which the fire is detected, strikes alarm-bells at the various distant stations a certain number of times, according to the number of the district in question. Such an arrangement has been used in Boston with great success.

926. THE HELIX, A MAGNET.—The helix, when traversed by a current of electricity, not only has high magnetizing powers, as we have seen, but is also itself a magnet. If

What is said of their claims? 923. To what has American ingenuity applied electro-magnetism? Show how an astronomical observation may be telegraphically recorded. 924. How may a number of clocks be made to keep uniform time by means of electro-magnetism? 925. For what has the principle of the telegraph been used? Show how an alarm of fire may be simultaneously raised at different stations. 926. What is the effect of an electric current traversing a helix on the helix itself?

suspended so as to allow it freedom of motion, it points north and south, and dips like the magnetic needle. So, like poles of two helices repel each other; unlike poles attract each other.

Even when not bent in the form of helices, two wires traversed by electric currents, if brought near each other in parallel lines and free to move, exhibit mutual attraction or repulsion. When their currents move in the same direction, they attract each other; when in contrary directions, they repel each other.

## Magneto-electricity.

927. Not only is magnetism developed by electric currents, but electric currents are produced by magnetism. That branch of science which treats of electric currents so produced is called Magneto-electricity.

The phenomena of magneto-electricity, like those of electro-magnetism, go far towards proving the intimate connection between electricity and magnetism, if not their actual identity.

928. *Experiments.*—Connect the ends of wire from a helix with a galvanometer. Then quickly thrust into the helix one of the poles of a bar magnet. The needle of the galvanometer is at once deflected, showing the passage of an electric current over the wire. If the opposite pole is introduced into the helix, a current passes in the contrary direction.

Within a helix place a soft iron bar of such length that each end may project a little. Over its ends bring the poles of a horse-shoe magnet, so suspended as to have freedom of revolution. On turning the magnet rapidly, the poles of the bar are reversed twice for each revolution, and an electric current is produced on the wire, as is shown by a galvanometer attached to it. This principle has been applied in different magneto-electric machines, with which water may be decomposed, platinum wire heated to redness, sparks produced, shocks given, and other experiments performed.

929. THE MAGNETO-ELECTRIC MACHINE.—Fig. 328 represents one form of the Magneto-electric Machine.

S is a compound horse-shoe magnet supported on three pillars. In front of its poles, and as near as it can be brought without touching, is a bar of soft iron bent at right angles, and surrounded with several coils of insulated copper wire. The ends of this wire are pressed by springs against a con-

Fig. 323.

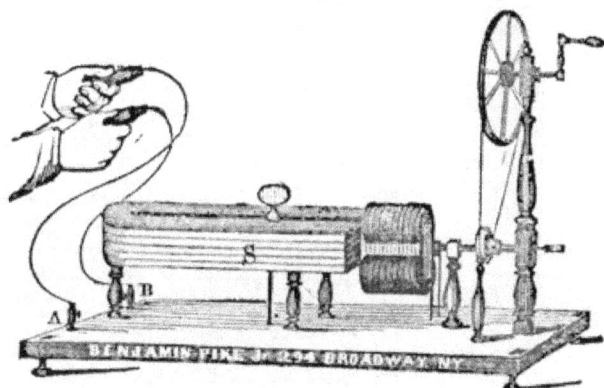

MAGNETO-ELECTRIC MACHINE.

ducting metallic plate, connected by wires passing under the stand with the screw-cups A, B. The soft iron armature just described is mounted on an axis which is made to revolve by a wheel turned by a handle. The handle being rapidly turned, each half-revolution of the armature brings its extremities near opposite poles of the magnet, thus reversing its polarity, and producing a strong electric current on the wire. If small copper cylinders attached to the wires are grasped one in each hand, as shown in the figure, a series of severe shocks are received, and the muscles are so contracted that it is almost impossible to open the hands and let go the conductors.

Machines of this kind, adapted to medical use, have been found efficacious in cases of rheumatism, dyspepsia, sprains, nervous diseases, &c., the current being made to pass through the diseased part.

## Diamagnetism.

930. Experiments with powerful electro-magnets show that almost all substances are susceptible of magnetic influence. Some are attracted by the magnet; others, repelled; while a few are not acted on at all, though when more powerful magnets shall be made they may perhaps be found to fall under one of the two previous classes.

Hence arises a three-fold division of bodies. 1. Magnetic bodies, or such as are attracted by an electro-magnet.

Machine represented in Fig. 323, and its mode of operation. What is the effect of such a machine on the human system? What use has been made of machines of this kind? 930. What has been shown by experiments with powerful electro-magnets? Name the three classes into which bodies are divided with reference to the influence

2. Diamagnetic, or such as are repelled.   3. Indifferent, or such as are not acted on at all.

Fig. 329.

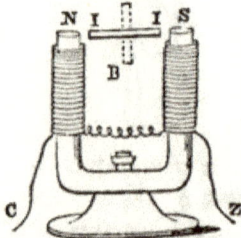

The difference between these three classes of bodies may be illustrated with the apparatus shown in Fig. 329. N, S, are the poles of an electro-magnet, which is connected by the wires C, Z, with a galvanic battery. A bar of iron, nickel, cobalt, manganese, or other magnetic substance, suspended between the poles so as to move freely, will come to rest with its ends as near them as possible, in the position I I. On the contrary, a bar of bismuth, phosphorus, zinc, tin, or other diamagnetic substance, similarly suspended, will be repelled and come to rest at right angles to the position just described, as shown by the dotted line,—with its sides opposite the poles of the axis and its ends as far from them as possible. Similar attraction and repulsion are exhibited if the substances are presented to either pole separately. An indifferent substance will remain in any position in which it is placed, being neither attracted like the iron nor repelled like the bismuth.

Similar experiments may be made on liquids and gases by enclosing them in tubes. It is thus found that oxygen is magnetic; water, alcohol, ether, and the oils, diamagnetic.

———— ••• ————

# CHAPTER XVIII.

## ASTRONOMY.

931. ASTRONOMY is the science that treats of the heavenly bodies,—their motions, size, distance, &c.

By the heavenly bodies are meant the sun, the moon, stars, planets, and comets.

932. Astronomy, as it is the most sublime, is also the oldest of sciences. The shepherds of the patriarchal age, tending their flocks by day and night beneath the canopy of heaven, naturally directed their gaze to the brilliant

exerted on them by electro-magnets. Define each. Illustrate the difference between these three classes with the apparatus represented in Fig. 329. How may similar experiments be made on liquids and gases? What gas is found to be magnetic? What liquids are diamagnetic?

931. What is Astronomy? What are meant by the heavenly bodies? 932. Who

orbs with which it is studded, observed their motions, and thus became the first astronomers. Chaldean observations are said to extend back to within a hundred years of the flood. The Chinese, also, paid great attention to this science in remote antiquity. We are told that more than 2,000 years before the birth of Christ, an emperor of China put to death his two chief astronomers for not predicting an eclipse of the sun.

Destitute of the admirable instruments which modern science has produced and used with signal success, the ancient astronomers of course erred in many of their conclusions. We can only wonder that they obtained as much knowledge as they did respecting the heavenly bodies.

933. To unfold the principles of astronomy at length would require a volume, and to understand them thoroughly, a knowledge of the higher mathematics is essential. We can here present only such leading facts as will serve to give a general view of the science.

934. FUNDAMENTAL FACTS.—The great facts established by the researches of astronomers are as follows:—

1. Space is filled with worlds.

Looking up into the heavens on a clear night, we see them all around us. The telescope reveals millions. There are no doubt millions more too remote to be seen at all, and others which from being non-luminous escape our vision. Powerful instruments reach to points from which light, travelling as it does with the enormous velocity of 192,000 miles in a second, would be 60,000 years in reaching us, and throughout the whole of this vast field worlds are everywhere scattered. We can but infer that the regions to which man's eye has never penetrated are similarly studded; and that, if an observer could be transported to the remotest star visible with his telescope, he would see spread before him in the same direction a firmament no less rich and splendid than that which he beheld from the earth.

2. These worlds are divided into systems, the members of which are bound together by mutual attraction. Each system has a central sun, round which the other members, called Planets, revolve. While this revolution is going on, the suns themselves with their respective planets move about a common fixed central point.

3. The stars that we see twinkling in the sky are suns.

were the first astronomers? How far back are Chaldean observations said to extend? What story shows the attention paid to astronomy by the Chinese in remote antiquity? What is said of the ancient astronomers? What is the first great fact established by astronomers? What facts are stated respecting the number of worlds? What inference is drawn respecting the regions of space unpenetrated by the eye of man? How are these worlds divided? What are the stars that wo see twinkling

16*

The planets that we suppose to revolve about them are non-luminous, and therefore invisible.

4. Some of these planets have satellites or moons moving around them, and with them around the sun of the system to which they belong.

5. The Earth, which we inhabit, is a planet belonging to what is known as the Solar System, of which the Sun is the centre. The Earth is attended by one satellite known as the Moon.

### The Solar System.

935. The Solar System, as at present known, consists of the sun, its centre; seventy planets revolving round it, of which sixty-two, on account of their small size, are called Asteroids (*starlike bodies*); twenty moons revolving round the planets; and many thousand comets, the exact number of which is unknown.

936. That the earth and other planets move round the sun, was taught by the philosopher Pythagoras about 500 B. C. Deceived by appearances, however, the ancients generally rejected this theory, and believed the earth to be the fixed centre of motion for all the heavenly bodies. Some made the planets revolve round the sun, and the sun carrying the planets with it to move round the earth. The Egyptian astronomer Ptolemy supposed the universe to consist of a number of hollow spheres arranged one within another, and appropriated respectively to the sun, the moon, the planets, and the stars. The earth, according to Ptolemy, was at the centre of these spheres, which turned round it from east to west every twenty-four hours, carrying the stars and planets with them; being of crystal, they were perfectly transparent, and the inner ones did not therefore obscure the more distant luminaries seen through them.

These theories, particularly Ptolemy's, prevailed till about the middle of the sixteenth century, when the Prussian philosopher Copernicus revived the teachings of Pythagoras, and established what is called from him the Copernican System, which is now acknowledged as the true theory of the universe. Fearing the prejudices of his fellow-men, Copernicus withheld his system from them for some years. His great work, in which his views were embod-

in the sky? Why are not their planets visible? By what are some of the planets attended? What is the Earth? By what is it attended? 935. Of what does the Solar System, as at present known, consist? 936. What was Pythagoras's theory of the universe? What was the belief of the ancients generally? Give an account of Ptolemy's theory. By whom and when was the true system revived? When was

fed, was finally published in 1543, just in time for a copy to be placed in his hands on his death-bed.

The Copernican system at first met with but moderate favor. Its truth, however, was established by Galileo, whose observations with the newly-invented telescope afforded him incontrovertible arguments in its favor. Yet the advocates of the old system were determined to close their eyes. On Galileo's announcing the discovery of four moons about the planet Jupiter, they denied the possibility of their existence; and when he urged them to look for themselves through his telescope, they refused to have anything to do with an instrument they despised. An astronomer of Florence gravely argued that as there were only *seven* apertures in the head—two eyes, two ears, two nostrils, and one mouth—and as there were only *seven* metals, and *seven* days in the week, so there could be only *seven* planets. As there were six principal planets and one moon then known, the number was complete, and Galileo's pretended planets must be impossibilities.—But such absurd arguments could not long obscure the light of truth.

937. THE SUN (☉).—The Sun, the great source of light and heat to the planets, is the centre of the solar system. It is an immense globe, five hundred times as large as all its planets put together. Its diameter is 882,000 miles. Placed where the earth is, it would fill the whole orbit of the moon, and extend 200,000 miles beyond it in all directions. Its volume is nearly a million and a half times as great as the earth's, and it contains more than 350,000 times as much matter.

938. *Solar Spots.*—Viewed through a telescope, the sun looks like a globe of fire. Its surface, however, is not always wholly luminous. A number of dark spots, surrounded by a lighter shadow, are at times scattered here and there within a zone extending 35 degrees on each side of the solar equator. The number and size of these spots differ at different times; for, while some last a couple of months or even longer, others change their form from day to day. They have been known to vanish almost instantly and to appear as suddenly. Some years none at all are visible; in others, as many as 200 are seen at once, cover-

the work of Copernicus relating to this subject published? By whom was the truth of the Copernican system established? What were the arguments with which Galileo was met? 937. What is the Sun? How great is its diameter? Placed where the earth is, how far would it extend? How does its volume compare with the earth's? Its matter? 938. How does the sun look, when viewed through a telescope? Describe the spots which are sometimes visible. What is said of their number and size? What

ing so much of the surface as materially to diminish the
quantity of light emitted.

By comparing a number of observations on the solar spots, we find that
they are subject to periodical increase and decrease. They become larger
and more numerous for a certain time till they reach a maximum, after which
they gradually diminish, till all disappear, or nearly so; new ones then become visible, and go on increasing during the same period as before. This
period seems to be a little over eleven years.

Spots have occasionally appeared of such size that they could be readily
discerned with the naked eye. One thus seen for a week in June, 1843, must
have been 77,000 miles across, or nearly ten times the size of the earth.

Astronomers have tried to account for the solar spots in various ways.
The prevailing opinion is that the light received from the sun does not come
from its surface, but from a luminous atmosphere of great depth with which
it is surrounded; and that the spots in question are simply portions of the
dark body of the sun, which become visible when the luminous atmosphere
is opened by upward currents from the surface or any other agency. The
disturbance of this atmosphere, by whatever it is caused, is most frequent
near the solar equator.—Peculiarly bright streaks of light, called *faculæ*, are
often found near the spots or where they have just disappeared. They are
supposed to be the ridges of vast waves in the luminous atmosphere just described.

939. *Constitution of the Sun.*—The sun's density is
about one-fourth that of the earth. Respecting its constitution little is known, nor are we any better informed as to
what produces its intense heat and light. It was formerly
supposed that the whole mass was in a state of combustion.
But how can such combustion be kept up without diminishing the material on which it feeds? The difficulty of
answering this question has led the later astronomers to
point to friction or electricity as the most probable source
of solar heat and light.

940. *Motions.*—The more permanent of the sun's spots,
if observed from time to time, are found to change their
position on its disk, or face. First becoming visible on the

is found by comparing a number of observations on the solar spots? What is the
length of the period? Of what size have spots occasionally appeared? What was
the diameter of one seen in June, 1843? What is the prevailing opinion of astronomers respecting these spots? What are *faculæ*? What are they supposed to be?
939. How does the sun's density compare with the earth's? What is known respecting its constitution and heat? To what have the later astronomers pointed as the
most probable source of solar heat and light? 940. How is it proved that the sun

east side, they gradually move towards the west, and in about thirteen days are lost from sight in that direction. After a similar period they reappear in the east. This phenomenon shows that the sun turns on its axis from west to east; its revolution is performed in about 25 days, 8 hours.

Besides turning on its axis, the sun, attended by its planets, moves at the rate of 8 miles a second in a circular path round a centre far off in the fields of space. So vast is this path that it will take the sun 18,200,000 years to get once completely round it.

941. *The Zodiacal Light.*—A faint light, shaped like a sugar-loaf, is sometimes seen stretching obliquely upward in the heavens, from 70 to 100 degrees, from that part of the horizon where the sun is about rising or has just set. This phenomenon is known as the Zo-di'-a-cal Light. It is brightest and most distinctly defined in tropical regions, where it is visible most of the time. In high latitudes it is seldom clearly seen, except during March and April just after sun-set, and in September and October immediately before dawn.

The cause of the zodiacal light is unknown. Some suppose it to be an expansion of the solar atmosphere; others, a thin vapor, charged with matter from the tails of comets, of which the sun's attraction has deprived them; others, again, have suggested that it is a remnant of the original matter of which both sun and planets were made. The latest theory is, that it is a nebulous ring, surrounding the Earth, like the ring of the planet Saturn.

### The Planets.

942. By the Planets of the solar system are meant those heavenly bodies that revolve directly about the sun in oblong curves, and shine by its reflected light.

The word *planetes* in Greek means "a wanderer", and the bodies in question are so called in contradistinction to the fixed stars, which keep the same

turns on its axis? What is the time of its revolution? What other motion has the sun? How large is the path it travels? 941. Describe the Zodiacal Light. Where is it brightest? When is it seen in high latitudes? What opinions have been advanced to account for the zodiacal light? 942. What are the Planets? What does the word *planetes* mean? From what are the planets to be distinguished? How

position in the heavens relatively to each other. The planets and the fixed stars are easily distinguished; the former shine with a steady light, the latter twinkle.

943. The moons are sometimes called Secondary Planets. In that case, the bodies that revolve directly about the sun are called Primary Planets.

944. The planets are also distinguished as Inferior and Superior. The Inferior Planets are those that are nearer to the sun than the earth is; the Superior Planets are those that are farther from the sun than the earth is.

945. ORBITS OF THE PLANETS.—The path of a planet round the sun is called its Orbit. The planets being at different distances from the sun, their orbits differ in length, though they are similar in shape.

946. The planetary orbits are not circles, but oblong curves called Ellipses. Hence a planet is nearer the sun in one part of its course than in another. That point of its orbit at which it is nearest the sun is called its perihelion (plural, *perihelia*); that in which it is farthest from the sun is its aphelion (plural, *aphelia*). When a planet's distance from the sun is spoken of, its *mean* distance is meant. This is obtained by adding its greatest and least distance together and dividing by 2.

Fig. 330.

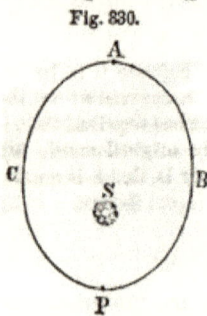

These definitions are illustrated in Fig. 330. A B P C represents an ellipse. S is the sun, situated not at the centre of the ellipse, but at one of two points within it called *foci*. P shows the position of a planet at its perihelion, and A at its aphelion.

The orbits of the planets lie in different planes, more or less inclined to each other.

947. Besides their revolution round the sun, the planets have another motion round their own axes. The time that

can the planets and the fixed stars be told apart? 943. What constitutes the difference between Primary and Secondary Planets? 944. Between Inferior and Superior Planets? 945. What is a planet's orbit? 946. What is the shape of the planetary orbits? What is a planet's Perihelion? Aphelion? When a planet's distance from the sun is spoken of, what is meant? How is the mean distance obtained? Illustrate these definitions with Fig. 330. What is said of the planes of the orbits? 947. What other motion have the planets besides their revolution round the sun?

it takes a planet to make one revolution on its axis is called its Day.

948. TABLE.—A Table of the planets follows, in the order of their distances from the sun, which are given in the second column. Their diameters in miles are given in the third column; the number of our days that it takes them to revolve round the sun, in the fourth; and the hours required for the revolution of each on its axis, in the fifth. The Tables in the Fifth Edition of Herschel's "Outlines of Astronomy" (1858) are here followed.

| Name. | Distance from Sun in miles. | Diameter in miles. | Year expressed in the Earth's days. | Day expressed in hours, &c. |
|---|---|---|---|---|
| Mercury . | 36,890,000 | 3,183 | 88 | 21$^h$ 5$^m$ |
| Venus . . | 68,770,000 | 8,108 | 225 | 23 21 |
| Earth . . | 95,298,260 | 7,926 | 365¼ | 24 |
| Mars . . | 145,205,000 | 4,546 | 687 | 24 37 |
| ASTEROIDS (62) | { from 210 to 301 millions | est. at from 100 to 1,000 | from 1,191 } to 2,051 } | unknown |
| Jupiter . | 495,815,500 | 90,734 | 4,333 | 9$^h$ 55$^m$ 27$^s$ |
| Saturn . . | 909,029,700 | 76,791 | 10,759 | 10 29 17 |
| Uranus . | 1,823,048,000 | 35,307 | 30,687 | 9 30 |
| Neptune . | 2,862,404,000 | 39,793 | 60,126 | unknown |

949. Mercury, Venus, Mars, Jupiter, and Saturn, being visible to the naked eye, were known to the ancients. Uranus was discovered in 1781 by Sir William Herschel, from whom it was first commonly called Herschel. Its discoverer gave it the name of Georgium Sidus, in honor of King George III. Both these names, however, were discarded for the mythological one by which it is at present known. The first of the asteroids, Ceres, was discovered in 1801 by the Sicilian astronomer Piazzi [pe-at'-ze]. Pallas was added to the list in 1804; Juno, in 1804; Vesta, in 1807; and the remainder, since 1844.

Neptune was discovered in 1846 by Dr. Galle [gal'-la], of Berlin. It was first called Le Verrier [luh va-re-a'], in honor of an eminent French astrono-

What is meant by a planet's day? 948. Referring to the Table, which of the planets do you find the smallest (the asteroids excepted), and which the largest? Which takes the shortest time to revolve around the sun, and which the longest? Which three have a day very nearly as long as the Earth's? Which three have days less than half as long as the Earth's? 949. Which of the planets were known to the ancients? Which was the next discovered? What other names has Uranus borne? When and by whom was the first asteroid discovered? When were the rest added to the list? When and by whom was Neptune discovered? What was it first called, and why?

mer, who by a series of calculations established the fact that there was a
more distant planet than Uranus, and instructed Dr. Galle in what part of the
heavens to look for it.

950. BODE'S LAW.—By comparing the distances of the
planets from the sun, Bode [bo'-da] arrived at the following
law :—Take the geometrical progression

          0    3    6    12    24    48    96    192    384,

each term of which (after the second) is obtained by doubling the preceding one.   To each term add 4, and we get

          4    7    10    16    28    52    100    196    388.

The distances of the nearer planets are approximately proportioned to these numbers.   That is, Mercury's distance being 36,890,000 miles, Venus's will be $\frac{7}{4}$ as much, the Earth's
$\frac{10}{4}$ ; &c.   Bode's law, however, does not apply to Saturn,
Uranus, and Neptune.   They are all, particularly the last,
much nearer the sun than this law would make them.

951. Fig. 331 shows the comparative size of the planets.
The asteroids are too small to appear on this scale.

Fig. 331.

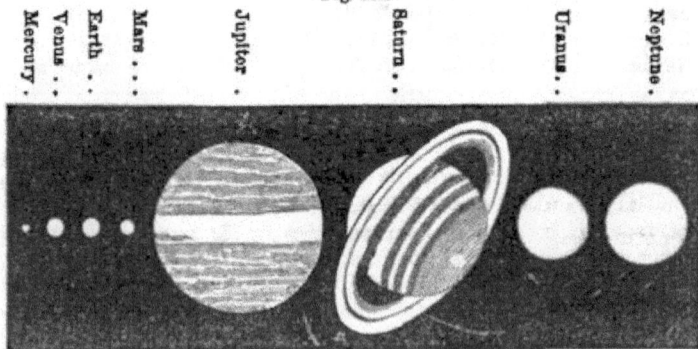

Herschel uses the following illustration to give an idea of the relative size
of the planets and their orbits :—"Choose any well levelled field or bowling-
green.   On it place a globe two feet in diameter; this will represent the Sun
Mercury will be represented by a grain of mustard-seed, on the circumference
of a circle 164 feet in diameter for its orbit; Venus a pea, on a circle of 284
feet in diameter; the Earth also a pea, on a circle of 430 feet; Mars a rather
large pin's head, on a circle of 654 feet; the Asteroids, grains of sand, in
orbits of from 1000 to 1200 feet; Jupiter, a moderate-sized orange, in a circle

nearly half a mile across; Saturn a small orange, on a circle of four-fifths of a mile; Uranus a full-sized cherry, or small plum, upon the circumference of a circle more than a mile and a half; and Neptune a good-sized plum, on a circle about two miles and a half in diameter."

952. KEPLER'S LAWS.—The laws that regulate the planetary motions were unknown till the commencement of the seventeenth century, when, after a long and careful comparison of numerous observations, they were discovered by John Kepler, a celebrated German astronomer, who thus won the title of "the Legislator of the Heavens". Kepler's laws apply to the moons in their revolutions about their primary planets, as well as to the latter.

953. *Kepler's First Law.—The orbits of the planets are ellipses having one focus in common, and in this common focus the sun is situated.*

The principal forces acting on the planets are the sun's attraction and the original force of projection. These forces alone would cause each planet to move about the sun in a perfect ellipse. The attraction of the other heavenly bodies, however, produces Perturbations, as they are called, and thus each orbit constantly deviates in a slight degree from an ellipse.

The ellipses described by the planets differ from circles in different degrees. The orbits of Mercury and several of the asteroids deviate most; those of Neptune and Venus are nearly circular.

954. *Kepler's Second Law.—The Radius Vector of a planet passes over equal areas in equal times.*

The Radius Vector is a line connecting the centre of a planet, as it traverses its orbit, with the centre of the sun.

Thus, in Fig. 332, the lines S A, S B, S C, &c., represent the radius vector of the planet there traversing its elliptical orbit. The whole space included within the orbit is divided into 12 equal triangles, 1, 2, 3, 4, &c.; and these,

Fig. 332.

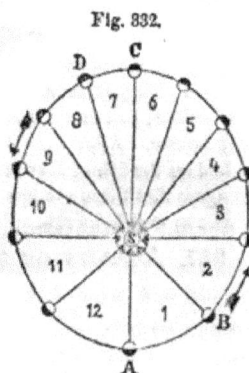

by Herschel to give an idea of the relative size of the planets. 952. When were the laws that regulate the planetary motions first known? By whom were they discovered? To what do Kepler's Laws apply? 953. Repeat Kepler's First Law. How is the elliptical form of the orbits accounted for? What is said of the ellipses described by the planets? Which of the orbits deviate most from a circle? Which deviate very little? 954. What is Kepler's Second Law? What is the Radius Vector? Illus-

according to the law just stated, must be traversed by the radius vector in equal times.

It follows from this law that the velocity of a planet differs at different points of its orbit, being greatest at its perihelion, and least at its aphelion. A B, C D, and the other arcs that form the bases of the twelve triangles, differ in length, but have to be traversed in the same time. The planet must therefore move fastest over the longest arcs, which are at its perihelion, and slowest over the shortest arcs, which are at its aphelion. In going from its aphelion to its perihelion, the arcs keep increasing, and the velocity of the planet is accelerated; from its perihelion to its aphelion, the arcs keep diminishing, and the velocity of the planet is retarded. Yet in going the whole distance from its aphelion to its perihelion a planet takes precisely the same time as in performing the opposite half of its course.

The cause of this difference of velocity is easily explained. In travelling towards its perihelion, a planet is constantly acted on by the sun's attraction in the same general direction as that in which it is moving, and this attraction becomes stronger and stronger as it approaches the sun. When returning to its aphelion, on the contrary, it is acted on by the sun's attraction in a direction opposite to that in which it is moving.

955. *Kepler's Third Law.—The squares of the planets' times of revolution round the sun are proportioned to the cubes of their distances from the latter.*

For example, Mercury's year consists of 88 days, Venus's of 225 days; Mercury is 36,890,000 miles from the sun, and Venus 68,770,000. Then the following proportion holds good, or nearly so :—

$$(88)^2 : (225)^2 :: (36,890,000)^3 : (68,770,000)^3$$

956. Kepler's laws have been verified by all the observations made since his time. They gave a wonderful impetus to the science, corrected many false notions, and enabled astronomers to arrive at new facts from facts already known. After many attempts and failures, the third law was finally reached on the 8th of May, 1618. "Perhaps", says Playfair, "philosophers will agree that there are few days in the scientific history of the world which deserve so well to be remembered."

957. ASPECTS OF THE PLANETS.—By the Aspects of the planets are meant their positions in their orbits relatively to each other. The aspects most frequently alluded to are Quadrature, Conjunction, and Opposition.

---

trate this law with Fig. 332. What follows from this law with respect to the velocity of a planet? In what part of its orbit does a planet move with accelerated velocity? In what, with retarded? Show the difference in the case of the Earth. What is the cause of this difference of velocity? 955. State Kepler's Third Law, and give an example. 956. What is said of Kepler's Laws and the estimation in which the third is held by philosophers? 957. What is meant by the Aspects of the planets? What

958. *Quadrature.*—Two heavenly bodies are said to be in Quadrature when they are 90 degrees apart; that is, when, if either were placed on the other's orbit at a point corresponding to its position on its own, the arc between them would subtend an angle of 90° at the focus. Thus, in Fig. 333, E represents the Earth, and Q Mars in quadrature. In almanacs and astronomical works, quadrature is denoted by the sign □.

Fig. 333.

959. *Conjunction.*—Heavenly bodies are said to be in Conjunction when they are seen in the same quarter of the heavens. Thus, in Fig. 333, Venus (V), the Sun (S), and Mars (N), are in conjunction, being in the same direction from the Earth (E). Conjunction is denoted by the sign ☌.

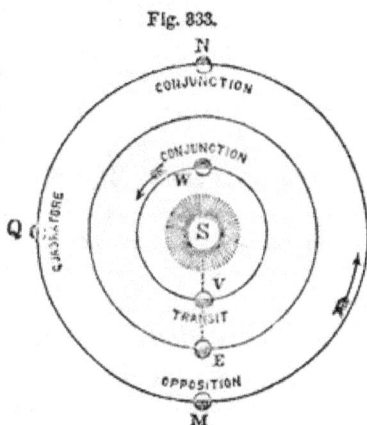

Conjunctions are of two kinds, Superior and Inferior. A planet is in Superior Conjunction when it is in conjunction on the opposite side of the sun from the Earth, as Venus at W, and Mars at N. A planet is in Inferior Conjunction when it is in conjunction on the same side of the Sun as the Earth is, as Venus at V. It is evident that the superior planets can never be in inferior conjunction.

960. *Opposition.*—Two heavenly bodies are said to be in Opposition when they are in directly opposite quarters of the heavens. Thus, in Fig. 333, Mars at M and the Sun (S) are in opposition, because relatively to the Earth (E) they lie in opposite directions. The inferior planets never appear in opposition. Opposition is denoted by the sign ☍.

**961.** *Transits.*—The passage of an inferior planet across the Sun's disk is called its Transit. In Fig. 333, Venus at V is making her transit.

A transit can take place only when a planet is in inferior conjunction. But, as the orbits of the planets are in different planes, there may be inferior conjunctions without any transit. Venus may be seen from the Earth in the same quarter as the Sun, and yet lie out of the plane which connects the centres of the Sun and the Earth.

**962.** *Occultation.*—When a planet or star is hid from the view of an observer on the Earth by the interposition of some other heavenly body, it is said *to suffer occultation.*

**963. REAL AND APPARENT MOTIONS.**—An observer at the Sun would see all the planets moving around him from west to east with perfect regularity and always in the same direction. He would see their Real Motions. An observer on the Earth sees only their Apparent Motions, and these are so irregular that one might almost fancy the bodies in question wandering through space without any fixed law to direct their course. They are seen at one time moving from west to east, at another stationary, and again pursuing a retrograde course from east to west.

The reasons of this are—1. We are 95,000,000 miles from their centre of motion. 2. We are ourselves moving, both round the sun and round the Earth's axis. Unconscious of these motions, we intuitively attribute the changes of direction produced by them to the motions of the orbs around us; just as a person on a boat, when it begins to move, seems to be at rest himself, and to see the wharf receding from him.

**964. ARE THE HEAVENLY BODIES INHABITED?**—This question is often asked, but can not be answered. No evidences of inhabitants have ever been discovered, even in the Moon, which is the nearest to us of all the heavenly bodies; nor can there be any till great improvements have been made in the telescope. Nothing, however, seems to be created without an object; and, humanly speaking, it would be strange if of all the orbs which Omnipotence has called into being our little world were the only one peopled by intelligent creatures.

If the planets are inhabited, it must be by creatures constituted very differently from the human race. Surrounded by entirely different circumstances as regards temperature, gravity, atmosphere, &c., the inhabitants of the different planets must be distinct races each from every other. Yet who can doubt that the same Infinite Wisdom that has adapted us to our sphere could as easily adapt them to theirs?

We proceed to consider the planets in turn. The character annexed to the name is the mark by which the planet is denoted.

965. MERCURY (☿).—The nearest planet to the Sun is Mercury. Under favorable circumstances, Mercury may be seen at certain times of the year for a few minutes after sun-set or before sun-rise. At other times it keeps so close to the Sun as to be invisible, being lost in the superior brightness of his rays in the daytime, and setting and rising so nearly at the same time with him as to afford no opportunity of observation.

To the naked eye Mercury looks like a star of the third magnitude, twinkling (unlike the other planets) with a pale rosy light. Viewed through the telescope, it exhibits similar phases or changes of appearance to those of the moon (from full to new); this is because we see more of its enlightened side at one time than another.

The solar heat received at Mercury is seven times as great as that of the Earth,—a temperature more than sufficient to make water boil. Mercury's light is also seven times as intense as ours, and the Sun seen from this planet would look seven times as large as it does to us. No permanent spots are visible either on Mercury or Venus, whence it has been supposed that we do not see the surfaces of these planets, but only their atmospheres loaded with clouds, which may serve to mitigate the otherwise intense glare of the sun. A German astronomer, however, at the commencement of the present century, observed what he regarded as a number of mountains on the surface of Mercury, one of which he computed to be over 10 miles in height.

respect to the heavenly bodies' being inhabited? 965. What is the nearest planet to the Sun? When is Mercury visible? What makes it invisible at other times? How does Mercury look to the naked eye? Viewed through the telescope, what phases does it present? How do the solar heat and light received at Mercury compare with ours? Are any permanent spots visible on Mercury or Venus? To what supposition has this fact led? What was observed on Mercury by a German astronomer?

Mercury's orbit deviates from a circle much more than that of any other
planet, the asteroids excepted. This circumstance, combined with the inclination of its axis to the plane of its orbit, must produce a great variety of
seasons, and extreme changes of temperature.

966. VENUS (♀).—The second planet from the Sun is
Venus. On account of its nearness, it appears larger and
more beautiful to us than any other member of our planetary system. So bright is Venus that it is sometimes visible at mid-day to the naked eye, and in the absence of the
Moon casts a perceptible shadow.

Being an inferior planet, Venus is never in opposition
to the sun, and is always below the horizon at midnight.
During part of the year, it rises before the Sun, and ushers
in, as it were, the day; when appearing at this time, the
ancients styled it Phosphor or Lucifer (*the light-bearer*), and
we call it the Morning Star. During the rest of the year,
it rises after the Sun; it was then styled Hesperus or Ves-
per by the ancients, and is distinguished by us as the Evening Star.

Venus is very nearly of the same size as the Earth. Its diameter has
generally been set down at 7,900 miles, somewhat less than the Earth's. In
Herschel's latest Tables, however, it is given as 8,108, which makes it a little larger.

Venus's heat and light are twice as great as ours. So intense is its
brightness that variations in its surface (if indeed its surface is not hid from
us by a cloudy atmosphere) for the most part escape detection, every portion
of the disk being flooded with light. Yet spots have occasionally been seen
on its surface, and mountains have been observed having an estimated height
of from 15 to 20 miles. Venus's phases, when viewed through the telescope,
are similar to those of Mercury and the Moon; but it never appears exactly full, being invisible at the time when this phase would otherwise be
presented.

967. THE EARTH (⊕).—The third planet from the Sun
is the Earth, which we inhabit.

The form of the Earth is that of an oblate spheroid,—

What is stated with respect to Mercury's seasons? 966. What is the second planet
from the Sun? How does it look to us, and why? What proofs have we of Venus's
brightness? When is Venus called the Morning, and when the Evening Star? How
does the size of Venus compare with that of the Earth? How do its heat and light
compare with ours? What have been observed on Venus's surface? What phases
does she present? 967. What is the third planet from the Sun? What is the form

that is, a sphere flattened at the poles like an orange. Its equatorial diameter is 7925.6 miles, and its polar diameter 26½ miles less. The circumference of a sphere is a little more than three times as great as its diameter; the distance round the earth, therefore, is about 25,000 miles.

The Earth is so large that its rotundity is not apparent to a person standing on its surface. We know it to be round, however, in several ways. 1. Navigators have sailed round it. By keeping the same general direction, east or west (as far as the land would allow), they have arrived at the place of starting. 2. The highest part of a vessel approaching in the distance is seen first, the lower part being obscured by the rotundity of the earth's surface. If the earth were a plain, we should see the hull as soon as the topmast.

968. *Motions.*—The Earth turns on its axis once in 24 hours. This is called its Diurnal Motion. Constantly bringing new points of the surface before the sun, and withdrawing others from its beams, this motion produces the succession of day and night.

The circumference of the earth being 25,000 miles, and a complete revolution being made in 24 hours, it follows that every point on the equator must revolve at the rate of a little over 1,000 miles an hour. As we go towards the poles, circles drawn round the earth parallel to the equator diminish in length, and points situated on them will consequently move with less velocity. At the poles there is no diurnal motion at all.

969. The Earth has also an Annual Motion,—about the Sun. Its orbit, like that of the other planets, is elliptical, but does not deviate much from a circle. Its perihelion is 3,000,000 miles nearer the Sun than its aphelion; consequently at the former point, other things being equal, it receives more heat than at any other part of its orbit.

The Earth reaches her perihelion on the 1st of January every year. Hence our winter is somewhat milder than that of the southern hemisphere; while the Sun at that period of a southern summer is perceptibly hotter than the summer sun at corresponding latitudes in the north. The heat in the inte-

of the Earth? What is its equatorial diameter? Its polar diameter? Its circumference? Why do we not see the roundness of the Earth? How do we know it to be round? 968. What is meant by the Earth's Diurnal Motion? What does it produce? What is the velocity of the diurnal motion at the equator? At the poles? At intermediate points? 969. What is meant by the Earth's Annual Motion? What is the shape of its orbit? When does the Earth receive the most solar heat, and why? How do the northern and southern winter and summer compare? Explain

rior of Australia at the time the Earth reaches her perihelion, is said to be more intense than any known even about the equator. Yet the difference of distance is so small compared with the whole, as not very materially to affect the Earth's temperature; nor has it anything to do with the change of seasons, as we shall presently see.

970. The Earth's orbit is nearly 600,000,000 miles in length; and to get round it in $365^{day}$ $5^{hrs}$ $48^m$ $48^s$ (which is the period of its revolution and constitutes our year), it must travel over 68,000 miles an hour.

Though we are constantly moving with this great velocity, we are unconscious of it. This is because we have never known what it is to be at absolute rest; and again, the motion is perfectly easy and regular, there being no obstructions in the way to make us sensible of it.

971. *The Earth in Space.*—Space extends infinitely on all sides of the Earth, studded with stars at different distances. To us, however, the stars appear equally distant, and seem to lie on the inner surface of a vast hollow sphere, at the centre of which we are placed. For purposes of definition and description, it is often convenient thus to allude to the firmament; and the expressions "celestial arch", "concave surface of the heavens", are used for the purpose,—not to denote any real objects, but the apparent arch or concave surface that we may conceive to be thrown around us.

972. *Horizon, Zenith, Nadir.*—The Sensible Horizon is the line that bounds the view,—that is, where earth and sky appear to meet. To an observer on the ocean, or on a vast plain where there is nothing to obstruct the view, this line is always a circle. The plane passing through the sensible horizon, and infinitely extended through space, is called the Plane of the Sensible Horizon.

The Rational Horizon is a plane passing through the Earth's centre, parallel to the plane of the sensible horizon.

At the Earth these planes are separated by the distance between the cen-

---

the cause. Is the Earth's temperature materially affected by this difference of distance? 970. With what velocity does the Earth travel round its orbit? Why are we not sensible of moving? 971. What is meant by the expressions, "celestial arch", "concave surface of the heavens"? 972. What is the Sensible Horizon? What is the Plane of the Sensible Horizon? What is the Rational Horizon? What

tre and the surface, or 4,000 miles; but so small is this distance compared with that at which the stars are situated that the two planes are regarded as striking the celestial arch at the same point. All heavenly bodies above the rational horizon at any given point are visible, and all below it invisible.

973. **The Poles of the Horizon** are two points in the heavens equally distant from the circle that bounds the view. One of these, the point directly overhead, is called the Ze'-nith; the opposite point, directly beneath us, is called the Na'-dir.

Every point on the Earth's surface has a horizon, zenith, and nadir of its own; and the horizon, zenith, and nadir of every point are constantly changing, owing to the revolution of the Earth on its axis. Hence, at night, new heavenly bodies are constantly coming into view in the east, while others are setting in the west.

974. *The Ecliptic.*—Seen from the Sun, the Earth would appear to describe a circle round that luminary, among the fixed stars on the concave surface of the heavens. This circle corresponds with the apparent path of the sun as seen from the Earth, and is called the Ecliptic.

The plane of the Earth's equator, extended till it meets the concave surface of the heavens, forms what is called the Celestial Equator, or the Equinoctial. The ecliptic and the equinoctial form an angle of 23° 28', and this angle is called the Obliquity of the Ecliptic. The axis of the Earth, therefore, instead of being perpendicular to the plane of its orbit, is inclined to it at an angle of (90° — 23° 28') 66° 32'.

975. The ecliptic cuts the equinoctial at two points, called Equinoxes, because when the sun appears at these points the days and nights are equal all over the world.

The equinoxes are distinguished as Vernal and Autumnal. The Vernal Equinox is that point at which the sun crosses the equinoctial from south to north, which takes place in our spring. The Autumnal Equinox is the point

is the distance between the two horizons at the Earth ? When they strike the celestial arch ? Which of the heavenly bodies are visible at any given point, and which invisible ? 973. What are the Poles of the Horizon ? What is the Zenith ? The Nadir ? What causes new heavenly bodies to keep coming into view at night and others to set ? 974. What is the Ecliptic ? What is the Celestial Equator, or Equinoctial ? What is the Obliquity of the Ecliptic ? 975. What are the Equinoxes ? Why are they so called ? How are they distinguished ? What is the Vernal Equi-

17

at which the sun crosses the equinoctial from north to south,—and this he does in our autumn.

976. *The Zodiac.*—The Zodiac is a belt on the concave surface of the heavens, sixteen degrees in width, eight of which lie on each side of the ecliptic. It is divided into twelve Signs, of 30 degrees each. The zodiac is peculiarly interesting to us, because it is the region within which the apparent motions of the Sun, the Moon, and all the greater planets, are performed.

The zodiac is so called from a Greek word signifying *animal*, because its signs were for the most part named after animals, of which the stars in each seemed to the ancients to be so grouped as to form rude outlines. Such groups of stars, which seem to be situated near each other because lying in the same direction from us, are called Constellations. Owing to what is known as the Precession of the Equinoxes,—that is, the sun's completing its revolution on the ecliptic every year before it reaches the same point of the heavens relatively to the fixed stars,—the signs of the zodiac do not now correspond in position with the constellations from which they were named. With the equinoxes, on which their position depends, they have retrograded 30 degrees towards the west. The signs of the zodiac and the constellations of the zodiac must therefore be distinguished from each other.

977. The names of the signs of the zodiac are given below in Latin and English, with the characters by which they are respectively denoted. They are given in their order, commencing at the vernal equinox.

| | | | | | |
|---|---|---|---|---|---|
| ♈ | *Aries,* | the ram. | ♎ | *Libra,* | the balance. |
| ♉ | *Taurus,* | the bull. | ♏ | *Scorpio,* | the scorpion. |
| ♊ | *Gemini,* | the twins. | ♐ | *Sagittarius,* | the archer. |
| ♋ | *Cancer,* | the crab. | ♑ | *Capricornus,* | the goat. |
| ♌ | *Leo,* | the lion. | ♒ | *Aquarius,* | the water-bearer. |
| ♍ | *Virgo,* | the virgin. | ♓ | *Pisces,* | the fishes. |

978. *The Change of Seasons.*—It has been stated that the Earth is nearer the Sun at one period of its revolution than at another. The change of seasons, however, is entirely independent of this fact, and is produced by the sun's rays falling on a given point of the Earth's surface with different degrees of obliquity at different parts of its orbit.

nox? What is the Autumnal Equinox? 976. What is the Zodiac? How is it divided? What makes it peculiarly interesting to us? From what is the zodiac so called? What are Constellations? How are the signs of the zodiac now situated relatively to the constellations from which they were named? To what is this owing? 977. Name the signs of the zodiac. 978. By what is the change of seasons pro-

When the Sun is vertical, or directly overhead, its heat is most intense; and the less its rays deviate from a vertical line in striking the surface, the more heat they impart to it.

The angle at which the Sun's rays strike a given part of the Earth's surface keeps constantly varying, in consequence of the Earth's revolving with its axis always pointing in the same direction, or, as it is generally expressed, everywhere parallel to itself. This will be understood from Fig. 334.

In Fig. 334 the Earth is represented as moving round the Sun, which is in one of the foci of her elliptical orbit. The dotted line is the zodiac, divided into its twelve signs. N S is the Earth's axis, which maintains the same direction in the four positions shown, and at every other part of the orbit.

Fig. 334.

At the vernal equinox (March 21), the equator is directly opposite the Sun; the solar rays fall at the same angle on the northern hemisphere as on the southern, and it is spring in the former, autumn in the latter. The Earth's axis is inclined neither to nor from the sun; consequently, half the surface, from pole to pole, is enlightened at a time, and day and night are of equal length all over the globe.

As the Earth moves eastward, the rays of the Sun no longer fall vertically on the equator, but on places north of it. This continues till June 21st, when the sun is vertical to places 23° 28' north (this being the obliquity of the ecliptic), and his rays extend over the same distance beyond the north pole. It is now summer in the north and winter in the south, for in proportion as the solar rays fall less obliquely on the former, they must fall more obliquely on the latter. It will be observed, also, that a space extending 23° 28' around the south pole is totally dark.

---

duced? When is the Sun's heat most intense? Why does the angle at which the Sun's rays strike a given part of the Earth's surface keep varying? What does Fig. 334 represent? Describe the position of the Earth and the circumstances attending

The Sun is never directly overhead to any place farther north of the equator than 23° 28'. As th· Earth continues her course eastward, it becomes vertical to places more and more to the south, and by the 22d of September, or thereabouts, it is vertical to the equator just as it was six months before. This is the period of the autumnal equinox. The Earth again presents a full side from pole to pole to the Sun, and the days and nights are once more equal. We have now the southern spring and the northern autumn.

From this point, the solar rays become more and more oblique in the north and fall vertically on places farther and farther south, till the same limit of 23° 28' is attained, which takes place about December 21, and marks the northern winter and the southern summer. Beyond this limit the Sun is never directly overhead. As the Earth keeps on to the east, his vertical rays fall on latitudes nearer and nearer to the equator, till finally on the 21st of March places on the equator have the Sun in their zenith as they had six and twelve months before.

979. The explanation just given shows that there are two points of the ecliptic in which the Sun is about 23½ degrees from the equator, and from which he seems to turn back towards that line. These points are called Solstices (*standing-points of the Sun*), because the Sun appears to stand still for several days at the same place in the heavens before taking an opposite direction. The solstice reached in June is called the Summer Solstice ; that in December, the Winter Solstice.

980. Circles on the Earth's surface about 23½ degrees north and south of the equator form the limits beyond which the Sun's rays are never vertical. These circles are called Tropics (from a Greek word meaning *to turn*), because on reaching them the vertical rays turn back towards the equator. The northern tropic is called the Tropic of Cancer, because when the Sun reaches this line he is seen from the Earth in the sign Cancer, as will be apparent from Fig. 334. For a similar reason the southern tropic is called the Tropic of Capricorn.

981. It appears from Fig. 334 that from March 21 to September 22 the north pole is constantly illuminated and the south pole in darkness, notwithstanding the revolution of the Earth on its axis; while from September 22

it, at March 21. At June 21. At September 22. At December 21. 979. What are the Solstices? Why are they so called? How are they distinguished? 980. What are the Tropics? Whence is their name derived? What is the northern tropic

to March 21, darkness reigns at the north pole and the south pole enjoys continual light. At the summer solstice there is a space of 23½ degrees about the north pole on which the Sun does not set, and at the winter solstice a corresponding space about the south pole. The lines that bound these regions are called the Polar Circles. The one near the north pole is called the Arctic Circle; that near the south pole, the Antarctic Circle.

982. If, instead of being inclined, the Earth's axis were perpendicular to the plane of its orbit, the regions on the equator would have the Sun constantly in their zenith, day and night would always be equal over the whole globe, there would be no variety of seasons, and a given place would have about the same temperature from one year's end to another. Something of this kind must be the case on the planet Jupiter, whose axis is nearly perpendicular to the plane of its orbit. On the other hand, the more the axis of a planet is inclined, the greater are the extremes of temperature incident to its several seasons.

983. THE MOON (☾).—The Earth is attended by one satellite called the Moon,—a beautiful orb which ' rules the night ' with its gentle brilliancy, produces in part the tides, and sensibly affects the Earth's motions by its attraction.

984. *Size.*—The Moon's diameter is 2,165 miles, but its apparent size is almost equal to the Sun's in consequence of its nearness to our planet. Its density is not much more than one-half that of the Earth, and it contains about one-eightieth as much matter.

985. *Motions.*—The Moon is 240,000 miles from the Earth, and revolves about the latter so as to reach the same point relatively to the fixed stars in 27 days, 8 hours. To reach the same point relatively to the Sun requires 29 days, 13 hours, since the Earth has itself meanwhile advanced in its orbit.—When nearest the Earth, the Moon is said to be in her Per'-i-gee, and when farthest from it in her Ap'-o-gee.

The terms *perigee* and *apogee* (which mean *near the Earth* and *away from the Earth*) are also applied to the apparent position of the Sun. When the Earth is at its perihelion, the Sun is said to be in perigee; and when the Earth is at its aphelion, the Sun is in apogee.

called, and why? The southern? 981. What are the Polar Circles? What is the one near the north pole called? That near the south pole? 982. If the Earth's axis were perpendicular to the plane of its orbit, what would follow? What is said of Jupiter? 983. By what is the Earth attended? 984. How great is the Moon's diameter? Its density? Its mass? 985. How far is the Moon from the Earth? What is the period of her revolution? When is the Moon said to be in perigee? In ap-

The Moon also turns on its axis in exactly the same time that it takes to revolve round the Earth, and in the same direction. The consequence is that she always presents the same side to the Earth. Nearly one-half of our fair attendant we never see, and to the inhabitants of half her surface, if she has any, we are invisible.

986. *Phases.*—The Moon is non-luminous, and shines only by the reflected light of the Sun; hence the hemisphere presented to the Sun is bright, while the opposite one is dark. As the Sun, Moon, and Earth are constantly taking different positions relatively to each other, the portion of illuminated lunar surface presented to us is as constantly changing. Hence arise what are called the Phases of the Moon.

When *new*, the Moon lies between the Earth and the Sun, near a line connecting their centres. Her dark side is then towards us, and she is invisible. Soon, however, she gets so far east of the Sun as to appear in the west shortly after his setting. A bright crescent then becomes visible on the side nearest the Sun, the rest of her circular disk being just discernible, not by sun-light directly received, but by sun-light reflected from the Earth to the Moon, and by her reflected back to us. The crescent gradually grows larger, until, when the Moon is 90 degrees from the Sun, or in quadrature, half her disk is illumined. She is then said to be in her First Quarter.

Each succeeding night now finds the enlightened portion larger and larger, and the Moon is said to be *gibbous*. At last she reaches a point at which she is again almost in a line with the Sun and the Earth, but this time the Earth is in the middle. The Moon rises in the east as the Sun sets in the west; the whole of her enlightened hemisphere is therefore turned towards us, and she is said to be *full*.

After this the Moon again becomes gibbous, and we see less and less of her enlightened surface, till at length half of her disk is dark, when she is said to be in her Third Quarter. Advancing beyond her third quarter, she wanes still further to a crescent, and at length on arriving in conjunction with the Sun disappears entirely,—to go through the same phases again as she makes another revolution in her orbit.

987. To the inhabitants of the Moon, if any there be, the Earth presents the same phases that the Moon does to us, but in reversed order. When the Moon is new to us, the Earth is full to them,—a splendid orb, thirteen times

as large as the full Moon. When she is in her first quarter, the Earth is in her third quarter, &c.

988. The Moon has either no atmosphere at all, or one exceedingly rare, and not extending more than a mile from its surface. Hence it must be destitute of water, for any liquid on its surface would long since have been dissipated by the heat of the lunar days, there being no atmospheric pressure to check evaporation. If there were any water on the surface of the Moon, clouds would certainly be observed at times dimming its face.

989. Viewed through a telescope, the surface of the Moon appears exceedingly rough, covered with isolated rocks, deep valleys, yawning chasms, craters of extinct volcanoes, in some cases more than 100 miles in width, and lofty mountains, several of which are from three to four miles high and cast their shadows a great distance over the rugged plains. Every thing is desolate in the extreme. Several of the earlier astronomers thought that they discerned volcanoes in a state of eruption; but later observers are of the contrary opinion, attributing the peculiar brightness of the supposed volcanic summits to phosphorescence, or superior reflective properties.

Names have been given to the various mountains and spots visible on the Moon, and a map has been prepared of the whole side presented to us, which has been pronounced "vastly more accurate than any map of the Earth we can yet produce."—The great telescope of the Earl of Rosse shows with distinctness every object on the lunar surface that is 100 feet in height. It has brought to light, however, no signs of life or habitation.

990. MARS ( ♂ ).—Mars, the fourth planet from the Sun, is 4,546 miles in diameter. Its day is of nearly the same length as ours, its year about twice as long. The inclination of its axis to the plane of its orbit does not differ much from the Earth's, and its seasons are therefore similar to ours. It is surrounded by an atmosphere of moderate density.

Mars is easily distinguished in the heavens by his red fiery light, which is supposed to owe its color to the soil from which it is reflected. The telescope distinctly shows continents of a dull red tinge, like that of sand-stone,

Moon? 988. What is said of the Moon's atmosphere? Why is the Moon supposed to be destitute of water? 989. How does the Moon look, when viewed through a telescope? What is now thought respecting the supposed volcanic eruptions formerly observed? How high objects does the Earl of Rosse's telescope distinctly show? 990. Which is the fourth planet from the Sun? What is the length of its diameter? Its day? Its year? How do its seasons compare with ours? How may Mars be distinguished? What does the telescope show? What are seen about the

washed by seas of a greenish hue. Bright white spots are seen about the poles, which are no doubt occasioned by the reflection of the sun's light from the snow and ice collected there. It is observed that as each pole is turned towards the sun the spots about it diminish in size, owing to the melting of the snow by the solar heat.

991. THE ASTEROIDS.—The Asteroids are so small that, with the exception of one or two which have been seen without a telescope, they are invisible to the naked eye. Their diameters have not yet been accurately determined; some exceed 100 miles, and others probably fall somewhat under that mark. A number of them are provided with extensive atmospheres. The Asteroids are supposed by some to be the wreck of one large planet, which they believe to have originally revolved between Mars and Jupiter, and by some tremendous catastrophe to have burst into fragments. Many similar bodies probably remain to be discovered in this region.

The Asteroids are comparatively so diminutive that the force of gravity on their surfaces must be very small. A man placed on one of them would spring with ease 60 feet high, and sustain no greater shock in his descent than he does on the earth from leaping a yard. On such planets giants may exist; and those enormous animals which here require the buoyant power of water to counteract their weight, may there inhabit the land.

992. JUPITER ($\mathrm{2\!\!\!\!\perp}$).—Next to the asteroids is Jupiter, the largest of the planets, which exceeds the Earth in bulk nearly 1,300 times. Its revolution round the Sun is performed in about 12 years, and that around its axis in less than 10 hours. Jupiter is attended by four satellites, which revolve about it from west to east.

All of these satellites but one exceed our Moon in size. The largest would sometimes be visible to the naked eye as a very faint star, were it not lost in the superior brightness of its planet. Three of them are totally eclipsed during every revolution by the long shadow which the planet casts, and the fourth is very often eclipsed. The relation between their orbits and motions is such that for many years to come Jupiter will never be deprived of the light of all four at the same time.

poles? By what are they supposed to be caused? 991. Are the Asteroids visible to the naked eye? What is the length of their diameters? What are the Asteroids thought by many to be? What is stated with respect to the force of gravity on their surface? 992. How does Jupiter rank in size? How does it compare in bulk with the Earth? What is the length of its year? Its day? By what is it attended?

SATURN. 393

So large is Jupiter, and so short a time is it in revolving on its axis, that every point on its equator must turn at the rate of 450 miles a minute. The result is an immense excess of centrifugal force at the equator; and this is seen to have operated before the mass of the planet became hard, by flattening it very much at the poles.—Jupiter's disk is always crossed with a number of dark parallel belts, as shown in Fig. 331. They vary in breadth and situation, but are always parallel to the equator of the planet; hence they appear to be connected with its rotation on its axis, and are no doubt produced by disturbances in its atmosphere.

992. SATURN ( ♄ ).—Saturn, which is next to Jupiter in distance from the Sun, is also next to it in size, having a diameter of 76,791 miles, and consequently a bulk nearly 1,000 times that of the Earth. Its day is not half so long as ours; but it is 29½ of our years in making one complete revolution in its orbit.

Saturn has eight moons, seven of which were known for sixty years before the eighth was discovered. The largest of them has a diameter about half as large again as our Moon. Saturn's disk, like Jupiter's, is frequently diversified with belts; spots are of rare occurrence. An atmosphere of considerable density is supposed to surround the planet.

Saturn has a remarkable appendage, consisting of three bright, flat, and exceedingly thin rings, encircling its equator, and revolving with it around its axis in about the same time in which the planet itself revolves. The whole breadth of these rings is 27,000 miles, while their thickness does not exceed 100 miles. They are supposed to consist of a mixture of gases and vapors, sufficiently substantial to cast a shadow. The three rings are detached from each other, and lie in the same plane very close together, while the inner one is 19,000 miles from the surface of the planet. They are prevented from falling in upon the planet by the centrifugal force generated by their rapid revolution.

993. URANUS ( ♅ ).—Uranus, the next planet to Saturn, revolves about the Sun in 84 of our years. There being no spots on its surface, we are unable to fix the period of its revolution on its axis. It is attended by six moons, which move from east to west (unlike the satellites of the other

What is the size of the largest of these moons? What relation subsists between their orbits and motions? What is the shape of Jupiter? What has caused the flattening at the poles? With what is Jupiter's disk crossed? To what are these belts to be attributed? 992. What is the next planet to Jupiter? What is Saturn's diameter? How does its bulk compare with the Earth's? Its day? Its year? How many moons has Saturn? How is its disk diversified? What remarkable appendage has Saturn? Describe its rings. 993. What is the next planet to Saturn? What is the length of the year of Uranus? Its day? By what is it attended? How do its light

17*

planets) in orbits nearly perpendicular to that of the planet. The solar heat and light of Uranus are only $\frac{1}{360}$ of ours.

994. NEPTUNE (♆).—Neptune, the most remote planet of the solar system, is invisible to the naked eye. Seen through the telescope, it looks like a star of the eighth magnitude. The diameter of Neptune is 39,800 miles, which is 4,500 more than that of Uranus. Its revolution around the Sun is performed in about 165 of our years. Neptune has at least one moon, distant from it about as far as ours is from us.

The discovery of Neptune is one of the greatest triumphs of which science can boast. Comparing observations on Uranus, while it was still thought to be the most distant member of the solar system, astronomers found certain *perturbations* or irregularities, in its motions, which could be accounted for only on the supposition that there was some unknown planet beyond it by whose attraction it was affected. Le Verrier thoroughly investigated the subject, and even went so far as to compute the size and distance of the suspected planet, and to predict in what part of the heavens it would be found at a given date. A letter from the French astronomer, embracing the results of his calculations, reached Berlin, September 13, 1846; and that very evening, sweeping the heavens with his powerful telescope, according to Le Verrier's instructions, Dr. Galle discovered what was apparently a star of the eighth magnitude not laid down on his chart, but was proved by its change of place on the following evening to be a planet.—It is just to add that Adams, an English astronomer, had, about the same time with Le Verrier, made similar calculations, and with nearly the same result.

995. REAL AND APPARENT POSITION OF THE HEAVENLY BODIES.—We seldom see the heavenly bodies in their real position. This is owing to two causes,—Refraction and Parallax.

996. *Effect of Refraction.*—Refraction, which has been explained in the chapter on Optics, bends rays of light entering our atmosphere from a rarer medium, and causes the body from which they proceed to appear higher than it really is. The Sun is thus made visible a few moments before he actually rises and after he sets. The effect of re-

and heat compare with ours? 994. What is the most remote planet of the solar system? How does Neptune look, when seen through the telescope? What is its diameter? What is the period of its revolution? How many moons has Neptune? Give an account of the circumstances under which Neptune was discovered. 995. Why do we not see the heavenly bodies in their real position? 996. What is the effect of re-

fraction is greatest when a body is on the horizon, and diminishes as it ascends towards the zenith, at which point it entirely disappears.

997. *Effect of Parallax.*—A planet seen from different points of the Earth's surface appears to lie in different positions. This is evident from Fig. 335.

The planet C to an observer at A seems to lie at F; to one at B it appears to lie at D. To avoid the inconsistencies which would otherwise exist in observations made at different places, the centre of the earth is taken as a standard point; and the true position of a heavenly body is that point of the celestial arch which would be cut by a line connecting the centre of the Earth with the centre of the body in question, infinitely produced.

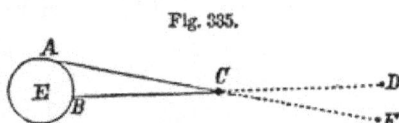

Fig. 335.

Parallax is the angle made by a line from a heavenly body to the Earth's centre and another line from the same body to the eye of an observer.

It is evident that, the nearer a heavenly body is, the greater is its parallax The fixed stars are so remote that they have no appreciable parallax. The Earth, if visible to them, would be nothing more than a minute point of light. —The parallax of a heavenly body is greatest when it is on the horizon. At the zenith it would be nothing, because from that point the lines to the observer's eye and the centre of the Earth would coincide.

998. ECLIPSES.—By an Eclipse of the Sun or Moon is meant its temporary obscuration by the interposition of some other body. An eclipse is called Total, when the whole disk is obscured; and Partial, when only a portion is darkened.

999. An eclipse of the Sun is caused by the Moon's getting between it and the Earth, and intercepting its rays. This can happen only at new Moon, because, when between us and the Sun, the Moon must present to us her unenlightened side.

fraction? 997. How does a planet seem to lie, when observed from different parts of the Earth's surface? Illustrate this with Fig. 335. What is the true position of a heavenly body? What is Parallax? What is said of the parallax of the fixed stars? What would be the effect of refraction and parallax on the apparent position of a body in our zenith? 998. What is an Eclipse? When is an eclipse called Total, and when Partial? 999. What causes an eclipse of the Sun? When alone can this hap-

If the Moon's orbit lay in the same plane as the Earth's, she would eclipse the Sun every time she became new; but, as her orbit is inclined to the ecliptic at an angle of more than 5 degrees, her shadow may fall above or below the Earth at the time of her change.

When the Moon intervenes between the Sun and the Earth at such a distance from the latter as to make her apparent diameter less than the Sun's, a singular phenomenon is exhibited. The whole disk of the Sun is obscured, except a narrow ring around the outside encircling the darkened centre. This is called an Annular Eclipse, from the Latin *annulus*, a ring.

1000. An eclipse of the Moon is caused by the Earth's getting between it and the Sun. This can take place only at full Moon, because when the Earth is between the Sun and the Moon the latter must present her enlightened side to the Earth.

Non-luminous itself, when cut off from the solar rays, the Moon must become invisible. There is this difference between an eclipse of the Sun and the Moon. In the former, the Sun shines the same as ever, and its brightness is undiminished to those who are out of the Moon's shadow. When the Moon is eclipsed, on the other hand, she diffuses no light, and is dark to all within whose range of vision she is situated.—Solar eclipses occur more frequently than lunar. The greatest number of both that can take place in a year, is seven; the smallest number, two; the usual number, four.

1001. When the Sun is totally eclipsed, the heavens are shrouded in darkness, the stars make their appearance, the birds go to roost, the animals by their uneasiness testify their alarm, and all nature seems to feel the unnatural deprivation of the light of day. It is not surprising that, when the cause of the phenomenon was unknown, it filled the minds of men with consternation. Even at the present day barbarous nations regard eclipses as indications of the displeasure of their gods. Columbus, on one occasion, when wrecked on the coast of Jamaica, and in imminent danger both of starvation and an attack from the Indians, saved himself and his men by taking advantage of this superstitious feeling. From his acquaintance with astronomy, he knew that an eclipse of the Moon was about to take place; and on the morning of the day, summoning the natives around him, he informed them that the Great Spirit was displeased because they had not treated the Spaniards better, and would shroud his face from them that night. When the Moon became dark, the Indians, convinced of the truth of his words, hastened to him with plentiful supplies, praying that he would beseech the Great Spirit to receive them again into favor.

pen? Why is not the Sun eclipsed every time the Moon becomes new? What is an Annular Eclipse? 1000. By what is an eclipse of the Moon caused? When can this take place? What difference is mentioned between an eclipse of the Sun and the Moon? Which occurs more frequently? What is the usual number in a year? 1001. Describe the appearance of things during a total eclipse of the Sun. How do barbarous nations regard eclipses? How did Columbus once save himself and his

1002. COMETS.—*Comet* is derived from a Greek word meaning *hair;* and the term is applied to a singular class of bodies belonging to the solar system, from which long trains of light, called *tails*, spread out like hair streaming on the wind. They differ very much in appearance; but, for the most part, they consist of a *nucleus*, which is a very bright spot, apparently denser than the other portions; an *envelope*, which is a luminous fog-like cover surrounding the nucleus; and a *tail*, which appears to be an expansion of the envelope produced by solar heat.

The tails of different comets differ greatly in shape and extent. In some this appendage is entirely wanting; in others it has been found to extend 120,000,000 miles. Several tails have been exhibited at the same time; the comet of 1744 threw out no less than six, like an enormous fan, over the heavens. Even in the same comet the tail keeps changing, being largest when near the Sun and diminishing as it recedes from that body.—The tail lies on the opposite side of the nucleus from the Sun,—behind it, when approaching its perihelion, and preceding it when retiring from that point.

1003. *Constitution.*—The matter of which comets are composed must be an exceedingly thin gas or vapor.

The nucleus is always bright, no matter what position in relation to the Earth it may occupy; no phases are presented, as in the case of the planets; this proves the nucleus to be so rare that the solar light (which alone renders it visible) can penetrate it and be seen on the side opposite to that which it strikes. Again, comets have on different occasions passed very near the planets, yet have never been found to cause any irregularities in their motions, while their own motions have been materially affected. The tail, in particular, must be exceedingly rare, perhaps not weighing more than a few ounces, even when most extensive.

1004. *Orbits, Velocity.*—The orbits of the comets are either ellipses, parabolas, or hyperbolas.

If ellipses, they generally deviate very much from a circle, being lengthened out an immense distance in proportion to their breadth. Comets that move in elliptical orbits return after a series of years; those that move in parabolas or hyperbolas never reappear, but after wheeling about the Sun dash off into the remote regions of space, perhaps to visit other systems.

Some comets at their perihelion pass very close to the Sun. The one that appeared in 1843 almost grazed its surface, approaching so near it that the solar disk must have appeared 47,000 times larger than it looks to us, and the heat received must have been twenty-five times greater than that required to melt rock-crystal.

1005. When near the Sun, comets move with incredible velocity,—sometimes at the rate of over a million miles an hour.

1006. *Number.*—The exact number of comets can not be determined. Over seven hundred have been seen and enumerated. Multitudes have visited our system without being seen from the Earth, in consequence of reaching their perihelion in the day-time, or when the heavens were obscured by mists and clouds. Arago estimated the number that have appeared or will appear within the orbit of Uranus at 7,000,000 ; the same calculation extended to Neptune's orbit would make the number 28,000,000.

1007. Comets were formerly regarded with superstitious terror as precursors of war, famine, and other misfortunes. In more modern times the fear of a collision made them formidable objects. This fear, however, has been dispelled by the discovery of their great rarity. A collision, however fatal it might be to the comet, would probably do little injury to a solid body like the Earth.

## The Fixed Stars.

1008. The Fixed Stars are so called in contradistinction to the planets, because they maintain the same position relatively to each other, not because they are absolutely at rest. They all move about some fixed point in immense orbits, which it will take millions of years for them to complete. Shining by their own light and not by reflection, they are suns, and are probably each the centre of a system of its own.

1009. *Magnitudes.*—Varying in size and situated at different distances from us, the stars are not all of the same brilliancy. They are divided into about twenty classes according to their brightness, and distinguished as stars

How near did the comet of 1843 pass to the Sun ? 1005. What is the velocity of comets, when near the Sun ? 1006. What is the number of the comets ? What prevents us from seeing many that visit our system ? What was Arago's estimate ? 1007. How were comets formerly regarded? How are they now looked upon ? 1008. Why are the Fixed Stars so called ? 1009. How are the fixed stars classified ? What are Tel-

of the First, Second, &c., Magnitude: The stars of the first six magnitudes are visible to the naked eye; the rest are called Telescopic Stars, because seen only with the telescope. There are about 24 stars of the first magnitude, 50 of the second, and 200 of the third; but the number in the lower classes increases so rapidly as to be almost beyond enumeration.

1010. *Constellations.*—For convenience of reference, the stars are divided into constellations, or groups, named after animals and other objects to which their outline bears some fancied resemblance. The twelve constellations of the zodiac have been already named; there were thirty-six more laid off by the ancients in other parts of the heavens. The whole number has been increased in modern times to ninety-three. The stars in each constellation are distinguished, according to their magnitude, first by the letters of the Greek alphabet, then by those of the Roman, and when both are exhausted, by figures.

1011. *Distance.*—The distance of the fixed stars is absolutely incredible. None of them can be less than 19,200,000,000,000 miles from the Earth, while the greater part are far more remote.

The recent improvements in telescopes have enabled astronomers to compute the distance of nine of the nearest stars. Sirius, the brightest of them, is found to be so far off that light, with a velocity of 192,000 miles a second, is fourteen years in reaching us; from the North Star it is over 48 years. The mind is lost in trying to comprehend such mighty distances; and yet it will be remembered these are among the nearest stars.

1012. Several remarkable facts are worthy of note in connection with the fixed stars. Some of them wane for a time, so as to be classed in a lower magnitude, and then resume their former brilliancy. Others, after vanishing entirely for a season, suddenly reappear; these are called Periodical Stars.

Many stars (at least several thousand), when viewed through a powerful telescope, are resolved into two stars, one of which is generally much fainter than the other. These are known as Binary or Double Stars. In some cases the faint one may only appear to be near the bright one from lying in the same direction, and really be millions of miles behind it; but there is generally reason for supposing that the fainter luminary revolves about the brighter one in obedience to that same great law of gravitation which prevails in our own system.—Some stars, apparently single, are resolved into three, four, and even six, by the telescope.

Many of the binary stars are tinged with complementary colors. The larger one is orange-colored, the smaller blue; or the one is red, and the

escopic Stars? How many stars are there of the first magnitude? Of the second? Of the third? 1010. How are the fixed stars divided? How many constellations were laid off by the ancients? How many have been added in modern times? How are the stars in each constellation distinguished? 1011. What is the distance of the fixed stars? What is the distance of Sirius? Of the North Star? 1012. What are Periodical Stars? What are Binary Stars? What relation seems to subsist between the brighter and fainter star? Into what are some stars resolved? With what are

other green. Some of the single stars look blood-red; but there are none that exhibit deep tinges of blue or green.

The size of several of the fixed stars has been calculated approximately. Their diameters are found to be enormous,—in one case not less than 200,000,000 miles. Sirius, "the dog-star", if set in the place of our Sun, would look 125 times as large as he, and give us 125 times as much light. Trillions of miles away, as it is, it dazzles the eye when seen through a powerful telescope.

1013. THE GALAXY.—The Galaxy, or Milky Way, is a broad zone of light which stretches across the sky from horizon to horizon, encircling the whole sphere and maintaining the same position relatively to the stars. Examined through a powerful telescope, it is found to consist entirely of stars, scattered by millions, like glittering dust, on the black ground of the heavens.

1014. NEBULÆ.—Nebulæ are clusters of stars so distant that they look like faint patches of cloud hardly discernible in the sky. They vary in shape, and are seen in different quarters of the heavens.

Lord Rosse's great telescope resolves some of the nebulæ into individual stars; it makes others appear bright, but not sufficiently so to be separated into the stars that compose them; and it calls up from the depths of space others which appear as faint even to its mighty magnifying power as those which it resolves appear to the unaided eye. The milky way is itself one of these nebulæ, more distinct than the others because nearer to us.

From the facts set forth we may conclude that the universe consists of a vast number of distinct clusters of worlds, separated from each other by immense intervals; that the fixed stars, the milky way, our Sun and its system, form one of these clusters; that the various nebulæ constitute other clusters, fainter or brighter according to their distance from us,—each composed of many different systems,—and having its members separated as widely as our Sun is from the brother suns about him.

How can the mind take in such mighty thoughts! How can the heart refuse its homage to the great Creator of all these worlds!

---

many of the binary stars tinged? What has been found with respect to the diameters of some of the fixed stars? How would Sirius look, if set in the place of our Sun? 1013. What is the Galaxy? How does it look through a powerful telescope? 1014. What are Nebulæ? How do nebulæ look through Lord Rosse's telescope? What may we conclude from the facts set forth?

# CHAPTER XIX.

## METEOROLOGY.

1015. METEOROLOGY is the science which treats of the phenomena of the atmosphere. Among these are winds, clouds, fog, dew, rain, snow, and hail.

Some of the phenomena of the atmosphere have been already described and explained in connection with the various subjects that have engaged our attention.

1016. WIND.—Wind is air put in motion.

The motion of the air is the result of changes constantly going on in the earth's temperature, in consequence of the alternation of day and night and the succession of the seasons. Those portions of the atmosphere that rest on the hotter regions of the earth become heated and rarefied, and rising leave a vacuum which is immediately filled by a rush of cooler air from the surrounding parts. Currents are thus produced, which we call *winds*.

The direction of the wind is determined by various local causes, modified by the revolution of the earth on its axis. The latter, operating alone, would make it appear to blow uniformly from the east; but the various projections on the earth's surface, and the unequal distribution of land and water (the latter of which is incapable of being heated to the same degree as the former), —these and other agencies constantly at work combine to give the wind different directions at different places, and to make it vary at the same place.

1017. *Velocity.*—The velocity of the wind is measured with an instrument called the An-e-mom'-e-ter.

There are several kinds of anemometers. One of the best consists of a small windmill, with an index attached for recording the number of revolutions made in a second.

It is found with the anemometer that a wind so slight as hardly to stir the leaves travels at the rate of 1 mile an hour; a gentle wind, 5 miles in the same time; a brisk gale, 15 miles; a high wind, 30; a storm, 50; a hurricane, 80; a violent hurricane, 100.

1018. *Kinds.*—There are three kinds of winds; Constant, Periodical, and Variable.

---

1015. What is Meteorology? Mention some of the phenomena of the atmosphere. 1016. What is Wind? What puts the air in motion? By what is the direction of the wind determined? 1017. How is the velocity of the wind measured? What is one of the best forms of the anemometer? How fast does a scarcely perceptible wind travel? A gentle wind? A brisk gale? A storm? A hurricane? 1018. How many

1019. Constant Winds are those that blow throughout the year in the same direction.

The most noted of these are the Trade Winds, which extend about 30 degrees on each side of the equator, a zone of 6 degrees near the centre known as the Region of Calms being excepted. They blow uninterruptedly on the ocean from north-east to south-west in the northern hemisphere, and from south-east to north-west in the southern. The regions on the equator being more heated than the surrounding parts, the air resting on them is rarefied, and rising flows over the cooler masses towards the poles, while cold air from the latter rushes in below to supply its place. Were the earth stationary, the trade winds would be due north on one side of the equator, and due south on the other. The earth's diurnal revolution, however, from west to east, modifies these directions so far as to make the north wind north-east and the south wind south-east.

The trade winds are of great service to mariners, enabling them to make certain voyages (for instance, from the Canaries to the northern coast of South America) with great rapidity, and almost without touching a sail. The zone in which they prevail is noted for its transparent atmosphere, its uniformity of temperature, and general peaceful aspect; whence it has been called by the Spaniards "the sea of the ladies".

1020. Periodical Winds are such as blow regularly in the same direction at a certain season of the year or hour of the day. The monsoon, the simoom, and the land and sea breezes, are periodical winds.

The monsoons are modifications of the trade winds, which sweep, sometimes with great violence, over the Indian Ocean and the whole of Hindostan. For six months they blow from a certain quarter, and for the next six months from the opposite one, owing to the change in the sun's position, and consequently in the heat received at a given point.

The simoom, originating in the deserts of Asia and Africa, is distinguished by its scorching heat and the fine sand it carries with it, raised from the parched surfaces it traverses. The simoom from the Desert of Sahara, sweeping over the intervening regions, finally reaches the northern shore of the Mediterranean, and is there called the Sirocco.—During the continuance of this hot and deadly wind, the animal and the vegetable creation droop with excessive exhaustion; travellers on the desert save their lives only by throwing themselves down with their faces in the sand.

Land and sea breezes are produced by the unequal heating of land and

kinds of winds are there? Name them. 1019. What are Constant Winds? What are the most noted constant winds? Where do the trade winds blow, and in what direction? Explain the origin of the trade winds. How do they benefit mariners? What do the Spaniards call the region in which they prevail, and why? 1020. What are Periodical Winds? Mention some. Describe the monsoons. Where does the simoom originate, and by what is it distinguished? What is the Sirocco? What is

water. During the day, the land receives more heat than the adjacent ocean, the rarefied air in contact with it rises, and a gentle breeze sets in from the sea about nine in the morning, which gradually increases to a brisk gale in the middle of the day. About 3 P. M. it begins to subside, and is followed in the evening by a land breeze, which blows freshly through the night: for after sunset the land rapidly parts with its heat by radiation, and the air resting on it, becoming cooler than that on the ocean, rushes to supply the place of the latter when it rises in consequence of being rarefied.

1021. Variable Winds are those which are irregular as to time, direction, and force, seldom continuing to blow for many days together. They prevail chiefly in the temperate and frigid zones, the winds of the torrid zone being for the most part constant or periodical.

1022. *Hurricanes.*—Hurricanes are storms that revolve on an axis, while at the same time they advance over the earth's surface.

Hurricanes are distinguished by their tremendous velocity and great extent. They are often 500 miles in diameter, and sometimes much more. In the southern hemisphere they always revolve in the same direction as the hands of a watch; in the northern hemisphere, in the opposite direction. There are three hurricane regions; the West Indies, the Indian Ocean, and the China Sea. In the last they are called Typhoons.

1023. *Tornadoes.*—Tornadoes, or Whirlwinds, are as violent as hurricanes, but more limited in extent. They are rarely more than a few hundred yards in breadth and twenty-five miles in length. Though lasting but a few seconds in a given place, they are frequently most disastrous in their effects, prostrating forests, overturning buildings, and ravaging the whole face of the country.

1024. *Water-spouts.*—A Water-spout is a phenomenon frequently observed at sea, consisting of a column of water raised sometimes to the height of a mile and tapering from each end towards the centre. It is supposed by some to be produced by a whirlwind of great intensity; by others it is attributed to electrical influences.

the effect of the simoom on the animal world? When do land and sea breezes blow, and how are they produced? 1021. What are Variable Winds? Where do they chiefly prevail? 1022. What are Hurricanes? By what are they distinguished? In what direction do they revolve? Name the three hurricane regions. 1023. What are Tornadoes? Describe their effects. 1024. What is a Water-spout? By what is it produced? Give an account of the way in which it is formed. 1025. What does

Water-spouts are formed as follows :—From a dark cloud a conical pillar is seen to descend with its point downward. As it approaches the water, the latter becomes violently agitated, and a similar column rises from it, point upward. The two finally unite, forming a continuous column from the cloud to the water. After remaining joined for a time, they again separate into two columns, one of which is drawn up into the cloud, while the other pours down in the form of heavy rain. Sometimes the two columns are dispersed before a junction is effected.

1025. ATMOSPHERIC MOISTURE. — The atmosphere always contains more or less moisture, derived from the earth's surface, particularly those portions of it that are covered with water, by the process of evaporation. When the air contains as much moisture as it is capable of holding at any given temperature, it is said to be *saturated.*

The higher the temperature of air, the more moisture it is capable of receiving. At 32° F., it will hold only $1/160$ of its own weight of watery vapor ; while at 113° it will receive eight times as much, or $1/20$ of its own weight.

1026. The earth gives out incredible quantities of moisture by evaporation. Experiments prove that an acre of ground apparently parched by the sun sends forth into the air over 3,000 gallons of water in 24 hours. Of course much greater quantities are evaporated from a moist soil and from surfaces covered with water.

1027. *The Hygrometer.*—The amount of moisture in the atmosphere is ascertained with an instrument called the Hygrometer. Hygrometers are made on different principles.

In some, the degree of humidity is indicated by the elongation of a hair, a fibre of whalebone, or some other animal substance which readily absorbs moisture and is increased in length by so doing. In others, it is shown by the increase of weight in some substance that absorbs moisture, such as sponge, cotton, or potash. In the more delicate instruments, the degree of moisture is shown by the greater or less facility with which it is condensed from the air in the form of dew on a cold surface. The more moisture in the air, the less cold will be required to condense it into dew.

1028. *Fog—Clouds.*—When the air is cooler than the earth, the moisture imparted to it in the manner just described is partially condensed and thus rendered visible, forming either *fog* or *clouds.* The only difference between the two is in their height. When the condensation takes place near the earth's surface, fog is the result; when in the upper regions of the atmosphere, clouds.

1029. *Kinds of Clouds.*—Clouds are divided into different classes, the principal of which are the Nimbus, the Cumulus, the Stratus, and the Cirrus.

The Nimbus, or rain-cloud, is a dense mass of vapor, of a leaden gray or blackish color, with a lighter tint on its edges.—The Cumulus has the appearance of many dense whitish clouds piled up one on another; or of a vast hemisphere with its base on the horizon, and peak rising above peak, looking like huge hills of snow when illumined by the sun. The cumulus may be called the cloud of day, and is an indication of fair weather.—The Stratus consists of a number of horizontal layers of cloud, not very far removed from the earth's surface. Forming at sunset and disappearing at sunrise, it may be called the cloud of night.—The Cirrus (called *cat's tail* by sailors) is a fleecy cloud, composed of thin feathery filaments disposed in every variety of form. The cirrus is the highest of all clouds, frequently reaching an altitude of from three to five miles. It is no doubt often composed of snowflakes, as the temperature of the regions in which it floats must be cold enough to freeze the watery particles.

1030. DEW.—When the moisture of the atmosphere comes in contact with an object colder than itself, it is condensed and deposited on the surface. This is the way in which Dew is formed.

A glass of ice-water on a warm day is almost immediately covered with a fine dew. So, in winter, when a number of persons are in a warm room, the moisture imparted to the air by their breath is condensed on the window-panes by the cold air without, and then sometimes frozen, giving them a beautiful frosted appearance.—Just so, in the evening, when objects on the earth's surface are cooled down by radiation, the moisture of the atmosphere is deposited on them in the form of dew.

1031. Dew is never abundant except during calm serene nights. It is generally more plentiful in spring and autumn than in summer, because the difference between the temperature of day and night is greater in those seasons. The quantity of dew precipitated on different bodies depends much upon their nature. Thus grass and leaves will frequently be found glistening with crystal drops at sunrise, when gravelled walks, stones, wood-work, and metallic surfaces, are comparatively dry—another striking proof of the wisdom with which Providence orders the economy of nature.

1032. Frost is nothing more than frozen dew.

formed? What is the difference between them? 1029. Name the different kinds of clouds. Describe the Nimbus. The Cumulus. The Stratus. The Cirrus. 1030. Under what circumstances is Dew formed? What familiar instances of the formation of dew are mentioned? 1031. When is dew most abundant? How does its precipi-

**1033. RAIN.**—Rain is water taken up by the air in the form of vapor and returned to the earth in drops.

When two masses of damp air differing considerably in temperature are mingled, they become incapable of retaining the same amount of moisture which they held while they remained apart. The excess is precipitated in the form of rain, the vesicles of vapor under the influence of mutual attraction blending together and forming drops.

Some parts of the earth never have any rain, vegetation, when it exists at all, being supported entirely by dew. This is the case with Peru, the Desert of Sahara, portions of Arabia and Egypt, and extensive districts in Central Asia. In other parts, for example Guiana, it rains almost constantly. The Island of Chiloe has a rather moist climate; the people there have a current saying, that it rains six days in the week and is cloudy the seventh.

**1034. SNOW.**—Snow consists of the watery particles of the atmosphere frozen for the most part in a crystalline form.

Viewed through a microscope, snow-flakes exhibit forms of great beauty and endless variety. Between six and seven hundred different forms have been distinguished, many of them belonging to the six-sided system of crystals.

Snow of a beautiful crimson color and a delicate green has been found in different parts of the world. These tints are due to minute plants or animalcules in different stages of development.

**1035. HAIL.**—Hail consists of globules of ice formed in the atmosphere by the congelation of its moisture and precipitated to the earth.

Hail is produced by an intense degree of cold in the atmosphere, and is generally accompanied with electrical phenomena. It is rare at the level of the sea within the tropics, and in high latitudes is totally unknown, being most abundant in temperate climates. Hail-storms seldom continue a quarter of an hour, but while they last large quantities of ice fall. The stones are generally pear-shaped, and frequently weigh ten or twelve ounces. Masses weighing 6, 8, and even 14 pounds, have been known to fall.

---

tation show the goodness of Providence? 1032. What is Frost? 1033. What is Rain? How is rain formed? What parts of the earth never have any rain? Where does it rain almost constantly? 1034. Of what does Snow consist? What is the form of snow-flakes? Of what color has snow sometimes been found? How is this accounted for? 1035. Of what does Hail consist? How is it produced? Where is it most frequent? What is the shape of hail-stones? How large have they been known to fall?

# FIGURES.

For the convenience of the pupil during recitation, the Figures to which reference is made by letters are here reproduced. The numbers correspond with those of the text.

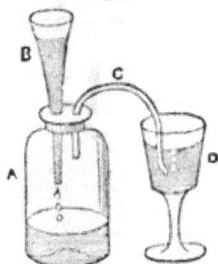

Fig. 1.

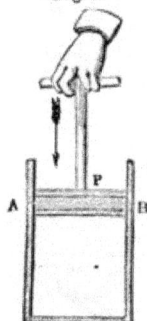

Fig. 8.

Fig. 13.

Fig. 14.

Fig. 15.

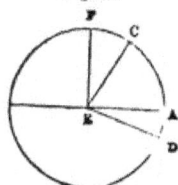

Fig. 16.

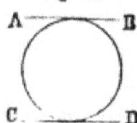

Fig. 17.

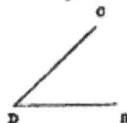

Fig. 18.

Fig. 20.

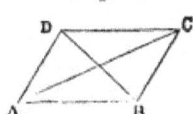

Fig. 21.

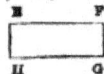

Fig. 22.

Fig. 23.

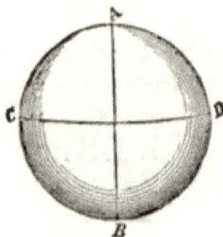

Fig. 24.

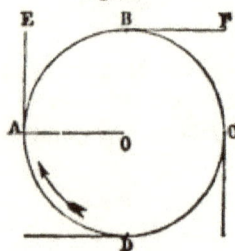

Fig. 25.

Fig. 29.

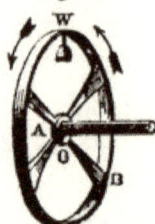

Fig. 31.

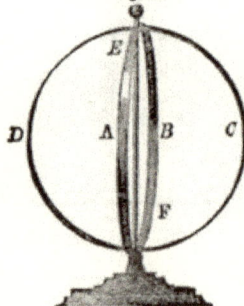

Fig. 32.

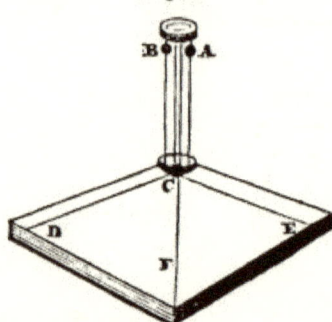

Fig. 33.

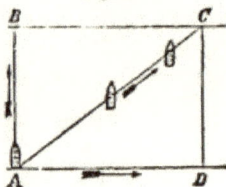

Fig. 34.

Fig. 36.

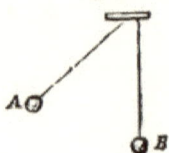

Fig. 37.

Fig. 38.

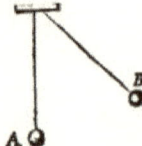

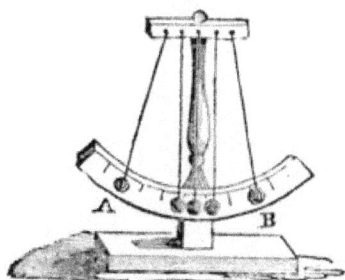

Fig. 39.

Fig. 49.

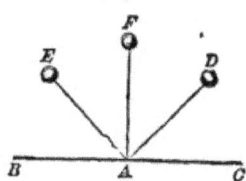

Fig. 40.

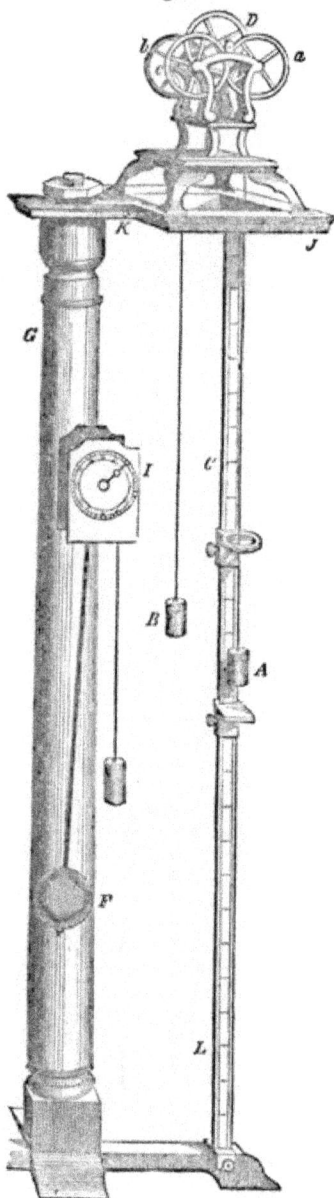

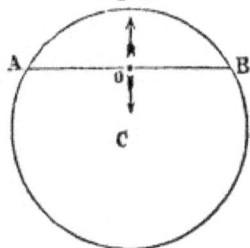

Fig. 43.

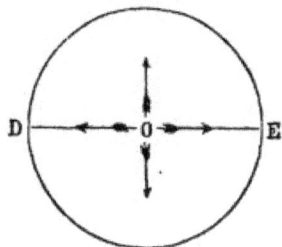

Fig. 44.

18

Fig. 50.

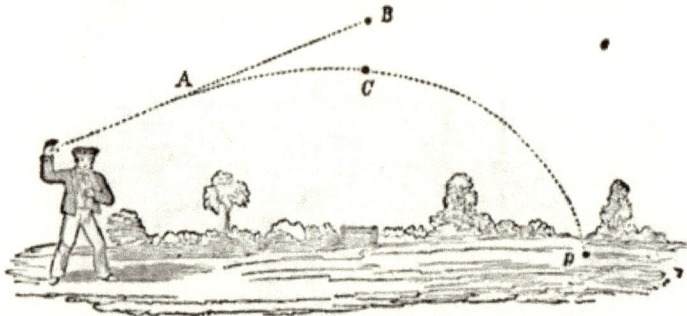

Fig. 51.

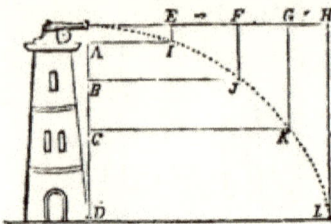

Fig. 53.

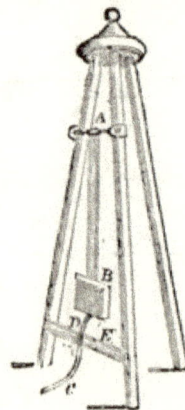

Fig. 54.

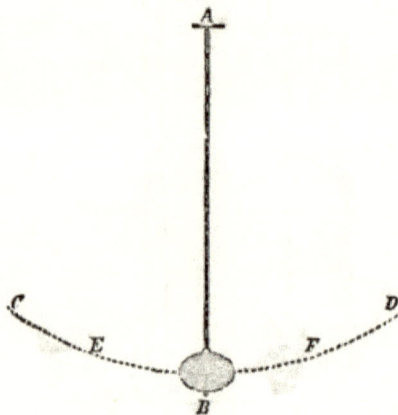

Fig. 56.

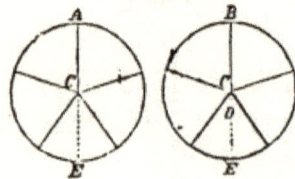

Fig. 58.

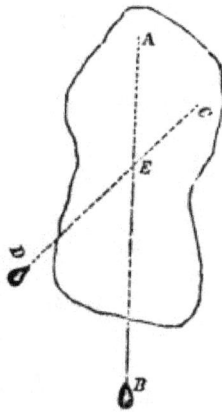

Fig. 60.

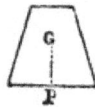

Fig. 61.

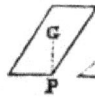

Fig. 62.

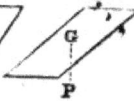

Fig. 74.

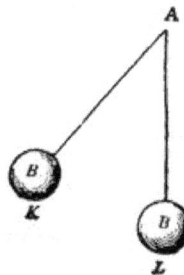

Fig. 75.

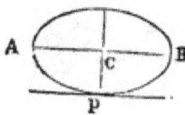

Fig. 76.

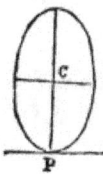

Fig. 78.

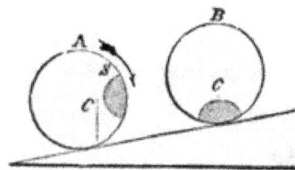

Fig. 81.

Fig. 85.

Fig. 86.

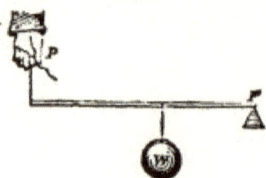

Fig. 87.

Fig. 91.

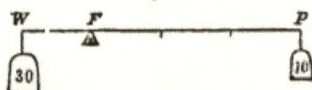

Fig. 95.

Fig. 93.

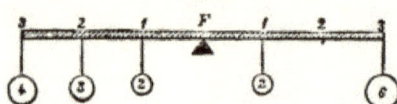

Fig. 96.

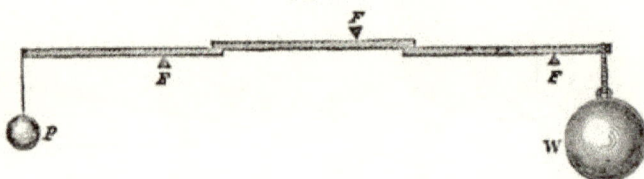

Fig. 98.

Fig. 99.

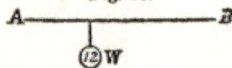

Fig. 100.

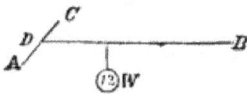

Fig. 103.

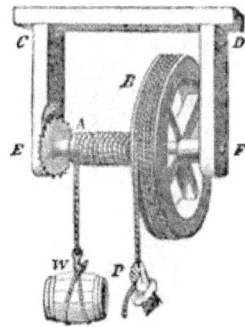

Fig. 102.

Fig. 104.

Fig. 105.

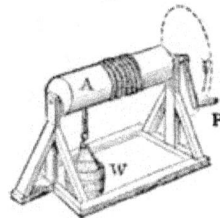

Fig. 106.

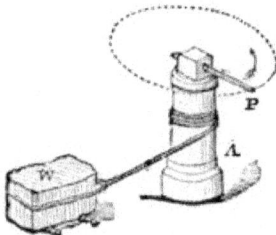

Fig. 108.

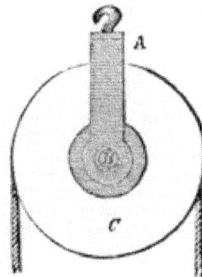

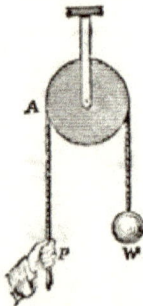

Fig. 109.

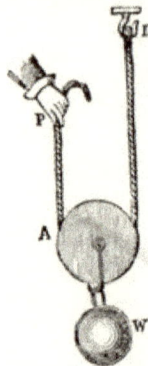

Fig. 111.

Fig. 112.

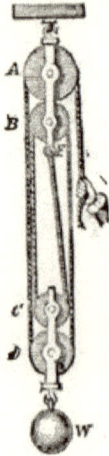

Fig. 113.

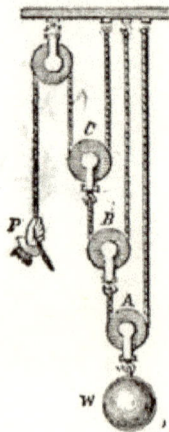

Fig. 115.

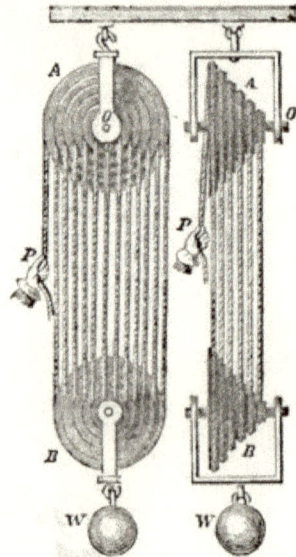

Fig. 114.

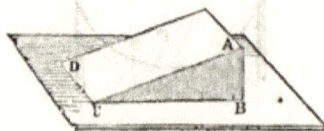

Fig. 116.

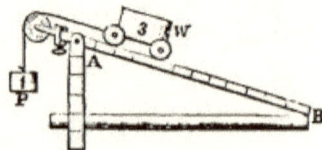

Fig. 117.

Fig. 118.

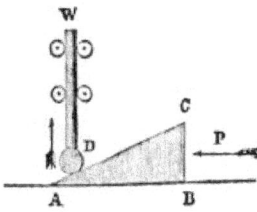

Fig. 119.

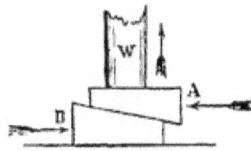

Fig. 123.

Fig. 122.

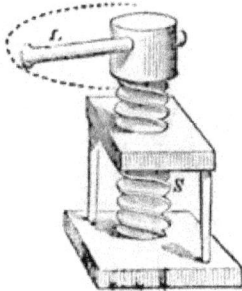

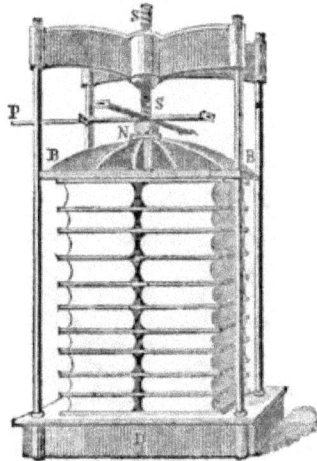

Fig. 124.

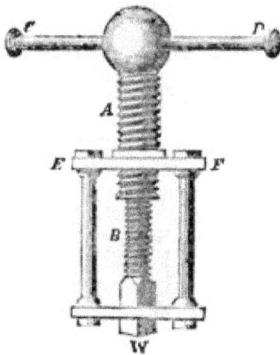

Fig. 126.

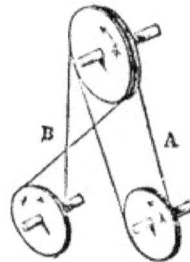

Fig. 127.

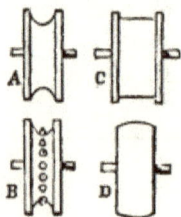

Fig. 129.

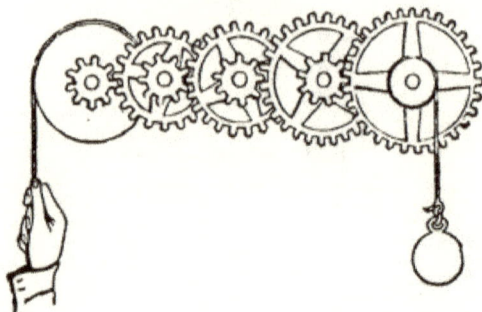

Fig. 180.

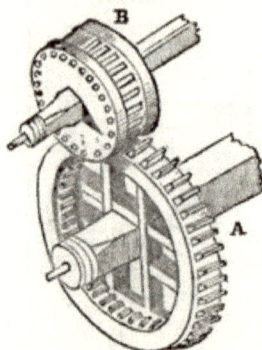

Fig. 131.

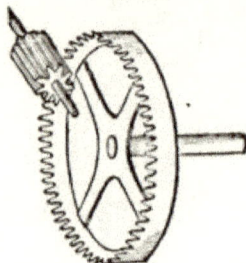

Fig. 134.

Fig. 136.

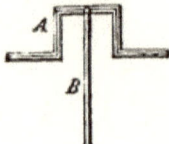

Fig. 137.

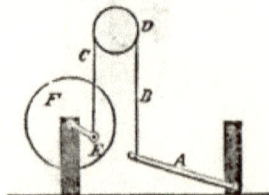

Fig. 139.

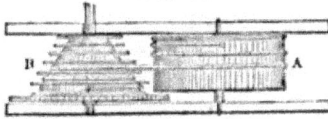

Fig. 139.

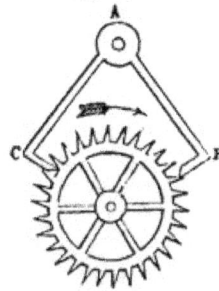

Fig. 140.

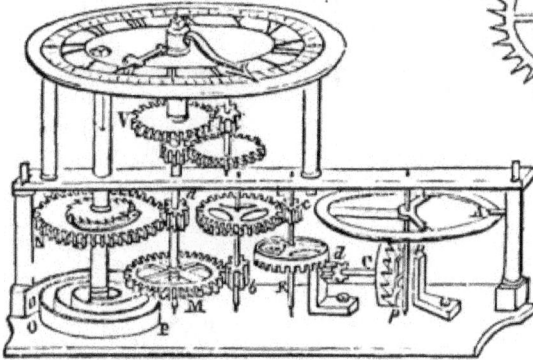

Fig. 142.

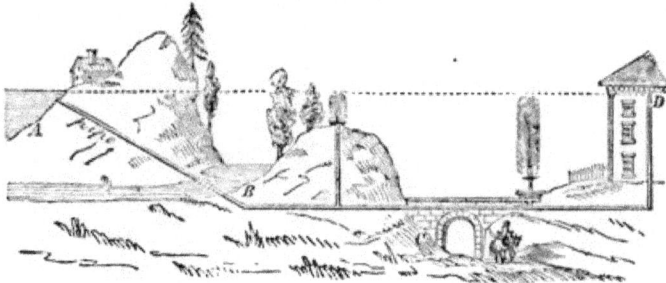

Fig. 143.

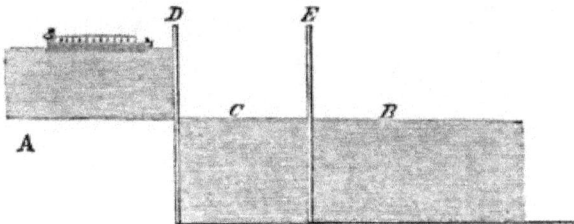

18*

Fig. 145.

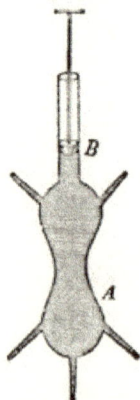

Fig. 146.

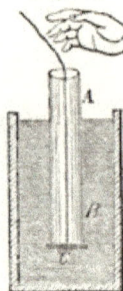

Fig. 149.

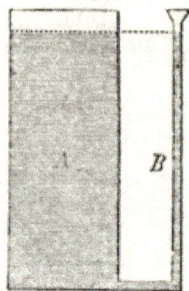

Fig. 151.

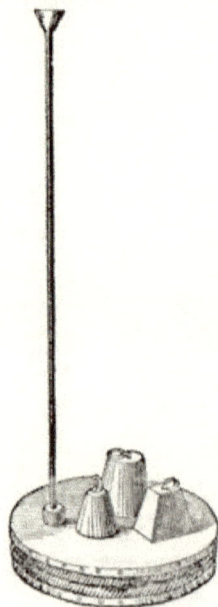

Fig. 152.

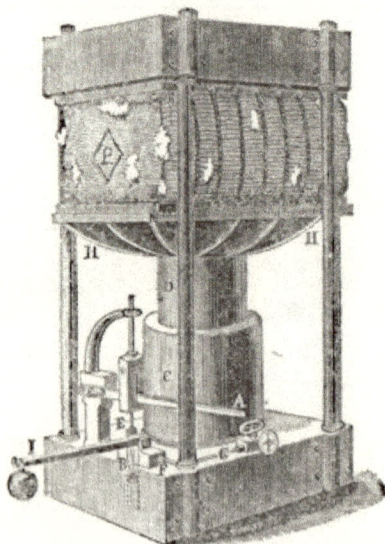

Fig. 153.

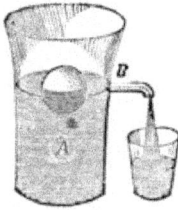

Fig. 155.

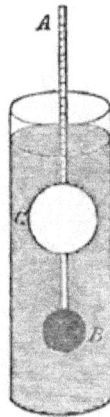

Fig. 160.

Fig 163.

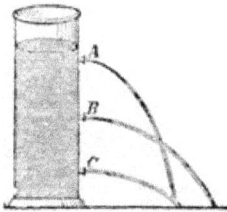

Fig. 161.

Fig. 102.

Fig. 164.

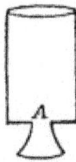

Fig. 165.

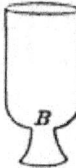

Fig. 167

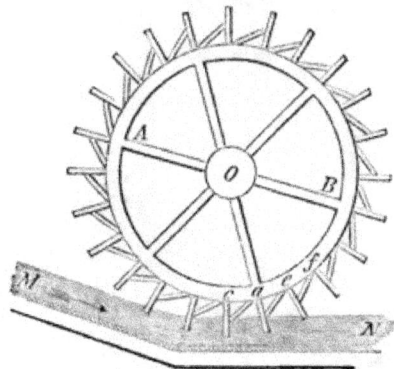

Fig. 166.

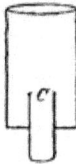

Fig. 168.

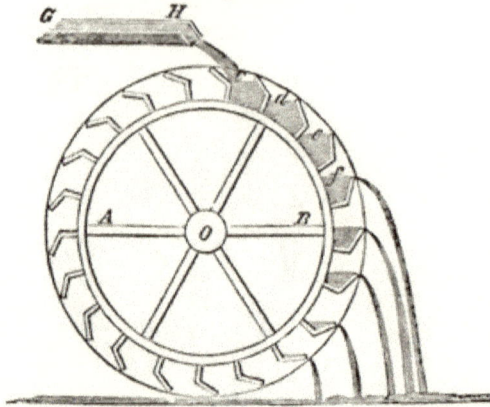

Fig. 170.

Fig. 171.

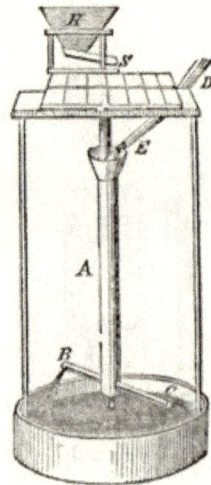

Fig. 172.

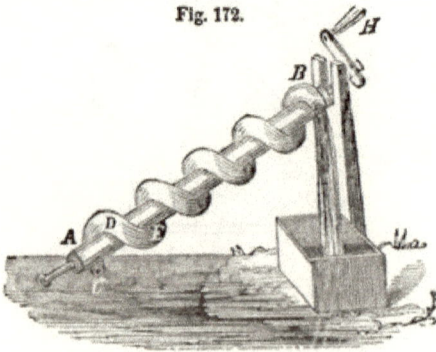

FIGURES.

Fig. 173.

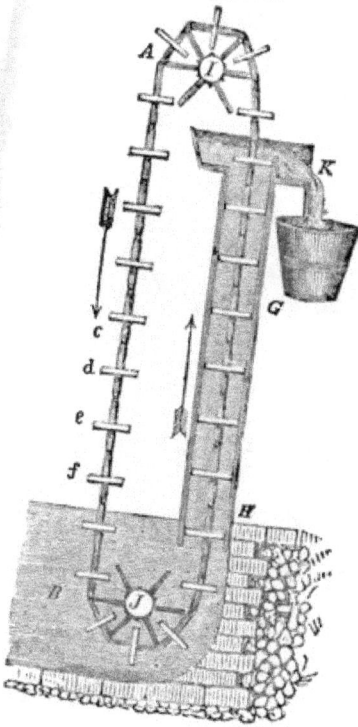

Fig. 175.

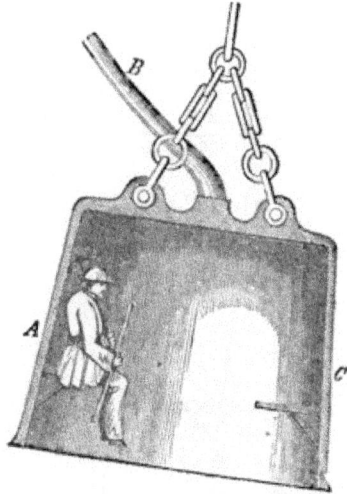

Fig. 174.

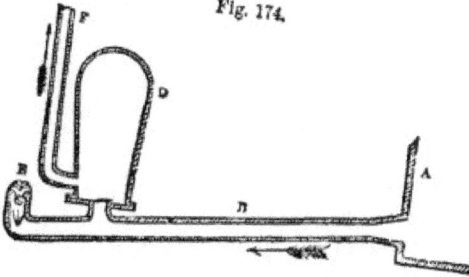

Fig. 177.

Fig. 179.

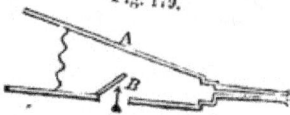

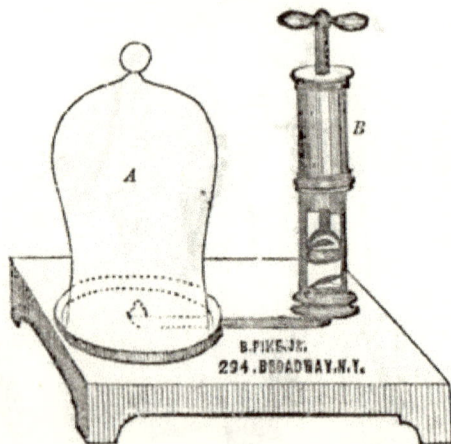

Fig. 187.

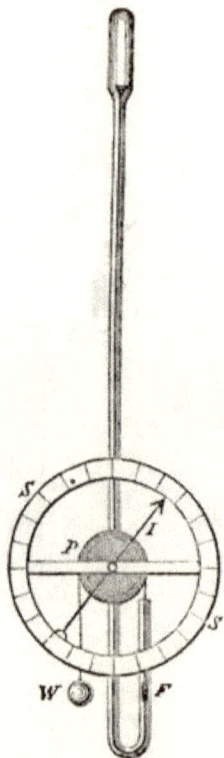

Fig. 182.

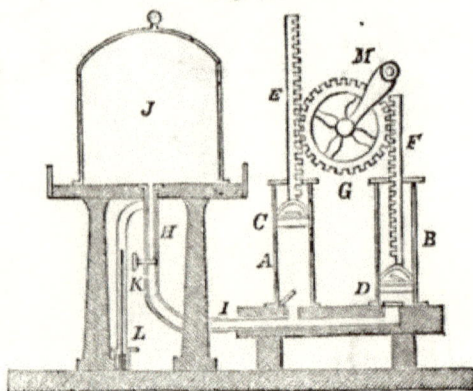

Fig. 189.

Fig. 190.

Fig. 192.

Fig. 193.

Fig. 197.

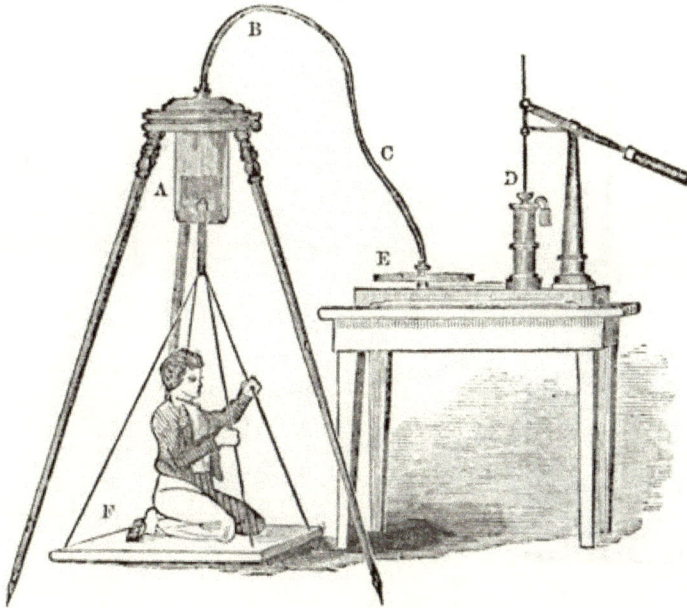

Fig. 198.

Fig. 196.

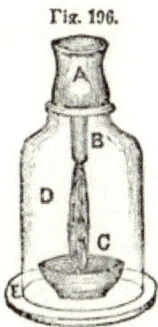

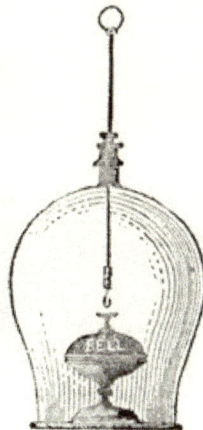

Fig. 200.

Fig. 201.

Fig. 203.

Fig. 204.

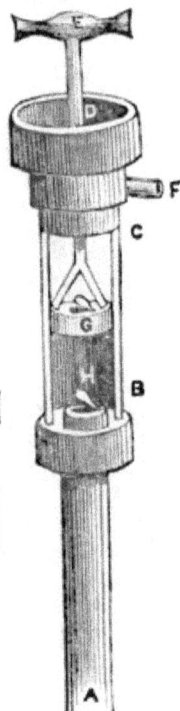

Fig. 205.

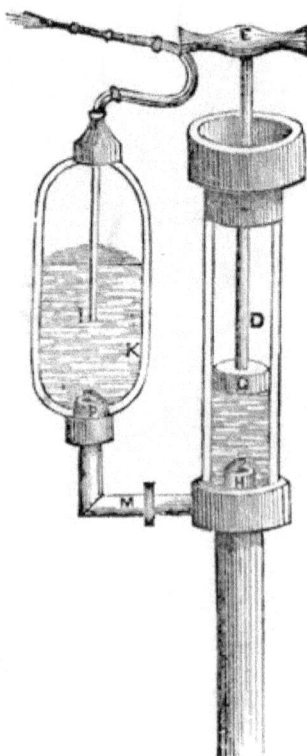

Fig. 202.

Fig. 207.

Fig. 206.

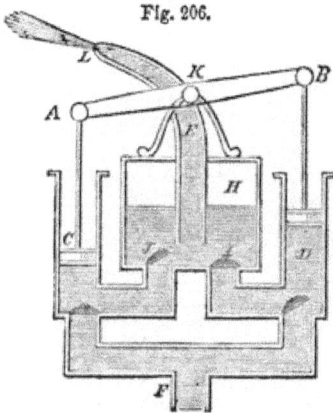

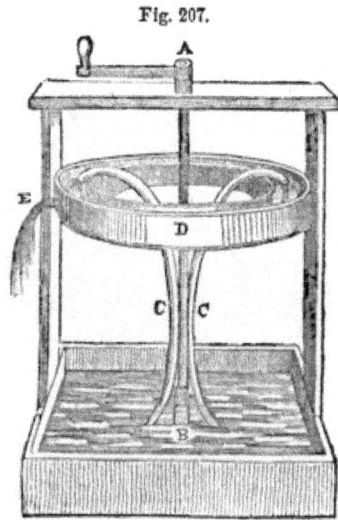

Fig. 208.

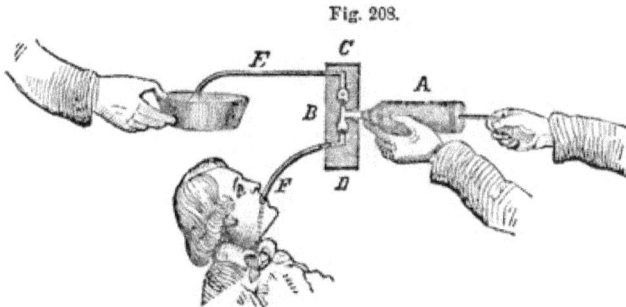

Fig. 213.

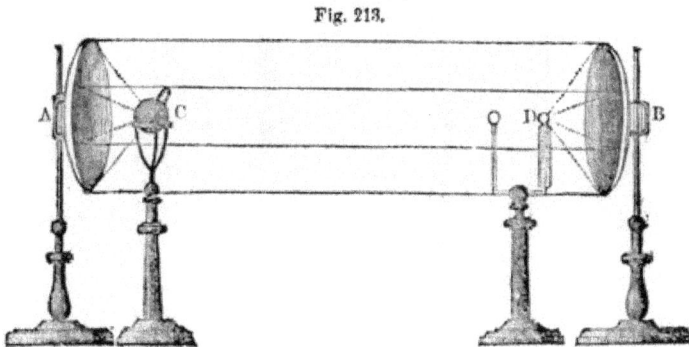

Fig. 215.

Fig. 219.

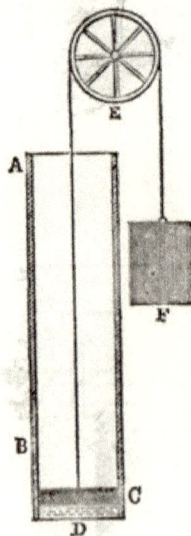

Fig. 221.

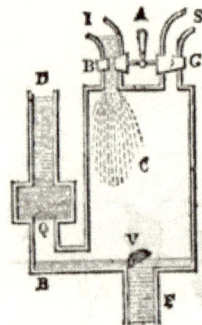

Fig. 222.

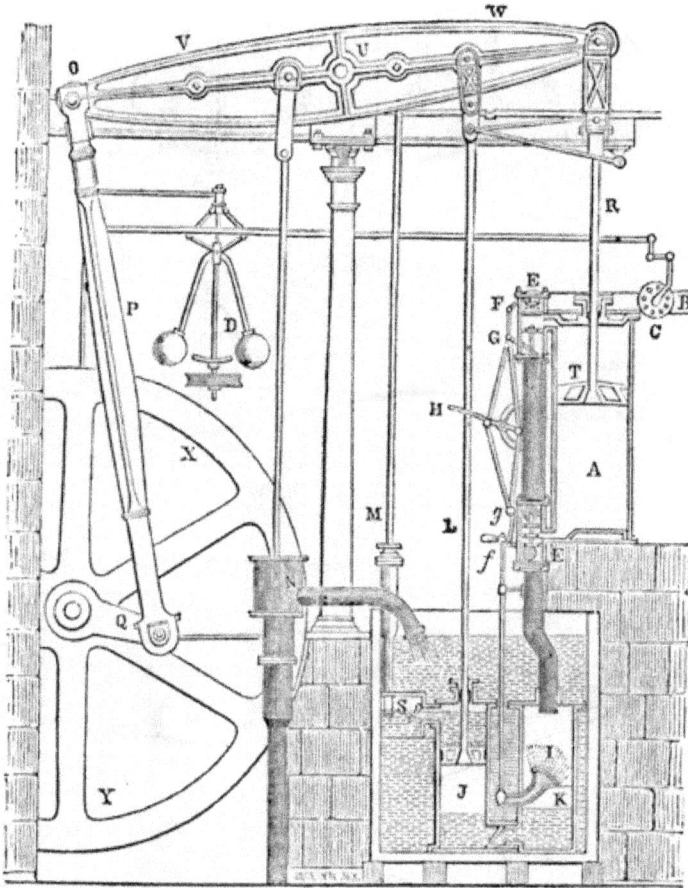

Fig. 222

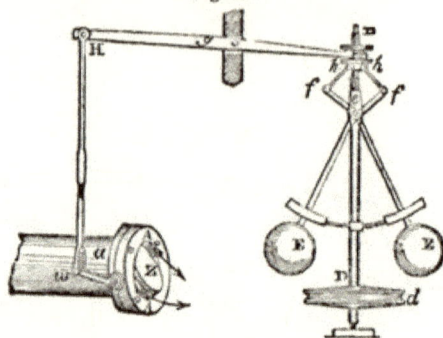

Fig. 224.

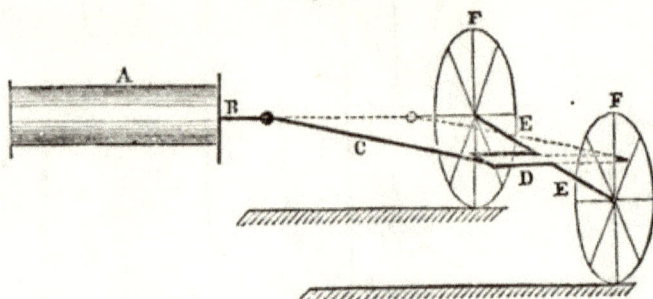

Fig. 228.

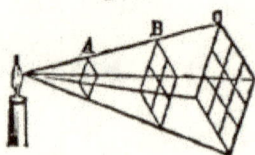

Fig. 229.

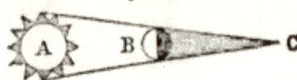

Fig. 230.

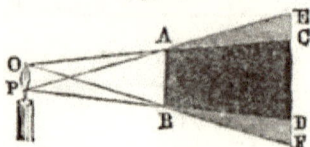

Fig. 231.

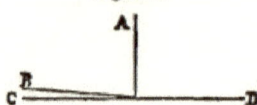

Fig. 232.

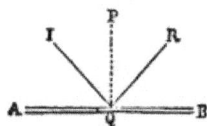

Fig. 233.

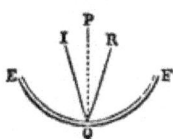

Fig. 234.

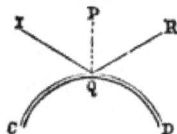

Fig. 235.

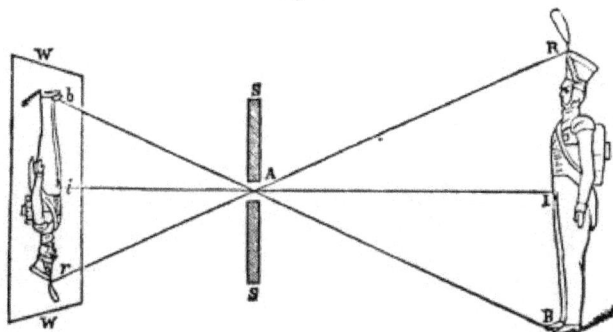

Fig. 236.

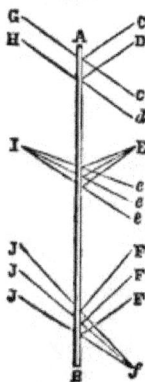

Fig. 238.

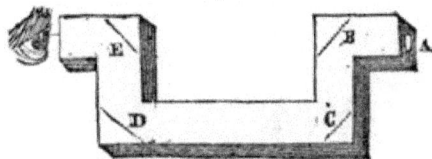

Fig. 237.

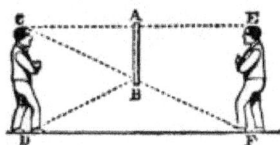

Fig. 239.

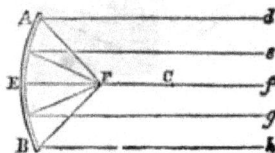

Fig. 240.

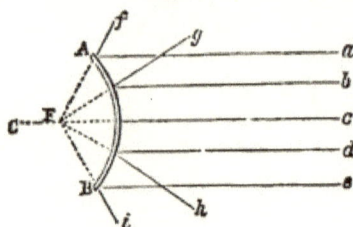

Fig. 241.

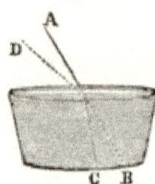

Fig. 242.

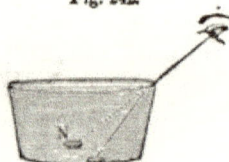

Fig. 244.

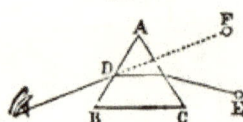

Fig. 243.

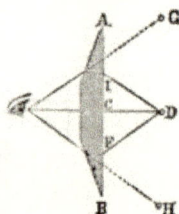

Fig. 250.

Fig. 253.

Fig. 254.

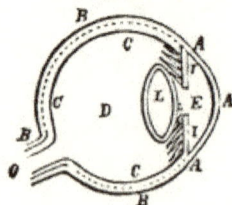

Fig. 255.

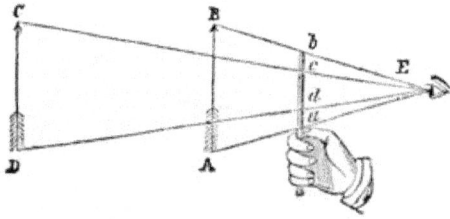

Fig. 256.

Fig. 258.

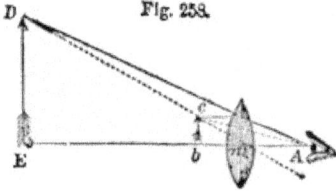

Fig. 257.

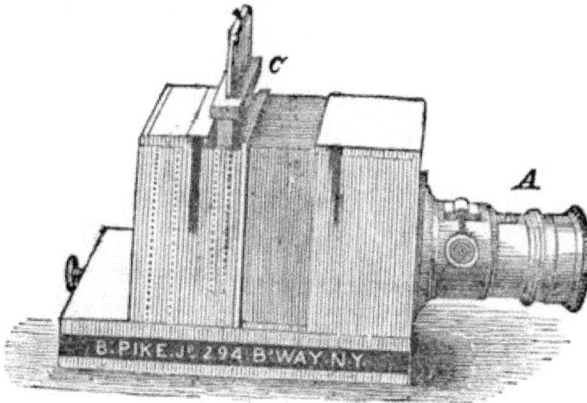

B. PIKE J. 294 B WAY N.Y.

Fig. 259.

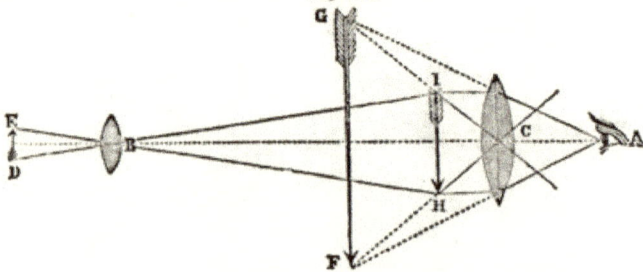

Fig. 260.

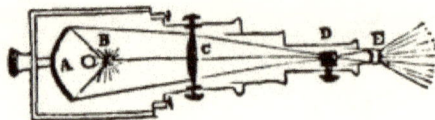

Fig. 261.

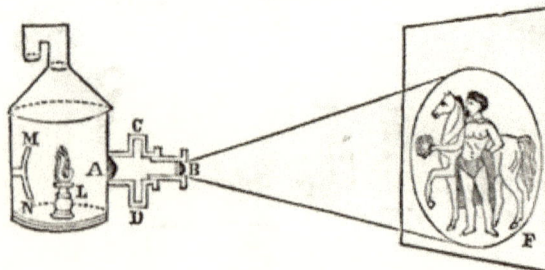

Fig. 262.

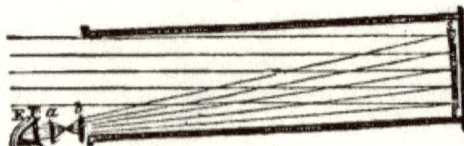

Fig. 264.

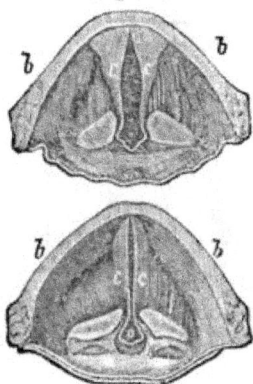

Fig. 265.

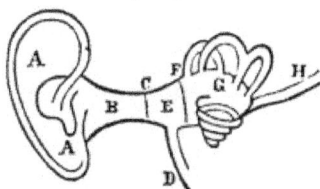

Fig. 266.

19

Fig. 269.

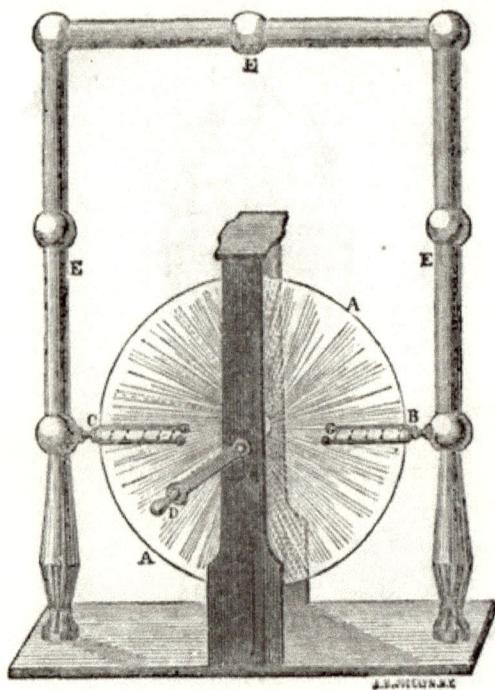

Fig. 274.

Fig. 285.

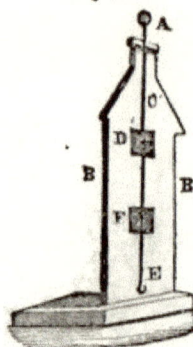

Fig. 286.

Fig. 287.

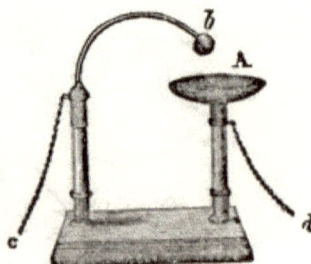

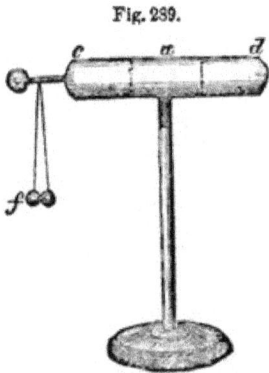
Fig. 289.

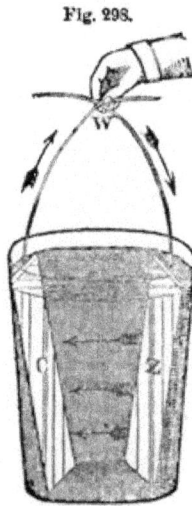
Fig. 298.

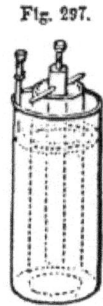
Fig. 297.

Fig. 299.

Fig. 300.

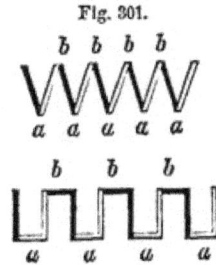
Fig. 301.

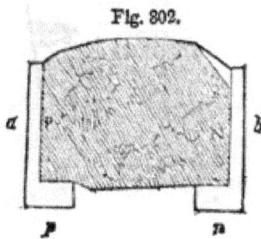
Fig. 302.

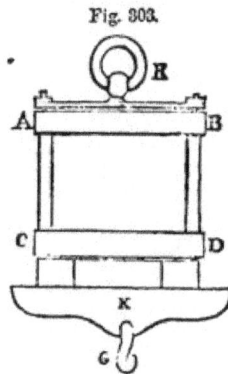
Fig. 303.

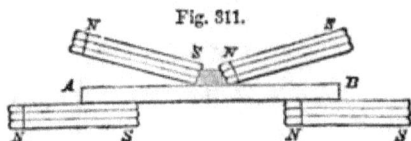

Fig. 311.

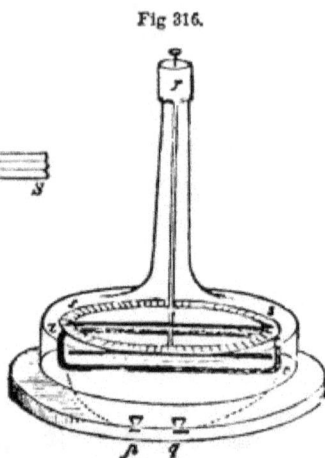

Fig 316.

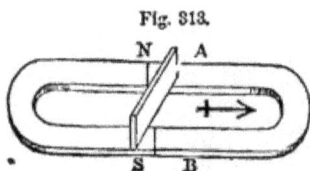

Fig. 313.

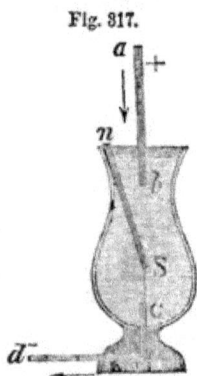

Fig. 317.

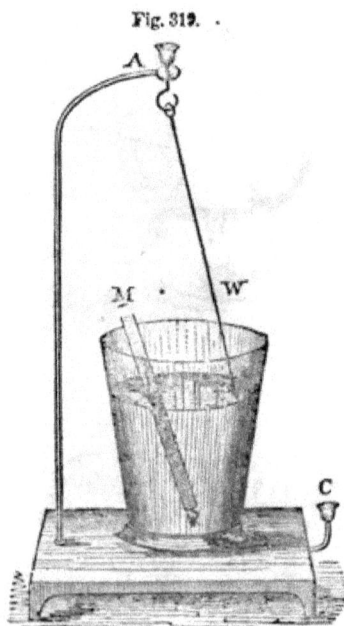

Fig. 319.

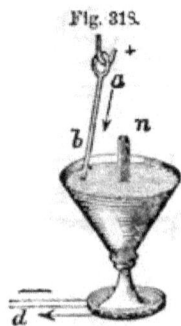

Fig. 318.

Fig. 323.

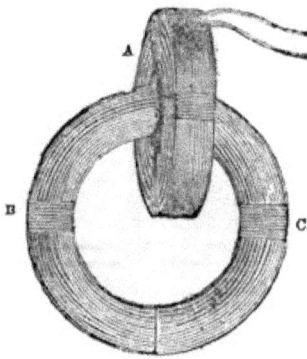

Fig. 324.

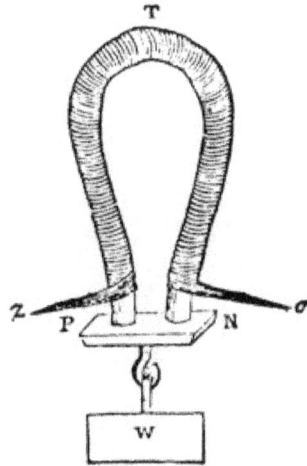

Fig. 326.

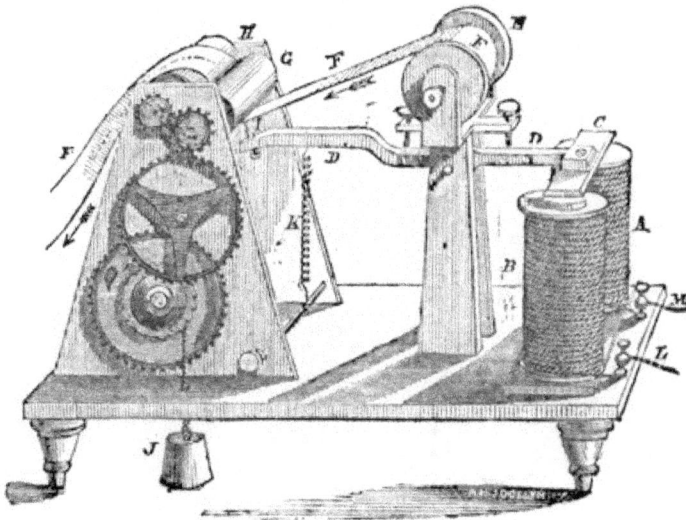

Fig. 328.

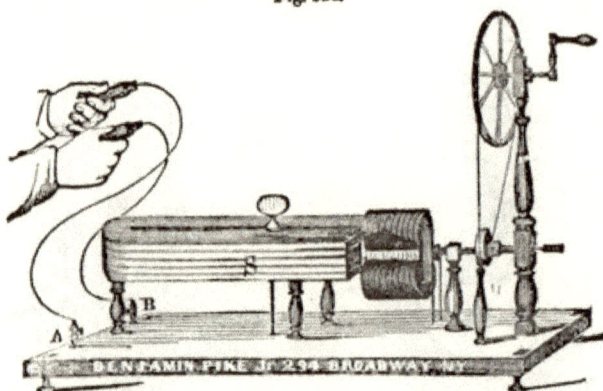

Fig. 329.

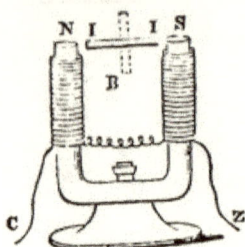

Fig. 330.

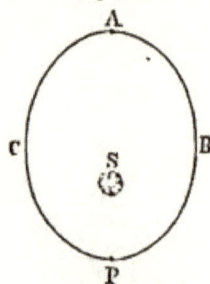

Fig. 332.

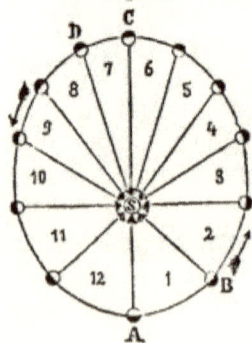

Fig. 333.

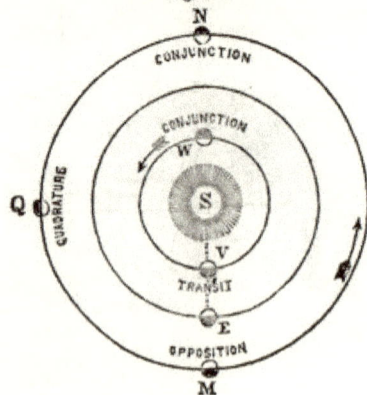

Fig. 884.

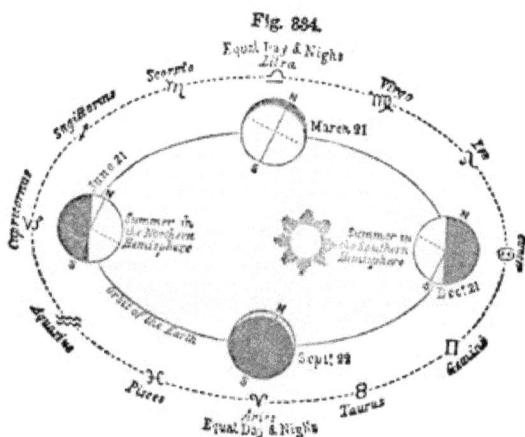

Fig. 885.

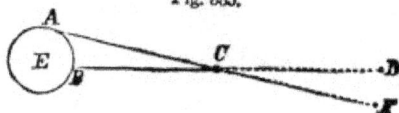

# INDEX.

[THE FIGURES REFER TO PAGES, NOT TO SECTIONS.]

## A.

*Aberration*, chromatic, 259.
*Acoustics*, defined, 274.
*Actinism*, 257.
*Action*, defined, 42. Equal to reaction, 43.
*Adhesion*, defined, 21. Experiments illustrative of, 21.
*Adjutage*, different forms of, 155.
*Aëriform bodies*, defined, 8.
*Affinity*, chemical, 9.
*Agents*, defined, 7.
*Air*, composition of, 9. Tends to stop motion, 36. Resistance of the, 54; effect of, 59, 62. Exists in every substance, 166. Is impenetrable, 166. Is compressible, 167. Is elastic, 167. Has weight, 168. Density of, at different levels, 174. Effects of its rarity, 175. Rarefied by heat, 175. Navigation of the, 177. Essential to life and combustion, 184. Supports a column of water from 32 to 34 feet high, 187. A non-conductor of heat, 201.
*Air-gun*, the, 168.
*Air-pump*, the, 177. Single-barrelled, 178. Double-barrelled, 179. Experiments with, 180.
*Alfred the Great*, his mode of measuring time, 126.
*Amazon*, the, its fall, 156. Its discharge, 156.
*Anemometer*, the, 401.
*Angle*, defined, 35. Vertex of, 35. Right, 35. Obtuse, 35. Acute, 35. Visual, 264.

*Anode*, defined, 324.
*Aphelion*, of a planet, 374.
*Apogee*, 359.
*Apple-cutter*, the, 180.
*Aqueducts*, of the ancient Romans, 132.
*Arc*, defined, 34.
*Arch*, the voltaic, 328. Celestial, 384.
*Archimedes*, reasoned by induction, 10. Explained the properties of the lever, 94. Discovered the leading principles of specific gravity, 146. His screw, 162. Fired the Roman fleet with concave mirrors, 194.
*Aristotle*, his doctrine respecting falling bodies, 54.
*Armature*, of magnets, 335.
*Ascending Bodies*, 59. Height reached by, 60.
*Asteroids*, the, 392.
*Astronomy*, defined, 363. Fundamental facts of, 369.
*Athermanous substances*, defined, 206.
*Atmosphere*, the, 166. Pressure of, 169, 174; proposed mode of transmitting mails by the, 182. How heated by the sun, 204. Moisture of, 404.
*Atomic Theory*, 17.
*Atoms*, what they are, 17.
*Attraction*, of gravitation, 20. Molecular, 21. Capillary, 146. Between floating bodies, 149. Electrical, 289, 290. Magnetic, 337.
*Atwood's Machine*, 56, 57.
*Aurora borealis*, 303.
*Axis*, of a sphere, 71.

**B.**

*Bacon, Roger,* invented spectacles, 263.
*Balance,* the, 96. Of a watch, 128.
*Ballistic Pendulum,* the, 64.
*Balloons,* why they rise, 53. Rose, why they lose their buoyancy, 150. Invention of, 176.
*Barometer,* the, 171. The wheel, 172. Its use as a weather guide, 172.
*Barometer gauge,* the, 180.
*Base,* of a body, 72.
*Battery,* the electrical, 300. The galvanic, 319; the trough, 320; Smee's, 320; Daniell's, 321; Grove's, 322; Bunsen's, 322; theory of, 323. The thermo-electric, 332. The relay, 361.
*Bay of Fundy,* tides of the, 157.
*Beam,* a, of light, 230.
*Bell,* vacuum, 183. Electrical, 301.
*Bellows,* hydrostatic, 137. Principle of the common, 170.
*Bladder-glass,* the, 181.
*Blindness,* cause of, 263.
*Boats,* propulsion of, 160. Shape of, 161.
*Bode,* his law, 376.
*Body,* a, what it is, 7. A simple, 8. A compound, 8. The simple bodies, 8. When it stands and falls, 73.
*Boilers,* how made, 225.
*Boiling,* process of, explained, 203.
*Bottle,* thousand grain, 141.
*Bottle Imps,* 167, 182.
*Breadth,* defined, 12.
*Breathing* process of, explained, 170.
*Breezes,* land and sea, 402.
*Brewster, Sir David,* invented the kaleidoscope, 241.
*British Channel,* tides of the, 157.
*Brittleness,* defined, 23.
*Buckets,* of a wheel, 158.
*Buffon,* his experiment with concave mirrors, 194.
*Burning,* process of, 195.
*Burning glasses,* 243, 251.

**C.**

*Calms,* region of, 402.
*Caloric,* 192.
*Camera obscura,* 266. Draughtsman's, 267. Daguerreotypist's, 267.

*Canals,* principle of their construction, 134.
*Cancer,* tropic of, 388.
*Capillary Attraction,* what it is, 146. Cause of, 146. Familiar examples of, 147. Laws of, 148. Interesting facts connected with, 149.
*Capricorn,* tropic of, 388.
*Capstan,* the, described, 105.
*Carbon,* combines with oxygen to produce animal heat, 196.
*Carbonic acid,* found at the bottom of wells, 140.
*Cathode,* defined, 324.
*Catoptrics,* 236.
*Centre,* of gravity, 70. Of magnitude, 71. Of motion, 71.
*Centre of gravity,* what it is, 70. How to find it, 71. In man, 77. Tends to get as low as possible, 73.
*Centrifugal Force,* defined, 37. Examples of, 38, 39. Law of the, 39. Its effect on revolving bodies, 39. Apparatus to illustrate the, 40. Gave its form to the earth, 41.
*Centripetal Force,* defined, 37.
*Ceres,* when discovered, 375.
*Chemical action,* a source of heat, 195.
*Chemical affinity,* 9.
*Chemistry,* defined, 9.
*Chinese,* the, early acquainted with gunpowder, 63. First used the magnet in navigation, 342.
*Chord,* what it is, 285.
*Chords,* vocal, 286.
*Chromatics,* 254.
*Chronometers,* how perfect, 127.
*Circle,* defined, 84. Simple galvanic, 319. The arctic, 389. The antarctic, 389.
*Circumference,* defined, 84. How divided, 85.
*Cirrus,* the, defined, 405.
*Clepsydra,* the, 126. Described, 153. Invented by Ctesibius, 187.
*Clocks,* how regulated, 68. History of, 126. Pendulum applied to, 126. Works of, 127. Electro-magnetic, 364.
*Clouds,* how formed, 404. Kinds of, 405.
*Coat,* choroid, 262. Sclerotic, 262.
*Cogs,* what they are, 123.
*Cohesion,* defined, 21.
*Cold,* what it is, 192.

*Colors*, the primary, 255. Difference of, explained, 256. Complementary, 257.

*Columbus*, his discovery respecting the variation of the compass, 343. Saved himself and his men by predicting an eclipse, 396.

*Combustion*, what it is, 195. Produces most of our artificial light, 231.

*Comets*, what they consist of, 397. Their orbits, 397. Their velocity, 397. Their number, 398.

*Compass*, land or surveyor's, 341. Mariner's, 342. Boxing the, 342.

*Compressibility*, 19, Of air, 20.

*Concord*, 255.

*Condensation*, 212. Of steam, 219.

*Condenser*, the, 184. Of the steam-engine, 222.

*Conductometer*, the, 199.

*Conductors*, of heat, 199. Of electricity, 293.

*Conjunction*, 379.

*Constellations*, 386, 399.

*Convection*, of heat, 202.

*Copernicus*, revived the true theory of the universe, 370.

*Cornea*, the, 261. Use of, 262.

*Couronne des tasses*, the, 320.

*Crank*, the, 124.

*Crown-wheel* and pinion, 123.

*Crutch*, the, of an escapement, 127.

*Ctesibius*, invented the lifting-pump, 186. Invented the clepsydra, 187. Supposed to have invented the water-organ, 284.

*Cumulus*, the, described, 405.

*Cup*, Tantalus's, 186. The phosphorus, 305.

*Cupping-glasses*, principle of, 175.

*Curb*, of a watch, 129.

*Curves*, magnetic, 338.

*Cylinder*, defined, 86.

**D.**

*Daguerreotype* process, the, 268.

*Dams*, should increase in strength at the base, 136.

*Dead-point*, the, of a crank, 125.

*Density*, 19. In optics, 246.

*Descartes*, advanced the undulatory theory of light, 229.

*Dew*, how formed, 405.

*Diagonal*, defined, 85.

*Diamagnetism*, 367.

*Diameter*, defined, 34.

*Diathermanous substances*, defined, 206.

*Dionysius*, ear of, 281.

*Dioptrics*, 246.

*Dip*, magnetic, 340.

*Direction*, line of, 71.

*Discharger*, the jointed, 298. The universal, 298.

*Discord*, 255.

*Dispersion*, of light, 259.

*Distillation*, process of, described, 212.

*Diving-bell*, the, 166.

*Divisibility*, defined, 17. Instances of, 18.

*Double cone*, may be made to roll up an inclined plane, 80.

*Draft*, how produced in a chimney, 176.

*Driver*, the, 120.

*Drum*, the, 283.

*Ductility*, defined, 26. Of platinum, 26. Of gold, 26. Of glass, 26.

*Du Fay*, his theory respecting electricity, 291.

**E.**

*Ear*, the human, 287.

*Earth*, the, owes its form to the centrifugal force, 41. Magnetic poles of, 344. Form of, 382. Motions of, 383. Orbit of, 384. Phases of, to the moon, 390. How it would look from the fixed stars, 393.

*Ear-trumpet*, the, 280.

*Echoes*, 279.

*Eclipse*, of the sun, how produced, 395. Annular, 396. Of the moon, 396.

*Ecliptic*, the, 385. Obliquity of, 385.

*Eel*, the Surinam, 316.

*Elasticity*, defined, 24. Perfect, 24. Belongs to hard solids, 24. Of steel, 24. A limit to, 25.

*Electricity*, a source of light, 232. What it is, 289. Sources of, 290. Developed by friction, 290. Vitreous, or positive, 291. Resinous, or negative, 291. Nature of, 291. Conduction of, 293. Path of, 294. Velocity of, 294. Machines for developing, 294; experiments with, 301. Mechanical effects of the passage of, 304. From steam, 310. Atmospheric, 310. Voltaic, 316. Difference between frictional and voltaic, 324. Developed by heat, 332. Connection between magnetism and, 352. Developed by magnetism, 366.

*Electrics*, 293.

*Electrodes*, what they are, 323.

*Electro-magnetism*, defined, 349. As a motive power, 357.

*Electro-magnets*, 356, 357.

*Electro-metallurgy*, 326.

*Electrometer*, the, 309. The quadrant, 309.

*Electrophorus*, the, 308.

*Electroscope*, the, 308.

*Electrotyping*, process of, described, 327.

*Elements*, sixty-two in number, 8. Divided into metals and non-metallic, 8. The non-metallic enumerated, 9.

*Endless Band*, 121.

*Endosmose*, 150.

*Engine*, defined, 88. Atmospheric, 219. Steam, 219. Hero's, 219. Marquis of Worcester's, 220. Savery's, 221. Newcomen's, 222. Watts', 222. The low pressure, 226. The high pressure, 226. The locomotive, 226.

*Equator*, the magnetic, 341. The celestial, 385.

*Equilibrium*, stable and unstable, 79.

*Equinoctial*, the, 385.

*Equinoxes*, 385. Precession of the, 386.

*Escapement*, of clocks, 127. Of watches, 128.

*Esquimaux*, why they thrive on fat, 196.

*Evaporation*, 211.

*Exosmose*, 150.

*Expansibility*, 19. Of air, 20.

*Expansion*, 207.

*Experiment*, what it consists in, 10.

*Extension*, defined, 12.

*Eye*, the, 260. Parts of, 261. Adaptation of, 265.

**F.**

*Fahrenheit*, his thermometrical scale, 214.

*Falling bodies*, velocity of, 54. Law of, 56. Rules relating to, 58.

*Faraday*, his theory respecting electricity, 292.

*Fata morgana*, 248.

*Figure*, defined, 12.

*Fire*, St. Elmo's, 311.

*Fire-alarm*, electro-magnetic, 365.

*Fire-balls*, 312.

*Fire-engine*, principle of the, 188

*Fire-escape*, the, 107.

*Fire-house*, the electrical, 307.

*Fish*, how they rise and sink in water, 145. Electrical, 316.

*Flame*, how produced, 193.

*Float-boards*, 158.

*Fluids*, embrace liquids and aëriform bodies, 8. Difference between them and solids, 8. Non-elastic, 25. Elastic, 25. Division of elastic, 165.

*Flute*, the, principle of, 283.

*Flyer*, the electrical, 305.

*Fly-wheel*, the, 125.

*Focus*, the principal, 242. The virtual, 244.

*Fog*, how formed, 404.

*Follower*, the, 120.

*Force*, defined, 26. Striking, 31. Centrifugal, 37. Centripetal, 37.

*Forge-hammer*, the, 124.

*Fountains*, how high they rise, 132. Vacuum, 181.

*Franklin*, his theory respecting electricity, 292. Proved lightning to be produced by an electric discharge, 312. Invented the lightning-rod, 315.

*Friction*, what it is, 37, 85. How it opposes motion, 85. Kinds of, 85. Sliding, converted into rolling, 86. Laws of, 86, 87. Modes of lessening, 87. Uses of, 88. Of one wheel on another, 120. Of water against the sides of pipes, 155. Of a stream against its banks, 156. Enables the wind to produce waves, 156. A source of heat, 197. A source of electricity, 290.

*Frost*, what it is, 406.

*Fulcrum*, what it is, 94.

*Furnace*, of a steam-engine, 226.

*Fusee*, of a watch, 128.

**G.**

*Galaxy*, the, 400.

*Galileo*, his doctrine respecting falling bodies, 54. Invented the pendulum, 67. First made a practical use of the telescope, 272. Established the truth of the Copernican system, 371.

*Galle, Dr.*, discovered Neptune, 394.

*Galleries*, whispering, 281.

*Galvani*, discovered voltaic electricity, 317. His experiment, 317.

*Galvanism*, 316. (*For particulars, see* Voltaic electricity.)

*Galvanometer*, the, 351. With astatic needle, 352.

*Gamut*, the, 284.

*Gases*, what they are, 165. Specific gravity of, how found, 143. Exhibit endosmose and exosmose, 150. Conducting power of, 201. Expansion of, 210.

*Gearing*, what it is, 121.

*Gioia, Flavio*, improved the compass, 342.

*Glottis*, the, 286.

*Governor*, the, 225.

*Gravitation*, defined, 20. Circumstances attending its discovery, 47. Facts established respecting it, 47. Direction of, 48. Laws of, 49.

*Gravity*, terrestrial, 46. Laws for the force of, 49. Sometimes causes bodies to rise, 53. Centre of, 70. Used as a motive power, 81. Specific, 139. Tables of specific, 144.

*Guericke*, invented the air-pump, 177. His famous experiment, 178. First contrived an electrical machine, 295.

*Gunnery*, 63.

*Gunpowder*, principle on which it acts, 63. Invention of, 63. Apparatus for firing, with electricity, 308.

*Gutta percha*, used for endless bands, 121.

## II.

*Hail*, its disastrous effects, 59. How formed, 406.

*Hair-spring*, the, of a watch, 128.

*Haloes*, what they are, 260.

*Hand-glass*, the, 180.

*Hardness*, defined, 22. Wanting in fluids, 22. Of various solids compared, 22.

*Harmony*, what it is, 285.

*Harp*, Æolian, 283.

*Heat*, what it is, 192. Free, or sensible, 192. Latent, 192. Theories respecting, 198. Has no weight, 193. Sources of, 193. The sun's, 194; how it may be increased, 194; how far it penetrates into the earth, 194. Below the earth's surface, 194. Produced by chemical action, 195. Animal, or vital, 196. Produced by mechanical action, 196. From friction, 197. From percussion, 197. Produced by electricity, 198. Diffusion of, 198; by conduction, 199; by convection, 202; by radiation, 203. Radiant, 204; law of, 204; reflection of, 205; absorption of, 206; transmission of, 206. Effects of, 207. Instruments for measuring, 213. Specific, 216.

*Helix*, the, 355. Magnetizing power of, 355. Itself a magnet, 365.

*Hemispheres*, the Magdeburg, 178.

*Hero*, his steam-engine, 219.

*Herschel*, his telescope, 273. Discovered Uranus, 375.

*Hiero*, golden crown of, 145.

*Hooke, Dr.*, added the hair-spring to the balance, 127.

*Horizon*, the sensible, 384. The rational, 384. Poles of the, 385.

*Horse*, the, strength of, 82.

*Horse-power*, defined, 84.

*Humor*, aqueous, 261. Vitreous, 261.

*Hurricanes*, 403.

*Huygens*, applied the pendulum to clockwork, 67. Unfolded the undulatory theory of light, 229.

*Hydraulics*, defined, 152.

*Hydraulicon*, the, 284.

*Hydrogen*, the lightest substance known, 144. Used for inflating balloons, 176. Produces musical sounds, 284.

*Hydrometer*, the, 142.

*Hydrostatics*, defined, 130. Law of, 131. Hydrostatic paradox, 137. Hydrostatic bellows, 137. Hydrostatic press, 138.

*Hygrometer*, the, 404.

## I.

*Ice*, process of its formation, 210.

*Iceland spar*, exhibits double refraction, 252.

*Image*, an, what it is, 239.

*Impenetrability*, defined, 13. Of air, 13. Instances of, 13.

*Incandescence*, 213.

*Incidence*, angle of, 46. Equal to angle of reflection, 46.

*Inclined Plane*, the, 110. Law of, 110. Practical applications of, 111. Law of bodies rolling down, 111.

*Indestructibility*, defined, 13. Instances of, 13. Anecdote illustrative of, 13.

*Induction*, electrical, 302. Magnetic, 346.

*Inertia*, defined, 15. Examples of, 15. Experiments illustrative of, 15, 16. Proportioned to a body's weight, 17.

*Instruments*, optical, 266. Stringed, 282. Wind, 283.

*Insulators*, 294.

*Iridium*, one of the hardest metals, 22. The heaviest known substance, 144.

*Iris*, the, 261. Use of, 262.

*Iron*, great tenacity of, 23.

## J.

*Jar*, the Leyden, 299.

*Juno*, when discovered, 375.

*Jupiter*, velocity of light ascertained from the eclipses of one of its moons, 234. Its seasons, 389. Details of the planet, 392.

## K.

*Kaleidoscope*, the, 241.

*Kepler*, his laws, 377.

## L.

*Lamp*, principle on which it burns, 147.

*Landes*, shepherds of, 78.

*Lantern*, a species of wheel, 123. Origin of the, 126. The magic, 271. Phantasmagoria, 271.

*Larynx*, the, 286.

*Lava*, discharge of, accounted for, 195.

*Leaves*, what they are, 122.

*Length*, defined, 12.

*Lens*, the crystalline, 261.

*Lenses*, what they are, 246. Classes of, 249. Refraction by convex, 250. Refraction by concave, 251. Achromatic, 259.

*Le Sage*, first attempted to transmit messages by electricity, 263.

*Level*, the spirit, 134. The water, 135.

*Lever*, what it is, 94. Of the first kind, 94, 95. Practical applications of the, 98, 100, 102. The bent, 99. The compound, 99. Of the second kind, 99. Of the third kind, 101. Perpetual, 104. Often combined with the screw, 115.

*Le Verrier*, his prediction verified by the discovery of Neptune, 394.

*Life-boats*, principle of, 144.

*Life-preservers*, principle of, 144.

*Light*, what it is, 229. Corpuscular theory of, 229. Undulatory theory of, 229. Sources of, 231. Of the sun, 232. Of the stars, 232. Propagation of, 232. Velocity of, 233. Intensity of, at different distances, 234. Reflection of, 236. Refraction of, 245. Laws of refracted, 246. Polarization of, 258. Dispersion of, 259. The electric, 325. The zodiacal, 373.

*Lightning*, 312. Effects of, 313.

*Lightning rod*, the, 313.

*Lights*, northern, 308.

*Line*, a right, defined, 34. Parallel lines defined, 34. A curve, 34. The neutral, of a magnet, 334.

*Liquefaction*, 210.

*Liquids*, defined, 8. How they differ from solids, 130. Have little cohesion, 131. Compressibility of, 131. Not devoid of elasticity, 131. Pressure of, 135. Rule for finding their pressure on the bottom of a vessel, 137. Specific gravity of, 141. Exhibit endosmose and exosmose, 150. Flow of, through orifices, 152. Flow of, in pipes, 155. Conducting power of, 201. Expansion of, 209. Converted into vapor by heat, 211. Good conductors of sound, 276.

*Living Force*, 81.

*Loadstone*, described, 334.

*Lock*, on a canal, 134.

*Locomotive*, the, 226.

*Lubricants*, 87.

*Lungs-glass*, the, 151.

## M.

*Machines*, what they are, 88. Can not create power, 88. Law of, 89. Advantages of using, 90. All, combinations of the six mechanical powers, 120. Must be regular in their motion, 125. For raising water, 161. Electrical, 294. Cylinder, 295, 296. Plate, 297. Hydroelectric, 310. Magneto-electric, 367.

*Magic lantern*, the, 271.

*Magnetism*, defined, 333. Theory of, 343. Terrestrial, 344; intensity of, 345. By induction, 346. By the sun's rays, 347. By contact with a magnet, 347. Developed by electricity, 349. Connection between electricity and, 352.

*Magneto-electricity*, 366. Medical use of, 367.

*Magnets*, what they are, 333. Natural,

334. Poles of, 334, 337. Power of natural, 335. Armed, 335. Artificial, 335. Bar, 336. Horse-shoe, 336. Compound, 336. Power of, how increased and diminished, 337. Attraction of, 337; law of, 338. Polarity of, 338. Production of artificial, 345.

*Magnifying glasses*, 251.

*Main-spring*, the, of a watch, 123.

*Malleability*, defined, 25. Of the metals, 25, 26.

*Mariotte's Law*, 163.

*Mars*, details of the planet, 391.

*Matter*, defined, 7. Ponderable, 7. Imponderable, 7. Forms of ponderable, 7. Properties of, 12.

*Mechanical Powers*, the, 94.

*Mechanics*, defined, 26.

*Medium*, a, what it is, 231. A uniform, 231. A dense, 246. A rare, 246.

*Melody*, what it is, 285.

*Meniscus*, what it is, 249.

*Mercury*, details of the planet, 381.

*Meridian*, the magnetic, 340.

*Metals*, the principal, 9. Specific gravity of various, 144. Precious, how tested, 145. Protection of, by voltaic electricity, 328.

*Meteorology*, defined, 401.

*Metius*, supposed to have invented the telescope, 272.

*Microscope*, wonders revealed by the, 18, 270. What it is, 268. The single, 268. The compound, 269. Solar, 270. Oxyhydrogen, 270.

*Milky way*, the, 400.

*Mill*, Barker's, 161.

*Mill-stones*, how made in France, 148.

*Mirage*, 248.

*Mirrors*, concave, Roman fleet fired with, 194. What they are, 237. Plane, 238. Concave, 238. Convex, 238. Reflection from plane, 240. Images formed by plane, 241. Reflection from concave, 242. Reflection from convex, 244.

*Mississippi*, the, its discharge, 156.

*Mixtures*, freezing, 211.

*Mobility*, defined, 20.

*Momentum*, what it is, 29. Rule for finding the, 30.

*Monsoons*, 402.

*Montgolfier* brothers, balloons invented by the, 176.

*Moon*, the, 389. Produces tides, 157. Size of, 389. Motions of, 389. Phases of, 390. New, 390. Gibbous, 390. Full, 390. How it appears through the telescope, 391. Eclipses of, 396.

*Moons*, of Jupiter, 392. Of Saturn, 393. Of Uranus, 393.

*Morse*, his telegraph, 358. His telegraphic alphabet, 361.

*Motion*, what it is, 27. Absolute, 27. Relative, 27. Kinds of, 28. Uniform, 28. Accelerated, 29. Retarded, 29. First law of, 36. Second law of, 41. Simple, 41. Resultant, 41. Parallelogram of, 42. Third law of, 43. Reflected, 45; law of, 46. Rotary, may keep a body from falling, 76. Perpetual, 89. Circular, how converted into rectilinear, 124. Alternate up-and-down, how produced, 124. Real and apparent, of the planets, 380.

*Motive Powers*, 81.

*Multiplying Glass*, the, 252.

*Muschenbroeck*, his electric shock, 300.

**N.**

*Nadir*, the, 385.

*Natural Philosophy*, defined, 9. Modes of investigation in, 10. Branches of, 11.

*Nebulæ*, 400.

*Needles*, magnetic, 336. Horizontal, 336. Dipping, 336, 341. Astatic, 339. How to magnetize, 347. Effects of electric currents on magnetic, 349.

*Neptune*, when discovered, 375. First called Le Verrier, 375. Details of the planet, 394.

*Nerve*, optic, 261, 262. Acoustic, 288.

*Newcomen*, his steam-engine, 222.

*Newton*, discovered the law of gravitation, 47. Held the corpuscular theory of light, 229.

*Nimbus*, the, described, 405.

*Non-conductors*, of heat, 199. Of electricity, 293.

*Non-electrics*, 293.

*Non-luminous bodies*, defined, 230.

*North star*, distance of the, 399.

**O.**

*Observation*, what it consists in, 10.

*Occultation*, 390.

*Ocean*, the surface of, spherical, 131. Pressure of, at great depths, 136.

*Octaves*, what they are, 284.

*Oersted*, discovered the phenomena of electro-magnetism, 349.

*Oil*, how extracted from seeds, 112.

*Opaque bodies*, defined, 231.

*Opera-glass*, the, 273.

*Opposition*, 379.

*Optics*, defined, 229.

*Organ*, the, 284.

*Orifices*, velocity of streams flowing through, 153. Course of streams flowing through, 154. Volume discharged from, 154.

*Oxygen*, promotes combustion, 195. Combines with carbon to produce animal heat, 196.

## P.

*Paddles*, of a wheel, 160.

*Pallas*, when discovered, 375.

*Pallets*, of an escapement, 127, 129.

*Parachute*, the, 55.

*Paradoxes*, 80. Hydrostatic Paradox, 137.

*Parallax*, 395.

*Parallelogram*, defined, 85. Of motion, 42.

*Pascal*, constructed the first barometer, 171.

*Pencil*, a, of light, 230. A diverging, 230. A converging, 230.

*Pendulum*, the, what it is, 65. Laws of its vibration, 65, 66. Application to clock-work, 67. Vibrates differently in different latitudes, 67. Effect of heat on its vibrations, 68. Compensation, 68. Gridiron, 68. Ballistic, 64. Its use in clock-work, 127.

*Penumbra*, the, 235.

*Percussion*, a source of heat, 197. A source of light, 232.

*Perigee*, 389.

*Perihelion*, of a planet, 374.

*Perspective*, the magic, 242.

*Perturbations*, 394.

*Phantasmagoria*, 272.

*Philosophy*, natural, 9. Meaning of the term, 9.

*Photographic* process, the, 238.

*Physics*, another name for Natural Philosophy, 9.

*Pile*, Volta's, 318. The dry, 322.

*Pinions*, defined, 122.

*Pipes*, flow of liquids in, 155.

*Pisa*, tower of, 75. Scene of an interesting experiment, 54.

*Pistol*, the electrical, 304.

*Planets*, the, 373. Secondary, 374. Primary, 374. Inferior, 374. Superior, 374. Orbits of, 374. Table of, 375. Aspects of, 378. Are they inhabited, 380.

*Plating*, 326.

*Pneumatics*, defined, 163.

*Points*, the cardinal, 341. Of the compass, 342.

*Polarity*, magnetic, 338.

*Polarization*, of light, 252.

*Poles*, of a galvanic battery, 323. Of natural magnets, 334. Of artificial magnets, 337. Magnetic, of the earth, 344. Of the horizon, 385.

*Pores*, what they are, 18.

*Porosity*, defined, 19. Of various substances, 19.

*Powers*, the mechanical, 94.

*Press*, book-binder's, 115. Bramah's hydrostatic (or hydraulic), 138.

*Pressure*, of liquids, 135. Of the atmosphere, 169.

*Prisms*, refraction by, 248. Decompose light, 255.

*Projectile*, a, what it is, 60. Forces by which it is acted on, 61. Path of, 61. Random of, 62.

*Propeller*, the screw, 160.

*Properties*, universal, 12. Accessory, 12.

*Ptolemy*, his system of the universe, 370.

*Pulley*, the, 106. The fixed, 106. The movable, 107. White's, 108. Much of its advantage lost by friction, 109.

*Pump*, the chain, 162. The lifting, 186. The forcing, 187. The centrifugal, 189. The stomach, 190.

*Pupil*, the, 261. Of beasts of prey, 262.

*Pyramids*, the most stable of figures, 74. Egyptian, 74.

*Pyrometer*, the, 215.

*Pyronomics*, defined, 192.

*Pythagoras*, the first to use the term *philosophy*, 9. Taught the true theory of the solar system, 370.

## Q.

*Quadrant*, defined, 85.

*Quadrature*, 879.
*Quadrilateral*, defined, 85.
*Quarter*, first, of the moon, 390. Third, of the moon, 390.

**R.**

*Race*, a, what it is, 158.
*Rack and pinion*, 124.
*Radiation*, of heat, 204.
*Radius*, defined, 34.
*Radius Vector*, the, 377.
*Rain*, 406.
*Rainbow*, the, 259. Primary and secondary, 260. Lunar, 260.
*Ram*, the hydraulic, 163.
*Random*, 62. At what angle it is greatest, 63.
*Rarity*, 19. In optics, 246.
*Rays*, what they are, 230. Incident, 236.
*Reaction*, defined, 43. Equal to action, 43. Examples of, 43. Often nullifies action, 43.
*Reasoning*, by induction, 10. By analogy, 10.
*Reaumur*, his thermometrical scale, 214.
*Receivers*, 177.
*Rectangle*, defined, 36.
*Reflection*, angle of, 46. Equal to angle of incidence, 46. Of light, 236; great law of, 238.
*Refraction*, 245. Atmospheric, 247. By convex lenses, 250. By concave lenses, 251. Double, 252. Its effect on the apparent position of the heavenly bodies, 394.
*Refractory substances*, defined, 210.
*Refrigerators*, what their sides are filled with, 209.
*Regulator*, of a watch, 129.
*Repulsion*, between the particles of aëriform bodies, 21. Between solids and liquids, 147. Electrical, 290.
*Resistance*, what it is, 27. Appears in various forms, 84.
*Rest*, what it is, 27. Absolute, 27. Relative, 27.
*Restitution*, force of, 24.
*Retina*, the, 261, 262. Images formed on, 264.
*Rhodium*, one of the hardest metals, 22.
*Rivers*, velocity of, how retarded, 156.
*Rocking Horse*, the, 76.

*Rocking Stones*, 79.
*Rocks*, how rent, 136.
*Roemer*, first used mercury in the thermometer, 214. Discovered the velocity of light, 234.
*Rope-dancers*, how they balance themselves, 78.
*Rosse, Earl of*, his telescope, 273.
*Rotation*, electro-magnetic, 353.
*Rubber*, the, 289.

**S.**

*Safes*, what the sides are filled with, 200.
*Sap*, how it ascends and descends in plants, 151.
*Saturn*, details of the planet, 393.
*Savery*, his steam-engine, 221.
*Scale*, diatonic, 285.
*Scape-wheel*, the, of a watch, 127.
*Screw*, the, what it consists of, 114. Kinds of, 114. Advantage gained by, 114. Practical uses of, 116. Hunter's, 116. The endless, 117. Archimedes', 162.
*Seasons*, the change of, 386.
*Sea-water*, heavier than fresh water, 144.
*See-saw*, the electrical, 301.
*Selenite*, polarizes light, 254.
*Self-luminous bodies*, defined, 230.
*Shadows*, 235.
*Shock*, the electric, 299; anecdote connected with, 300.
*Shower*, the mercury, 182.
*Signal key* the, 360.
*Silurus electricus*, the, 316.
*Simoom*, the, 402.
*Siphon*, the, 185.
*Sirius*, its light compared with the sun's, 232. Distance of, 339.
*Sirocco*, the, 402.
*Sling*, the principle on which it acts, 38.
*Smoke*, why it rises, 176.
*Snow*, protects vegetation, 202. How formed, 406. Colored, 406.
*Solids*, defined, 7. Difference between them and fluids, 8. Specific gravity of, 142. Porosity of, proved with the air-pump, 184. Expansion of, 207. Melted by heat, 210.
*Solstices*, the, 388.
*Sonorous bodies*, defined, 275.
*Sound*, nature of, 274. Transmission of, 275. Velocity of, 277. Distance to

which it is transmitted, 278. Interference of, 279. Reflection of, 279. A musical, how produced, 281; loudness of, 281; pitch of, 281; quality of, 281.
*Spark*, the electric, 297. Color of, 306. Length of, 307. Ignition by, 307.
*Speaking-trumpet*, the, 279.
*Specific gravity*, 139. Of liquids, 141. Tables of, 144. How to ascertain the weight of a body from its, 145.
*Spectacles*, 263.
*Spectrum*, the solar, 255. Properties of the, 257. Dark lines in the, 253.
*Speculum*, a, what it is, 237.
*Sphere*, defined, 36. Axis of, 36, 71. Poles of, 36. Equator of, 36.
*Spheroid*, oblate, 36. Prolate, 36.
*Spirit-level*, the, 135.
*Spots*, solar, 371.
*Springs*, used as a motive power, 82.
*Springs*, origin of, 133. Hot, accounted for, 195.
*Square*, defined, 36.
*Stability*, of bodies, 72. Depends on the position of the centre of gravity, 75. How increased, 76. Of a sphere, how increased, 79.
*Stammering*, 287.
*Stars*, the, a source of light, 232. Magnitudes of, 398. Distance of, 399. Periodical, 399. Binary, 399. Telescopic, 399.
*Staves*, of wheels, 123.
*Steam*, the most effective of motive powers, 83. Generation of, 216. Temperature of, 217. Properties of, 217. Condensation of, 218. Electricity from, 310.
*Steam-engine*, Hero's, 219. De Garay's, 220. Of De Caus, 220. Branca's, 220. Marquis of Worcester's, 220. Papin's, 221. Savery's, 221. Newcomen's, 222. Watts', 222. Parts of the, 223, 224. The low pressure, 226. The high pressure, 226. The locomotive, 226; history of, 227.
*Steel*, elasticity of, 24.
*Steelyard*, the, 97.
*Stephenson*, improved the locomotive, 228.
*Stethoscope*, the, principle of, 278.
*St. Helena*, tides at, 157.
*Still*, the, described, 212.
*Stilts*, used by French shepherds, 78.
*Stool*, the insulating, 298.

*Stratus*, the, defined, 405.
*Strength*, of men and animals, used as a motive power, 82. Of materials, 91. Of rods and beams, 91.
*Striking Force*, 81. Difference between it and momentum, 31. Rule for finding the, 31.
*Strings*, of musical instruments, 282.
*Sucker*, the, principle of, 170.
*Sun*, the, a source of heat, 193. A source of light, 232. Size of, 371. Constitution of, 372. Motions of, 372. Eclipses of, 395.
*Sun-dial*, the, 126.
*Syringe*, the fire, 197.
*System*, the Solar, 370. True theory of, taught by Pythagoras, 370; revived by Copernicus, 370.

**T.**

*Tangent*, defined, 34.
*Teeth*, connect wheels, 122.
*Telegraph*, electro-magnetic, 358. Morse's, 358. House's, 362. Bain's, 362. The sub-marine, 362. The Atlantic, 363. History of the, 363.
*Telescope*, the, 272. Refracting, 272. Astronomical, 273. Terrestrial, 273. Reflecting, 273. Herschel's, 273. Earl of Rosse's, 273.
*Temperature*, what it is, 192.
*Tempering*, how effected, 24.
*Tenacity*, defined, 22. Distinguished from hardness, 22. Belongs to the metals, 22. Of different substances compared, 23. Of liquids, 23.
*Thermo-electricity*, 332.
*Thermometer*, the, 213. Invention of, 214. The differential, 214.
*Thickness*, defined, 12.
*Thunder*, 312.
*Thunder house*, the, 306.
*Tides*, what they are, 157. How produced, 157. Spring, 157. Neap, 157. Height of, 157.
*Tools*, defined, 88.
*Top*, why it does not fall when spinning, 76.
*Tornadoes*, 403.
*Torpedo*, the, 316.
*Torricelli*, proved the pressure of the atmosphere, 171.
*Tourmaline*, polarizes light, 254.

*Train*, of wheels, 120. Of wheels and pinions, 122.
*Transit*, of a planet, 380.
*Translucent bodies*, defined, 231.
*Transparent bodies*, defined, 231.
*Treadle*, the, 125.
*Trevithick*, constructed the first practical locomotive, 227.
*Triangle*, defined, 35.
*Tripoli*, formed of fossilized animalcules, 18.
*Tropics*, the, 388.
*Trundle*, a, what it is, 123.
*Tubes*, acoustic, 278. Aurora, 302.
*Turbine*, the, 159.
*Tympanum*, the, 289.
*Typhoons*, 403.

**U.**

*Uranus*, when discovered, 375. Its former names, 375. Details of the planet, 393.

**V.**

*Vacuum*, what it is, 166. Torricellian, 172. Fountain, 181. Bell, 183.
*Valve*, the safety, 226.
*Vaporization*, 211.
*Vapors*, what they are, 165. Conducting power of, 201. Expansion of, 210.
*Variation*, magnetic, 340. Lines of no, 340.
*Velocity*, what it is, 27. Rule for finding the, 28. Of various moving objects, 28.
*Ventriloquism*, 287.
*Venus*, details of the planet, 382.
*Verge*, of a watch, 129.
*Vesta*, when discovered, 375.
*Veta*, the, 175.
*Views*, dissolving, 272.
*Vision*, 260. Defects of, 263.
*Voice*, the human, 285; when said *to change*, 286. Of the inferior animals, 287.
*Volta*, his theory respecting galvanism, 318. His pile, 318. Invented the *couronne des tasses*, 320.
*Voltaic electricity*, 316. Effects of, 324. Decomposes, 325. Luminous effects of, 328. Heating effects of, 329. Physiological effects of, 330. Medically applied, 331.

**W.**

*Watches*, history of, 126. Works of, 128. How regulated, 128. Parts of, 129.

*Water*, composition of, 9. Used as a motive power, 82. Quantity of, on the earth's surface, 130. Finds its level, 131. Conveyed in pipes, 132. How conveyed by the ancient Romans, 132. Its weight compared with air, 144. Wheels moved by, 158. Machines for raising, 161. Expansion of, in freezing, 209. Decomposed by the galvanic battery, 325.
*Water-clock*, the, 126, 153.
*Water-organ*, the, 284.
*Water-spouts*, 403.
*Watts*, his steam-engine, 222.
*Waves*, how produced, 156. Height of, 157.
*Wedge*, the, 112. Used for raising weights, 112. Familiar applications of, 113. Advantage gained by, 113.
*Weighing*, double, 97.
*Weight*, what it is, 50. Above and below the earth's surface, 50. Law of, 52. At different parts of the earth's surface, 53.
*Weight-lifter*, the, 182.
*Wells*, Artesian, 133.
*Wheel and Axle*, the, 103. Simply a revolving lever, 103. Law of, 104. Different forms of, 104.
*Wheels*, friction, 88. Enter largely into machinery, 120. Modes of connecting, 120. Different forms of the circumferences of, 121. Toothed, 122. Varieties of toothed, 122. Spur, 122. Cog, 123. Mill, 123. Mortice, 123. Crown, 123. Bevel, 123. How arranged in watches, 123. Undershot, 158. Overshot, 158. Breast, 159.
*Whirlwinds*, 403.
*Width*, defined, 12.
*Wind*, used as a motive power, 82. How produced, 401. Velocity of, how measured, 401. Constant winds, 402. Trade winds, 402. Periodical winds, 402. Variable winds, 403.
*Windlass*, the, described, 105.
*Wind-mills*, 82.
*Worcester*, his steam-engine, 220.
*Work*, unit of, 84.
*Wrapping connector*, 121.

**Z.**

*Zenith*, the, 385.
*Zodiac*, the, 386. Signs of, 386.